Mobile Ad Hoc Robots and Wireless Robotic Systems:

Design and Implementation

Raul Aquino Santos
University of Colima, Mexico

Omar Lengerke
Universidad Autónoma de Bucaramanga, Colombia

Arthur Edwards-Block
University of Colima, Mexico

A volume in the Advances in
Computational Intelligence and Robotics
(ACIR) Book Series

Managing Director:	Lindsay Johnston
Editorial Director:	Joel Gamon
Book Production Manager:	Jennifer Romanchak
Publishing Systems Analyst:	Adrienne Freeland
Assistant Acquisitions Editor:	Kayla Wolfe
Typesetter:	Nicole Sparano
Cover Design:	Nick Newcomer

Published in the United States of America by
Information Science Reference (an imprint of IGI Global)
701 E. Chocolate Avenue
Hershey PA 17033
Tel: 717-533-8845
Fax: 717-533-8661
E-mail: cust@igi-global.com
Web site: http://www.igi-global.com

Copyright © 2013 by IGI Global. All rights reserved. No part of this publication may be reproduced, stored or distributed in any form or by any means, electronic or mechanical, including photocopying, without written permission from the publisher. Product or company names used in this set are for identification purposes only. Inclusion of the names of the products or companies does not indicate a claim of ownership by IGI Global of the trademark or registered trademark.

Library of Congress Cataloging-in-Publication Data

Mobile ad hoc robots and wireless robotic systems: design and implementation / Raul Aquino Santos, Omar Lengerke, and Arthur Edwards Block, editors.
    pages cm
Includes bibliographical references and index.
  Summary: "This book introduces robotic theories, wireless technologies and routing applications involved in the development of mobile ad hoc robots"-- Provided by publisher.
  ISBN 978-1-4666-2658-4 (hardcover) -- ISBN 978-1-4666-2689-8 (ebook) -- ISBN (invalid) 978-1-4666-2720-8 (print & perpetual access) 1. Mobile robots. 2. Ad hoc networks (Computer networks) I. Aquino Santos, Raul, 1965- II. Lengerke, Omar, 1975- III. Block, Arthur Edwards, 1957-
  TJ211.415.M59 2013
  629.8'932--dc23
                                  2012029182

This book is published in the IGI Global book series Advances in Computational Intelligence and Robotics (ACIR) Book Series (ISSN: 2327-0411; eISSN: 2327-042X)

British Cataloguing in Publication Data
A Cataloguing in Publication record for this book is available from the British Library.

All work contributed to this book is new, previously-unpublished material. The views expressed in this book are those of the authors, but not necessarily of the publisher.

Advances in Computational Intelligence and Robotics (ACIR) Book Series

ISSN: 2327-0411
EISSN: 2327-042X

MISSION

While intelligence is traditionally a term applied to humans and human cognition, technology has progressed in such a way to allow for the development of intelligent systems able to simulate many human traits. With this new era of simulated and artificial intelligence, much research is needed in order to continue to advance the field and also to evaluate the ethical and societal concerns of the existence of artificial life and machine learning.

The **Advances in Computational Intelligence and Robotics (ACIR) Book Series** encourages scholarly discourse on all topics pertaining to evolutionary computing, artificial life, computational intelligence, machine learning, and robotics. ACIR presents the latest research being conducted on diverse topics in intelligence technologies with the goal of advancing knowledge and applications in this rapidly evolving field.

COVERAGE

- Adaptive & Complex Systems
- Agent Technologies
- Artificial Intelligence
- Cognitive Informatics
- Computational Intelligence
- Natural Language Processing
- Neural Networks
- Pattern Recognition
- Robotics
- Synthetic Emotions

IGI Global is currently accepting manuscripts for publication within this series. To submit a proposal for a volume in this series, please contact our Acquisition Editors at Acquisitions@igi-global.com or visit: http://www.igi-global.com/publish/.

The Advances in Computational Intelligence and Robotics (ACIR) Book Series (ISSN 2327-0411) is published by IGI Global, 701 E. Chocolate Avenue, Hershey, PA 17033-1240, USA, www.igi-global.com. This series is composed of titles available for purchase individually; each title is edited to be contextually exclusive from any other title within the series. For pricing and ordering information please visit http://www.igi-global.com/book-series/advances-computational-intelligence-robotics-acir/73674. Postmaster: Send all address changes to above address. Copyright © 2013 IGI Global. All rights, including translation in other languages reserved by the publisher. No part of this series may be reproduced or used in any form or by any means – graphics, electronic, or mechanical, including photocopying, recording, taping, or information and retrieval systems – without written permission from the publisher, except for non commercial, educational use, including classroom teaching purposes. The views expressed in this series are those of the authors, but not necessarily of IGI Global.

Titles in this Series

For a list of additional titles in this series, please visit: www.igi-global.com

Intelligent Technologies and Techniques for Pervasive Computing
Kostas Kolomvatsos (University of Athens, Greece) Christos Anagnostopoulos (Ionian University, Greece) and Stathes Hadjiefthymiades (University of Athens, Greece)
Information Science Reference • copyright 2013 • 349pp • H/C (ISBN: 9781466640382) • US $195.00 (our price)

Mobile Ad Hoc Robots and Wireless Robotic Systems Design and Implementation
Raul Aquino Santos (University of Colima, Mexico) Omar Lengerke (Universidad Autónoma de Bucaramanga, Colombia) and Arthur Edwards-Block (University of Colima, Mexico)
Information Science Reference • copyright 2013 • 347pp • H/C (ISBN: 9781466626584) • US $190.00 (our price)

Intelligent Planning for Mobile Robotics Algorithmic Approaches
Ritu Tiwari (ABV – Indian Institute of Information, India) Anupam Shukla (ABV – Indian Institute of Information, India) and Rahul Kala (School of Systems Engineering, University of Reading, UK)
Information Science Reference • copyright 2013 • 320pp • H/C (ISBN: 9781466620742) • US $195.00 (our price)

Simultaneous Localization and Mapping for Mobile Robots Introduction and Methods
Juan-Antonio Fernández-Madrigal (Universidad de Málaga, Spain) and José Luis Blanco Claraco (Universidad de Málaga, Spain)
Information Science Reference • copyright 2013 • 497pp • H/C (ISBN: 9781466621046) • US $195.00 (our price)

Prototyping of Robotic Systems Applications of Design and Implementation
Tarek Sobh (University of Bridgeport, USA) and Xingguo Xiong (University of Bridgeport, USA)
Information Science Reference • copyright 2012 • 321pp • H/C (ISBN: 9781466601765) • US $195.00 (our price)

Cross-Disciplinary Applications of Artificial Intelligence and Pattern Recognition Advancing Technologies
Vijay Kumar Mago (Simon Fraser University, Canada) and Nitin Bhatia (DAV College, India)
Information Science Reference • copyright 2012 • 784pp • H/C (ISBN: 9781613504291) • US $195.00 (our price)

Handbook of Research on Ambient Intelligence and Smart Environments Trends and Perspectives
Nak-Young Chong (Japan Advanced Institute of Science and Technology, Japan) and Fulvio Mastrogiovanni (University of Genova, Italy)
Information Science Reference • copyright 2011 • 770pp • H/C (ISBN: 9781616928575) • US $265.00 (our price)

Particle Swarm Optimization and Intelligence Advances and Applications
Konstantinos E. Parsopoulos (University of Ioannina, Greece) and Michael N. Vrahatis (University of Patras, Greece)
Information Science Reference • copyright 2010 • 328pp • H/C (ISBN: 9781615206667) • US $180.00 (our price)

www.igi-global.com

701 E. Chocolate Ave., Hershey, PA 17033
Order online at www.igi-global.com or call 717-533-8845 x100
To place a standing order for titles released in this series, contact: cust@igi-global.com
Mon-Fri 8:00 am - 5:00 pm (est) or fax 24 hours a day 717-533-8661

Table of Contents

Section 1
Robot Tracking Strategies

Stefano Panzieri, University Roma Tre, Italy
Federica Pascucci, University Roma Tre, Italy
Lorenzo Sciavicco, University Roma Tre, Italy
Roberto Setola, University Campus Bio-Medico, Italy

Anis Koubaa, Al-Imam Mohamed bin Saud University, Saudi Arabia & Polytechnic
Institute of Porto (ISEP/IPP), Portugal
Sahar Trigui, National School of Engineering, Tunisia
Imen Chaari, National School of Engineering, Tunisia

Lluis Pacheco, University of Girona, Spain
Ningsu Luo, University of Girona, Spain

Amina Waqar, National University of Computers and Emerging Sciences, Pakistan

Alejandro Hossian, Universidad Tecnológica Nacional, Argentina
Gustavo Monte, Universidad Tecnológica Nacional, Argentina
Verónica Olivera, Universidad Tecnológica Nacional, Argentina

Detailed Table of Contents

Section 1
Robot Tracking Strategies

Stefano Panzieri, University Roma Tre, Italy
Federica Pascucci, University Roma Tre, Italy
Lorenzo Sciavicco, University Roma Tre, Italy
Roberto Setola, University Campus Bio-Medico, Italy

In this chapter, the design of a completely decentralized and distributed multi-robot localization algorithm is presented. The issue is approached using an Interlaced Extended Kalman Filter (IEKF) algorithm. The proposed solution allows the dynamic correction of the position computed by any single robot through information shared during the random rendezvous of robots. The agents are supposed to carry short-range antennas to enable data communication when they have a "visual" contact. The information exchange is limited to the pose variables and the associated covariance matrix. The algorithm combines the robustness of a full-state EKF with the effortlessness of its interlaced implementation. The proposed unsupervised method provides great flexibility by using exteroceptive sensors, even if it does not guarantee the same position estimate accuracy for each agent. However, it can be effective in case of connectivity loss among team robots. Moreover, it does not need synchronization between agents.

Chapter 2

Anis Koubaa, Al-Imam Mohamed bin Saud University, Saudi Arabia & Polytechnic Institute of Porto (ISEP/IPP), Portugal
Sahar Trigui, National School of Engineering, Tunisia
Imen Chaari, National School of Engineering, Tunisia

Mobile robots and Wireless Sensor Networks (WSNs) are enabling technologies of ubiquitous and pervasive applications. Surveillance is one typical example of such applications for which the literature proposes several solutions using mobile robots and/or WSNs. However, robotics and WSNs have mostly been considered as separate research fields, and little work has investigated the marriage of these two technologies. In this chapter, the authors propose an indoor surveillance application, SURV-TRACK, which controls a team of multiple cooperative robots supported by a WSN infrastructure. They propose a system model for SURV-TRACK to demonstrate how robots and WSNs can complement each other to efficiently accomplish the surveillance task in a distributed manner. Furthermore, the authors investigate two typical underlying problems: (1) Multi-Robot Task Allocation (MRTA) for target tracking and capturing and (2) robot path planning. The novelty of the solutions lies in incorporating a WSN in the problems' models. The authors believe that this work advances the literature by demonstrating a concrete ubiquitous application that couples robotic and WSNs and proposes new solutions for path planning and MRTA problems.

Chapter 3

Lluis Pacheco, University of Girona, Spain
Ningsu Luo, University of Girona, Spain

Accurate path following is an important mobile robot research topic. In many cases, radio controlled robots are not able to work properly due to the lack of a good communication system. This problem can cause many difficulties when robot positioning is regarded. In this context, gaining automatic abilities becomes essential to achieving a major number of mission successes. This chapter presents a suitable control methodology used to achieve accurate path following and positioning of nonholonomic robots by using PID controllers. An important goal is to present the obtained experimental results by using the available mobile robot platform that consists of a differential driven one.

Chapter 4

Amina Waqar, National University of Computers and Emerging Sciences, Pakistan

Sensor-based localization has been found to be one of the most preliminary issues in the world of Mobile/Wireless Robotics. One can easily track a mobile robot using a Kalman Filter, which uses a Phase Locked Loop for tracing via averaging the values. Tracking has now become very easy, but one wants to proceed to navigation. The reason behind this is that tracking does not help one determine where one is going. One would like to use a more precise "Navigation" like Monte Carlo Localization. It is a more efficient and precise way than a feedback loop because the feedback loops are more sensitive to noise, making one modify the external loop filter according to the variation in the processing. In this case, the robot updates its belief in the form of a probability density function (pdf). The supposition is considered to be one meter square. This probability density function expands over the entire supposition. A door in a wall can be identified as peak/rise in the probability function or the belief of the robot. The mobile updates a window of 1 meter square (area depends on the sensors) as its belief. One starts with a uniform probability density function, and then the sensors update it. The authors use Monte Carlo Localization for updating the belief, which is an efficient method and requires less space. It is an efficient method because it can be applied to continuous data input, unlike the feedback loop. It requires less space. The robot does not need to store the map and, hence, can delete the previous belief without any hesitation.

Chapter 5

Alejandro Hossian, Universidad Tecnológica Nacional, Argentina

Gustavo Monte, Universidad Tecnológica Nacional, Argentina

Verónica Olivera, Universidad Tecnológica Nacional, Argentina

Robotic navigation applies to multiple disciplines and industrial environments. Coupled with the application of Artificial Intelligence (AI) with intelligent technologies, it has become significant in the field of cognitive robotics. The capacity of reaction of a robot in unexpected situations is one of the main qualities needed to function effectively in the environment where it should operate, indicating its degree of autonomy. This leads to improved performance in structured environments with obstacles identified by evaluating the performance of the reactive paradigm under the application of the technology of neural networks with supervised learning. The methodology implemented a simulation environment to train different robot trajectories and analyze its behavior in navigation and performance in the operation phase, highlighting the characteristics of the trajectories of training used and its operating environment, the scope and limitations of paradigm applied, and future research.

Chapter 6

Daniel S. F. Alves, Instituto de Matemática, UFRJ, Brazil

E. Elael M. Soares, Escola Politécnica, UFRJ, Brazil

Guilherme C. Strachan, Escola Politécnica, UFRJ, Brazil

Guilherme P. S. Carvalho, Escola Politécnica, UFRJ, Brazil

Marco F. S. Xaud, Escola Politécnica, UFRJ, Brazil

Marcos V. B. Couto, Escola Politécnica, UFRJ, Brazil

Rafael M. Mendonça, UERJ, Brazil

Renan S. Freitas, Escola Politécnica, UFRJ, Brazil

Thiago M. Santos, Escola Politécnica, UFRJ, Brazil

Vanessa C. F. Gonçalves, UFRJ, Brazil

Luiza M. Mourelle, UERJ, Brazil

Nadia Nedjah, UERJ, Brazil

Nelson Maculan, UFRJ, Brazil

Priscila M. V. Lima, Instituto de Ciências Exactas, UFRRJ, Brazil

Felipe M. G. França, UFRJ, Brazil

Many interesting and difficult practical problems need to be tackled in the areas of firefighting, biological and/or chemical decontamination, tactical and/or rescue searches, and Web spamming, among others. These problems, however, can be mapped onto the graph decontamination problem, also called the graph search problem. Once the target space is mapped onto a graph $G(N,E)$, where N is the set of G nodes and E the set of G edges, one initially considers all nodes in N to be contaminated. When a guard, i.e., a decontaminating agent, is placed in a node $i \in N$, i becomes (clean and) guarded. In case such a guard leaves node i, it can only be guaranteed that i will remain clean if all its neighboring nodes are either clean or clean and guarded. The graph decontamination/search problem consists of determining a sequence of guard movements, requiring the minimum number of guards needed for the decontamination of G. This chapter presents a novel swarm robotics approach to firefighting, a conflagration in a hypothetical apartment ground floor. The mechanism has been successfully simulated on the Webots platform, depicting a firefighting swarm of e-puck robots.

This chapter presents the development and implementation of three approaches that contribute to solving the mobile robot path planning problems in dynamic and static environments. The algorithms include some items regarding the implementation of on-line and off-line situations in an environment with static and mobile obstacles. A first technique involves the use of genetic algorithms where a fitness function and the emulation of the natural evolution are used to find a free-collision path. The second and third techniques consider the use of potential fields for path planning using two different ways. Brief descriptions of the techniques and experimental setup used to test the algorithms are also included. Finally, the results applying the algorithms using different obstacle configurations are included.

Chapter 8

One solution for trajectory tracking in a non-holonomic vehicle, like a mobile robot, is proposed in this chapter. Using the boundary values, a desired route is converted into a polynomial using a point-to-point algorithm. With the properties of Differential Flatness, the system is driven along this route, finding the necessary input values so that the system can perform the desired movement.

Section 2
Wireless Robotic Applications

Using a Single-Human Multiple-Robot System (SHMRS) to deploy rescue robots in Urban Search and Rescue (USAR) can induce high levels of cognitive workload and poor situation awareness. Yet, the provision of autonomous coordination between robots to alleviate cognitive workload and promote situation awareness must be made with careful management of limited robot computational and communication resources. Therefore, a technique for autonomous coordination using a hierarchically structured collective of robots has been devised to address these concerns. The technique calls for an Apex robot to perform most of the computation required for coordination, allowing Subordinate robots to be simpler computationally and to communicate with only the Apex robot instead of with many robots. This method has been integrated into a physical implementation of the SHMRS. As such, this chapter also presents practical components of the SHMRS including the robots used, the control station, and the graphical user interface.

Ítalo Jáder Loiola Batista, Federal Institute of Education, Science, and Technology of Ceará, Brazil

Antonio Themoteo Varela, Federal Institute of Education, Science, and Technology of Ceará, Brazil

Edicarla Pereira Andrade, Federal Institute of Education, Science, and Technology of Ceará, Brazil

José Victor Cavalcante Azevedo, Federal Institute of Education, Science, and Technology of Ceará, Brazil

Tiago Lessa Garcia, Federal Institute of Education, Science, and Technology of Ceará, Brazil

Daniel Henrique da Silva, Federal Institute of Education, Science, and Technology of Ceará, Brazil

Epitácio Kleber Franco Neto, Federal Institute of Education, Science, and Technology of Ceará, Brazil

Auzuir Ripardo Alexandria, Federal Institute of Education, Science, and Technology of Ceará, Brazil

André Luiz Carneiro Araújo, Federal Institute of Education, Science, and Technology of Ceará, Brazil

Driven by the rising demand for underwater operations concerning dam structure monitoring, Hydropower Plant (HPP), reservoir, and lake ecosystem inspection, and mining and oil exploration, underwater robotics applications are increasing rapidly. The increase in exploration, prospecting, monitoring, and security in lakes, rivers, and the sea in commercial applications has led large companies and research centers to invest underwater vehicle development. The purpose of this work is to present the design of an Autonomous Underwater Vehicle (AUV), focusing efforts on dimensioning structural elements and machinery and elaborating the sensory part, which includes navigation sensors and environmental conditions sensors. The integration of these sensors in an intelligent platform provides a satisfactory control of the vehicle, allowing the movement of the submarine on the three spatial axes. Because of the satisfactory fast response of the sensors, one can determine the acceleration and inclination as well as the attitude in relation to the trajectory instantaneously taken. This vehicle will be able to monitor the physical integrity of dams, making acquisition and storage of environmental parameters such as temperature, dissolved oxygen, pH, and conductivity, as well as document images of the biota from reservoir lake HPPs, with minimal cost, high availability, and low dependence on a skilled workforce to operate it.

Chapter 11

Leonimer Flávio de Melo, State University of Londrina, Brazil

Silvia Galvão de Souza Cervantes, State University of Londrina, Brazil

João Maurício Rosário, State University of Campinas, Brazil

This chapter presents a virtual environment implementation for embedded design, simulation, and conception of supervision and control systems for mobile robots, which are capable of operating and adapting to different environments and conditions. The purpose of this virtual system is to facilitate the development of embedded architecture systems, emphasizing the implementation of tools that allow the simulation of the kinematic, dynamic, and control conditions, in real time monitoring of all important system points. To achieve this, an open control architecture is proposed, integrating the two main techniques of robotic control implementation at the hardware level: systems microprocessors and reconfigurable hardware devices. The utilization of a hierarchic and open architecture, distributing the diverse actions of control in increasing levels of complexity and the use of resources of reconfigurable computation are made in a virtual simulator for mobile robots. The validation of this environment is made in a nonholonomic mobile robot and in a wheelchair; both of them used an embedded control rapid prototyping technique for the best navigation strategy implementation. After being tested and validated in the simulator, the control system is programmed in the control board memory of the mobile robot or wheelchair. Thus, the use of time and material is optimized, first validating the entire model virtually and then operating the physical implementation of the navigation system.

Chapter 12

Hernán González Acuña, Universidad Autónoma de Bucaramanga, Colombia

Alfonso René Quintero Lara, Universidad Autónoma de Bucaramanga, Colombia

Ricardo Ortiz Guerrero, Universidad Autónoma de Bucaramanga, Colombia

Jairo de Jesús Montes Alvarez, Universidad Autónoma de Bucaramanga, Colombia

Hernando González Acevedo, Universidad Autónoma de Bucaramanga, Colombia

Elkin Yesid Veslin Diaz, Universidad de Boyacá, Colombia

This chapter describes a Mechatronics Design methodology applied to the design of a mobile robot to climb vertical surfaces. The first part of this chapter reviews different ways to adhere to vertical surfaces and shows some examples developed by different research groups. The second part presents the stages of Mechatronics design methodology used in the design, including mechanical design, electronics design, and control design. These stages describe the most important topics for optimally successful design. The final part provides results that were obtained in the design process and construction of the robot. Finally, the conclusions of this research work are presented.

Chapter 13

John Blankenship, RobotBASIC, USA

Samuel Mishal, RobotBASIC, USA

Unlike most chapters in this book, this chapter does not introduce new methods or algorithms related to robotic navigation systems. Instead, it provides an overview of a simulation tool that, in some situations, can be useful for quickly evaluating the overall appropriateness of a wide variety of alternatives before focusing more advanced development activities on a chosen design. In addition, since the tool described herein is totally free, it can be used to help students and others new to robotics understand the value of utilizing a design-simulate-deploy approach to developing robotic behaviors. Robot Simulators can emulate nearly all aspects of a robot's functionality. Unfortunately, many programming environments that support simulation have steep learning curves and are difficult to use because of their ability to handle complex attributes such as 3D renderings and bearing friction. Fortunately, there are many situations where advanced attributes are unnecessary. When the primary goal is to quickly test the feasibility of a variety of algorithms for robotic behaviors, RobotBASIC provides an easy-to-use, economical alternative to more complex systems without sacrificing the features necessary to implement a complete design-simulate-deploy cycle. RobotBASIC's ability to simulate a variety of sensors makes it easy to quickly test the performance of various configurations in an assortment of environments. Once algorithm development is complete, the same programs used during the simulation phase of development can immediately control a real robot.

Chapter 14

Lafaete Creomar Lima Junior, Federal University of Rio de Janeiro, Brazil

Armando Carlos de Pina Filho, Federal University of Rio de Janeiro, Brazil

Aloísio Carlos de Pina, Federal University of Rio de Janeiro, Brazil

The chapter describes the stages of an autonomous mobile robot project, in this case, an underwater cleaning robot. First, the authors analyze the products already available for costumers, mainly focusing on the tasks they can perform (instead of the systems they use), in order to define the requirements of their project. Then, they build some models, based in the literature available. Based on them, the authors dimension the parts and systems by evaluating the results of these models. Finally, the authors use all information gathered to create a prototype, modeled with a CAE system.

Chapter 15

Gustavo Ramírez Torres, Siteldi Solutions, Mexico
Pedro Magaña Espinoza, Siteldi Solutions, Mexico
Guillermo Adrián Rodríguez Barragán, Siteldi Solutions, Mexico

With the educational mobile robot Worm-Type Mobile Educational Robot (Robot Móvil Educativo tipo Oruga, or ROMEO, by its Spanish acronym), the authors offer three hierarchical levels of experimental learning, where the operator can develop as far as his/her ability or imagination permits, gaining knowledge about the basics of sensors, communications, and mechanical and robot programming. Due the lack of learning focused on robotics in Mexican educational institutions, the authors present this chapter, where an early stimulation to this topic could trigger curiosity to research that leads to technological advancement. ROMEO is a mobile wireless communication platform with different types of sensors: moisture, brightness, temperature, etc., as well as a compass and accelerometers with similar characteristics to industrial and commercial applications that allow us to experiment with communication algorithms, sampling, and autonomous and semiautonomous navigation.

Chapter 16

Laura Victoria Escamilla Del Río, University of Colima, Mexico
Juan Michel García Díaz, University of Colima, Mexico

This chapter presents a theoretical and experimental comparison of electromagnetic propagation models for indoor robot communication using mobile ad-hoc IEEE802.11 and IEEE802.15.4. The analysis includes the behavior of the electromagnetic signal using the abovementioned standards in two scenarios, both located inside the building of the College of Telematics of the University of Colima. The results of the propagation of the electromagnetic signals in the two scenarios were then compared with the mathematical model.

Preface

It was Leonardo da Vinci (1452-1519) who first studied robotics in great depth. His great ability to project and develop mechanisms intended to reproduce natural characteristics he was interested in are published in his notebooks, *Codex Atlanticus*, Ms.B, Ms.I, today in the museum of science of Florence, Italy, and in private collections like the *Codex Leicester*, belonging to Bill Gates, the founder of Microsoft (Rosheim, 2006). The first robots were truly complex, automatic pieces of art capable of carrying out repetitive tasks. These robots gave rise to today's fixed-base, manipulating arms, widely used in industry (e.g. car-making industry). Mobile robots were developed recently, and their main feature is their guided, semi-autonomous, or fully autonomous motion capability.

The evolution of mobile robots has contributed to a number of solutions in the area of robot control. For example, the distributed control of robots requires wireless communication networks connecting the robots to other devices in their surroundings. This book intends to, among other things, provide examples of solutions related to the distributed control of mobile robots. Wifi and Bluetooth technologies are examples of such networks. Ad Hoc networks can be readily installed in a setting to offer coverage of a wide region by using a great number of nodes. Besides offering communication for the robots, a sensor network can aid their motion and location, as well as enhance their sensing capability. This book presents different platforms and studies that permit the integration of Ad Hoc networks into mobile robot applications with the goal of providing control of and communication among mobile robots through wireless technologies.

The last decade's fast advance in technology made mobility a reality. Each day witnesses the appearance of mobile and portable devices with ever-larger storage and processing capabilities. Likewise, wireless communication technologies have also been developed, searching for optimal relation of energy-unit-transmitted data. Perhaps the greatest advances have been in the area of wireless sensor networks.

Wireless Sensor Networks (WSNs) are made up of devices traditionally known as sensor nodes. These nodes are compact, autonomous, possess limited storage capacity, sense physical characteristics in their environment, and employ Radio Frequency (RF) signals over short distances, thus facilitating communication (usually Ad Hoc). Mobile robots are computerized mechanisms capable of moving and reacting to their surroundings by means of a combination of direction and operation established through a trajectory, defined plan, or action transmitted in real time. Sensor-led information measures both the robot's internal status and its surroundings. Mobile robots are classified based on the application type or locomotion type. Limbed robots can move across more irregular terrain than wheeled robots, but they demand more energy and are also slower (Pfeiffer, et al., 1998). Robot mobility can also be airborne, as is the case with robots used in aerial rescue, surveillance, and space exploration (Kaestner, et al., 2006). Some robots move quickly on the soil or on service lines (ducts, wiring, etc.) while others use magnetic levitation, which requires especially prepared surfaces. Lastly, there are sub-aquatic robots, which share stability problems with their space counterparts.

Mobile robotics is among the great variety of areas presently using wireless sensor networks. The combined use of these technologies is quite interesting from the perspective that they supplement each other. Mobile robots can be equipped with nodes and they can be interpreted as mobile nodes as part of a sensor network, or as they are utilized as a communication channel between isolated networks. On the other hand, a sensor network can be seen as an extension of the sensing and communication capabilities of mobile robots. There is, for example, extensive literature exploring two of the abovementioned characteristics.

The study of mobile robot location and control stands as an interesting challenge, given their great mobility and the constant topology changes characteristic of Ad Hoc Mobile Networks (MANET). These features call for efficient communications regarding the agent's individual status and rapid algorithms capable of calculating the relative positions of the systems involved.

Wireless mobile networks are communication systems where those stations constituting mobile nodes communicate by means of links (radio). These network types can be classified into two kinds: networks with infrastructure and Ad Hoc networks, or without infrastructure. The main feature of Ad Hoc networks is that they do not have any type of previously established infrastructure. Therefore, this kind of network can be quickly constructed, either in complex surroundings, or wherever previous installation of an infrastructure is difficult. In Ad Hoc networks, the nodes are responsible for carrying out those functions previously carried out by the access point; then, each network node becomes a router responsible for sending, forwarding, or receiving the messages with other nodes that must communicate among themselves to reach any defined area. Given that nodes can move randomly and dynamically, thus altering the topology of the network, the routing protocols used in Ad Hoc networks must be adaptive and capable of finding routes in this rapidly changing scenario.

Substantial work in the area of Ad Hoc networks has been presented in the last few years. The characteristic feature of these types of networks is the use of radio channels to promote communication between their nodes and the absence of fixed devices centralizing network communication, meaning that two nodes can communicate directly once they are within each other's transmission area. Whenever two nodes are quite distant from each other, intermediate nodes in the network can participate as intermediaries in this communication, carrying out routing functions of the messages between origin and destination nodes.

Several protocols have already been proposed in the literature to guarantee the routing of messages between all of the network nodes and enhance communication efficiency. Examples include DSDV (Perkins & Bhagwat, 1994), DSR (Johnson & Maltz, 1996), TORA (Park & Corson, 1997), and AODV (Perkins & Royer, 1999). These features allow nodes to move within a given area, dynamically altering the network topology while maintaining connectivity. All of this expands the number of possible applications of Ad Hoc networks.

The vision of Leonardo da Vinci is quickly becoming a reality. Robots can now emulate many natural movements in many areas of endeavor, and their application is sure to grow in the future as technology evolves and makes further advances possible. Robots and their intercommunication will one day make human life more enjoyable, productive, safer, and more entertaining. *Mobile Ad Hoc Robots and Wireless Robotic Systems: Design and Implementation* is a book whose objective is to present state-of-the-art work being done in the area of robotics and communications so that its readers can one day advance science and technology to our ultimate goal: produce robotic systems that can one day encompass every aspect of our lives, thus freeing the human species to advance in many other fields of endeavor.

Acknowledgment

Raul Aquino Santos would like to thank his wife, Tania, for her deep love and for giving him a wonderful family and his lovely daughters, Tania Iritzi and Dafne, for providing him with many years of happiness. In addition, he would like to thank his parents, Teodoro and Herlinda, for offering him the opportunity to study and to believe in God. Finally, he wants to thank his brother, Carlos, and his sisters, Teresa and Leticia, for their support.

Omar Lengerke would like to thank his wife, Magda. This book would not have been possible without her support and encouragement. Words cannot express his gratitude to Editor Prof. Aquino for his professional advice and assistance in polishing this manuscript. Special thanks to the researchers for their invaluable assistance in producing this volume.

Arthur Edwards Block would like to endlessly thank his wife, Marilú, for selflessly dedicating her life to making his life worth living. He would like to thank her for giving him the most precious children, Elisa and David, and for always firmly standing beside him. Thank you for walking life's path with him.

Raul Aquino Santos
University of Colima, Mexico

Omar Lengerke
University of Colima, Mexico

Arthur Edwards Block
University of Colima, Mexico

Section 1
Robot Tracking Strategies

Chapter 1
Distributed Multi–Robot Localization

Stefano Panzieri
University Roma Tre, Italy

Federica Pascucci
University Roma Tre, Italy

Lorenzo Sciavicco
University Roma Tre, Italy

Roberto Setola
University Campus Bio-Medico, Italy

ABSTRACT

In this chapter, the design of a completely decentralized and distributed multi-robot localization algorithm is presented. The issue is approached using an Interlaced Extended Kalman Filter (IEKF) algorithm. The proposed solution allows the dynamic correction of the position computed by any single robot through information shared during the random rendezvous of robots. The agents are supposed to carry short-range antennas to enable data communication when they have a "visual" contact. The information exchange is limited to the pose variables and the associated covariance matrix. The algorithm combines the robustness of a full-state EKF with the effortlessness of its interlaced implementation. The proposed unsupervised method provides great flexibility by using exteroceptive sensors, even if it does not guarantee the same position estimate accuracy for each agent. However, it can be effective in case of connectivity loss among team robots. Moreover, it does not need synchronization between agents.

1. INTRODUCTION

A fundamental issue in any mobile robot ad-hoc network is related to the ability of each agent to localize itself with respect to both the workspace and the other agents. In mobile robotics, this issue is referred to as localization when the environment is a known priori, while it is called Simultaneous Localization and Mapping (SLAM) when robots concurrently create a map of the environment and locate themselves on that map. Here, only the localization problem is addressed, although the proposed algorithm can be suitably extended to solve SLAM problems.

DOI: 10.4018/978-1-4666-2658-4.ch001

Copyright © 2013, IGI Global. Copying or distributing in print or electronic forms without written permission of IGI Global is prohibited.

In the literature, localization is approached using supervised and unsupervised communication paradigms. In the first case, a centralized supervisor (i.e., a base station) collects all the data coming from the robots and provides an estimate for the pose of the whole team. In the second approach, each robot runs a local algorithm to estimate its pose, using only its own sensors and shares data with its neighbors following the Mobile Ad hoc NETwork (MANET) model.

The complexity of a centralized algorithm grows in a nonlinear way with the number of robots. Moreover, this solution forces all mobile robots to communicate continuously with the supervisor: thereafter, robots need either to move closely to the supervisor location, or to set up a hierarchical network (Antonelli, Arricchiello, Chiaverini, & Setola, 2006). In both cases, some constraints over team mobility are imposed to guarantee that, at each moment, at least one communication-path from any robot to the supervisor exists.

According to the decentralized approach, each robot in the team collaborates to solve the localization problem. The easiest solution is achieved by decomposing the localization problem of the team into autonomous and independent localization tasks. Therefore, each robot implements its own localization algorithm, ignoring the presence of the other robots in the workspace. In other words, the localization problem is solved by multiplexing single robot localization, without exploiting additional information extracted from neighbors. Moreover, no relative measures are used, even if this information could be very important (i.e., formation control, collaborative exploration tasks). For these reasons, in the literature, the multi-robot localization problem is referred to as collaborative/cooperative localization, when the team exploits information sharing.

The relevance of information sharing during localization is particularly useful when it faces heterogeneous robot teams. In this case, some robots are equipped with expensive, high accuracy sensors (i.e., GPS, laser rangefinders, smart camera), while the others are equipped with low-cost sensors (i.e., sonar, Web-cam). Hence, information sharing enhances sensor data, since collaborative multi-robot localization exploits high-accuracy sensors across teams.

This chapter proposes an efficient probabilistic algorithm, the Interlaced Extended Kalman Filter (EKF) (Panzieri, Pascucci, & Setola, 2006), for collaborative multi-robot localization. It is based on the well-known prediction-update scheme of Bayesian filter and represents a sub-optimal solution for probabilistic state estimate. According to this approach, robots are able to compute their poses, exploiting the detection of other robots. Indeed, each robot implements an EKF to localize itself, using proprioceptive sensors and measurements retrieved from the environment. When two robots are visible to each other, they exchange information about their estimated poses and their measurements. By doing so, each robot takes advantage of its neighbors, using them as further virtual sensors. To cope with the noise and the ambiguity arising in real-world scenarios, observation models are set up in a probabilistic framework. Indeed, the covariance matrices related to these virtual devices are suitably manipulated to accommodate the different levels of uncertainties typical of pose estimates. The whole team localization problem is split over the robot network decomposing the whole filter into N sub-filters, where N is the cardinality of the team.

IEKF allows cooperative localization under severe communication and computational constraints. In fact, the communications are performed only during rendezvous and only limited data are broadcast. The computational load is reduced by the decomposition of the estimate procedure. Although reducing computational complexity and communication represents a great advantage, the main novelty of the proposed approach is related to the synchronization of the estimate. Several multi-robot localization algorithms based on Extended Kalman Filter decomposition rely on the assumption that the robotic network is completely synchronized. On the contrary, IEKF does not need any agent synchronization, since only the last estimate and measurements, together with their associated uncertainties, are exchanged on robots rendezvous. Moreover, there is no need for robots to use the same sampling frequency.

The remaining sections of this chapter are organized as follows: Literature about multi-robot localization is reviewed in Section 2. The problem of multi-robot localization is set as a nonlinear filtering problem and solving methods are introduced and explained in Section 3. Technical details about the Interlaced Kalman Filter (IKF) and its extended version (IEKF) are provided in Section 4, and Section 5 introduces the IEKF for multi-robot localization. Even though it is applicable to any kind of robot and sensor, here we focus our presentation on robots having unicycle kinematics and laser rangefinders for perceiving both robots and landmarks. Extensive simulations have been carried out to prove the effectiveness of the approach in indoor environment and are reported in Section 6. Future research developments are outlined in Section 7, followed by a discussion over the advantages and the limitations of the current approach.

2. RELATED WORK

A large number of studies related to localization have been carried out for single autonomous robots. Localization techniques developed for single robots have been extended and applied to multi-robot teams by adopting a centralized model. More recently, researchers have proposed *ad hoc* algorithms exploiting distributed computation and cooperative paradigms, due to the computational complexity and heavy communication overhead.

Multi-robot cooperative localization is a challenging task in mobile robotics, and it is the basis for both multi-robot navigation and exploration. Mobile platforms in the team are able to cooperate and estimate their poses in the environment. Robots in a cooperative localization system are able to make full use of their sensor information to refine their internal estimation and to improve their localization accuracy. Furthermore, the ability to exchange information during localization is particularly attractive, because each sight of another robot reduces the uncertainty of the estimated poses. This means that teams have to be designed to build a network infrastructure in order to perform cooperative tasks.

2.1 Ad Hoc Network in Multi-Robot Applications

Successfully controlling and coordinating a group of wireless-networked robots relies on the effectiveness of inter-robot communication. Early studies were carried out using a client-server approach and later evolved into teams of robots organized in ad hoc networks. Nowadays most robot communication applications choose ad hoc network configurations to reduce the vulnerability of centralized infrastructures. Ad hoc networks, indeed, are infrastructureless networks, where nodes are in charge of routing packets. There is no stationary infrastructure and the network topology can be highly dynamic.

To our knowledge, the first wireless communication adopted was based on infrared technology (Kahn & Barry, 1997), due to the low cost of infrared communication devices. Radio Frequency (RF) rapidly replaced infrared technology, due to the increased use of Internet wireless Local Access Networks (LANs) (IEEE 802.11). In robotics, Wi-Fi Networks serve as a backbone for data concentration and networking. They are used in conjunction with short-range, low-power devices in a wireless field network to collect data from the gateway and forward it to the data collection point (Dzung, Apneseth, Endersen, & Frey, 2005). Gerkey et al. (2001) developed *Player*, a network server that provides "transparent network access to all sensing and control" of multiple robots (p. 1226). Nguyen, Pezeshkian, Raymond, Gupta, and Spector (2003) developed a system aimed at increasing the range of a wirelessly controlled robot by using small slave robots that followed the main robot and relayed the signal onward.

Most of the robotic applications proposed in the literature are based on Wireless Personal Area Networks (IEEE 802.15) and mainly exploit Bluetooth and Zigbee transmission technologies. In Barnhard, McClain, Wimpey, and Potter (2004), Bluetooth communication is adopted to coordinate "search and navigation" tasks.

The technology defined by Zigbee is chosen (Song, Tian, Li, Zhou, & Liu, 2011) to coordinate a robot swarm using a low information flow. The Zigbee specification is intended to be simpler and less expensive than other WPANs, such as Bluetooth. Zigbee is targeted at RF applications that require a low data rate, long battery life, and secure networking.

2.2 Cooperative and Collaborative Localization

To the best of our knowledge, one of the first works about collaborative localization is in Kurazume, Nagata, and Hirose (1994). Authors consider a team of a mobile robots moving in an unknown environment, without the help of beacons. They introduce the concept of mobile landmark, since the exploration is performed using the robots themselves as landmarks. A data fusion algorithm collects data to improve the estimate, while each robot repeats move-and-stops to fix landmarks for the team. Robots are equipped with markers to aid detection.

A new sensing approach, named *robot tracker*, is used to improve the accuracy of the position estimation of each robot of the team in Rekleitis, Dudek, and Milios (2000), Rekleitis, Dudek, and Milios (2002), and Rekleitis, Dudek, and Milios (2003). A team of two robots is considered during exploration. Each platform is equipped with a robot tracker sensor to retrieve relative positions between robots. A particle filter is exploited to update pose estimates and the attached uncertainty.

A Bayesian approach is adopted by Fox, Burgard, Kruppa, and Thrun (2000). Here, two robots improve their global localization exchanging their belief (i.e., the posterior probability density over the state space conditioned to measurements) on rendezvous. A particle filter is at the basis of the algorithm, giving the possibility of managing a non-Gaussian shaped belief, and an easier global localization.

A more promising approach is developed in Roumeliotis and Bekey (2002, 2000) and examined in Martinelli, Pont, and Siegwart (2005) and Martinelli and Siegwart (2005), where a Kalman-based algorithm is used to achieve collaborative localization. Robots in the team collect data from their proprioceptive sensors and perform a prediction step of Kalman filter. They share exteroceptive data during the update, thus improving the estimate. Authors propose a distributed algorithm based on singular value decomposition of the covariance matrix. In this way, the complete filter is decomposed into a number of smaller communicating filter. When inter robot communications are not available, cross-correlation terms cannot be maintained, thus the algorithm suffers for application in large-scale environments.

In Trawny, Roumeliotis, and Giannakis (2005), the problem of cooperative localization under severe communication constraints is addressed. Specifically, both a minimum mean square error and maximum a posteriori estimators are considered. The filters are able to cope with quantized process measurements, since during navigation, each robot quantizes and broadcasts its measurements and receives the quantized observations of its teammates.

To eliminate the marker used for robot detection, a Metric-Based Iterative Closest Point (MbICP) algorithm is used in Zhuang, Gu, Wang, and Yu (2010). It derives the real-time optimal relative observations between any two robots equipped with laser range finders. In the same work, an adaptive localization period selection algorithm is proposed, based on autonomous motion state estimation.

Most approaches presented in the literature have been developed for homogenous teams in a probabilistic framework. To compute a good estimate, robots are assumed to be synchronized and to have the same sampling rate. Here we propose a probabilistic localization algorithm, which enables us to solve localization problems for a heterogeneous team. Time synchronization is not a requirement for such a procedure.

3. SETTING AND APPROACH

Let us consider a set of mobile robots moving in a known environment. The multi-robot localization problem can be solved by discovering the pose (i.e., position and heading direction) of each single mobile platform.

To mathematically describe our approach, we need to restore some assumptions. In the remainder of the chapter, we consider a set of N robots, supplied with a model of the environment (e.g., a map). They are equipped with both proprioceptive and exteroceptive sensors. The former are used to perform dead reckoning while the latter are exploited to enable robots both to relate their own position with respect to the map and to detect each others. Specifically, each robot continuously records its wheel encoders (dead reckoning) to generate odometric measurements at regular intervals. These measurements state the relative change of position according to the wheel encoders. Robots also poll their sensors (e.g., range finders, cameras) at regular time intervals as they generate measurements by means of robot-environment interaction. Additionally, each robot polls its sensors to discover the presence or absence of other robots.

Since all of these perceptions are typically confounded by noise, localization can be regarded as a nonlinear filtering problem, i.e., the problem of estimating the state of a dynamic system using a sequence of noisy measurements made on the system.

Throughout this chapter, the state space approach is chosen to model dynamic systems, together with a discreet-time formulation of the problem. In multi-robot localization, the state vector represents the poses of the robots comprising the team. In order to inspect inferences about a dynamic system, at least two models are required: first, a model describing the time evolution of the state, i.e. the system dynamic model or state transition model, and second, a model describing the relation between the noisy measurements and the state, i.e. the measurement or observation model.

To define the problem of nonlinear filtering (Ristic, Arulampalam, & Gordon, 2004), let us consider the state vector $x_k \in R^n$, where n is the dimension of the state vector and $k \in N$. Here the index k is attributed to a continuous-time instant t_k. The state vector evolves according to the following discrete-time stochastic model:

$$x_k = f_k(x_{k-1}, u_{k-1}, v_{k-1}) \tag{1}$$

where f_k is a known, possibly nonlinear function of the state, x_k, v_k is referred to as a process noise sequence, and u_k is the deterministic input of the system. In mobile robot localization, f_k represents the kinematic model of the robot. In multi-robot localization, the state transition model f_k is comprised of the kinematic models of the each robot and, eventually, some formation constraints, while the inputs of the system are represented by the proprioceptive data. Since process noise takes into account any mis-modeling effects or unpredictable disturbances in the system model, it describes both some dynamic interactions not considered in the kinematic models and noise arising in addition to odometric data.

The objective of the nonlinear filtering is to recursively estimate x_k from the measurements $z_k \in R^m$. The measurements are related to the state vector by means of the observation model:

$$z_k = h_k(x_k, w_k) \qquad (2)$$

where h_k is a known, possibly nonlinear function and w_k is a measurement noise sequence. In multi robot localization, h_k represents the stacked relation between the poses of the different robots and their exteroceptive measurements.

We suppose that these measurements are available in a probabilistic form: in this way, the probabilistic state-space formulation and the requirement for updating the information on receipt of new measurements are well suited for the Bayesian approach, which provides a rigorous general framework for dynamic state estimation problems. The Bayesian approach to dynamic state estimation's goal is to restore the *posterior probability density function* (pdf or density) of the state, based on all available information, including all the sequences of received measurements. Since either the system or measurement model should be nonlinear, the posterior pdf is, in general, non-Gaussian. The pdf encompasses all available statistics and represents the complete solution to the estimation problem. In principle, an optimal (with respect to any criterion) estimate of the state may be obtained from the posterior pdf, as well as a measure of the accuracy of the estimate.

To set up an effective multi-robot localization procedure, an online estimation is mandatory. In this case a recursive filter is a convenient solution, since received data can be processed sequentially rather than as a batch and there is no need to store the complete data set or to reprocess existing data as new measurements become available. From a Bayesian perspective, the problem is to recursively quantify the belief in the state x_k at time k at different values, given data Z_k up to time k, where $Z_k = \{z_i, i = 1 \ldots k\}$ is the sequence of all available measurements up to time k, i.e., it is required to compute the posterior pdf $p(x_k \mid Z_k)$. Such filters consist of essentially two stages: prediction and update.

Let us suppose that the posterior pdf $p(x_{k-1} \mid Z_{k-1})$ at time $k-1$ is available and let us assume that the noise sequences v_k and w_k are white, with known probability density functions and mutually independent. Moreover, let us assume that the initial state vector x_0 has a known pdf $p(x_0)$ and is also independent of noise sequences. The prediction stage evolves using the system model to obtain the prediction density of the state at time k via the Chapman-Kolmogorov equation:

$$p(x_k \mid Z_{k-1}) = \int p(x_k \mid x_{k-1}) p(x_{k-1} \mid Z_{k-1}) dx_{k-1}$$
$$(3)$$

The prediction stage exploits the system model for predicting the state pdf forward from one measurement time to the next one. Since the state is usually subject to unknown disturbances (modeled by random noise), prediction generally translates, deforms, and enlarges the state pdf. In multi-robot localization, during the prediction, the kinematic model of the robots is exploited together with proprioceptive measurements. At this stage the uncertainties grow, since the interaction between robots and environment are not considered.

At time step k, when a measurement z_k becomes available, the update stage is performed. This involves an update of the prior pdf via Bayes' rule:

$$p(x_k \mid Z_k) = p(x_k \mid z_k, Z_{k-1})$$
$$= \frac{p(z_k \mid x_k, Z_{k-1})p(x_k \mid Z_{k-1})}{p(z_k \mid Z_{k-1})}$$
$$= \frac{p(z_k \mid x_k)p(x_k \mid Z_{k-1})}{p(z_k \mid Z_{k-1})}$$

(4)

where $p(z_k \mid Z_{k-1})$ is a normalizing constant that depends on the likelihood function $p(z_k \mid x_k)$, defined by the measurement model and the known statistics of w_k. In the update, the measurement is used to reshape the prior density, obtaining the required posterior density of the current state. The update step harnesses the latest measurement to tighten the prediction pdf, since Bayes theorem allows updating knowledge about the state vector by means of including extra information from new data. In the multi-robot localization problem, exteroceptive measurements are involved in the update step, thus reducing the estimate uncertainty.

Equations (3) and (4) form the basis for the optimal Bayesian solution. Once the posterior density $p(x_k \mid Z_k)$ is known, an optimal state estimate with respect to any criterion can also be computed. For example, the Minimum Mean Square Error (MMSE) estimate is the conditional mean of x_k:

$$\hat{x}_{k|k} = \varepsilon\{x_k \mid Z_k\} = \int x_{k|k}p(x_{k|k} \mid Z_k)dx_k.$$

(5)

Similarly, a measure of accuracy of a state estimate may also be obtained.

The multi-robot localization solution given by Equations (3) and (4) is only conceptual, since, in general, it cannot be resolved analytically. First of all, the implementation of Equations (3) and (4) requires the storage of non-Gaussian pdf, which is, in general terms, equivalent to an infinite dimensional vector. Moreover, the dimensionality of the state vector grows with the number of robots. Distributions over x_k, hence, are exponential to the number of robots. Furthermore, each robot position is described by at least three values, thus modeling the joint distribution of the poses of all robots results infeasible for even a small N value. Since in most practical situations the analytic solution is intractable, one has to use approximations or suboptimal Bayesian filter. Adopting a decentralized and cooperative approach, distributed filters also have to be taken into account in order to split the computational load of the localization procedure over whole team.

4. BACKGROUND: THE INTERLACED EXTENDED KALMAN FILTER

In the linear-Gaussian case, a conceptual solution for Equations (3) and (4) is given by Kalman Filter. In this case, the posterior pdf at each time step is Gaussian. As a result, it can be completely described by its mean value and covariance. Glielmo, Setola, and Vasca (1999) proposed using the Interlaced Kalman Filter (IKF) to reduce the computational complexity of the estimation process for a class of nonlinear systems. This idea comes from the multi-player dynamic game theory, where the game solution is achieved when each player selects his strategy by maximizing its gain as the optimal response to the strategy chosen by the other players.

Briefly, the IKF is built on m parallel implementations of Kalman Filters (KF). Each KF works independently from the others. Each filter is designed to produce an estimate for a subset of the state vector, considering the remaining parts as deterministic time varying parameters.

To partially reduce the error introduced by decoupling, the noise covariance matrices are increased, as explained in the following.

Let us consider a system partitioned into only two subsets:

$$x_k^{(1)} = [A_{k-1}^{(1)} + F^{(12)}(x_{k-1}^{(2)})]x_{k-1}^{(1)} + f^{(1)}(x_{k-1}^{(2)}) + v_k^{(1)} \tag{6}$$

$$x_k^{(2)} = [A_{k-1}^{(2)} + F^{(21)}(x_{k-1}^{(1)})]x_{k-1}^{(2)} + f^{(2)}(x_{k-1}^{(1)}) + v_k^{(2)} \tag{7}$$

where the state vector $x \in R^n$ has been partitioned into $x^{(1)} \in R^{n_1}$ and $x^{(2)} \in R^{n_2}$ (with $n = n_1 + n_2$), $f_k^i(\cdot), i = 1, 2$ are differentiable functions, and $w_k^{(i)} \in R^{n_i}, i = 1, 2$ are zero-mean uncorrelated white process noise vectors characterized by the covariance matrices $Q_k^{(i)}, i = 1, 2$.

The output of the system can be decoupled into the following equivalent forms:

$$z_k = C^{(1)}(x_k^{(2)})x_k^{(1)} + d^{(1)}(x_k^{(2)}) + w_k^{(1)} \tag{8}$$

$$z_k = C^{(2)}(x_k^{(1)})x_k^{(2)} + d^{(2)}(x_k^{(1)}) + w_k^{(2)} \tag{9}$$

where $d^i(\cdot), i = 1, 2$ are differentiable functions, $w_k^{(i)} \in R^{m_i}, i = 1, 2$, are zero-mean uncorrelated white measurement noise vectors having covariance matrices R_k.

Under these hypotheses, the first filter of the IKF estimates $\hat{x}_{k|k}^{(1)}$ are obtained by exploiting $\hat{x}_{k|k-1}^{(2)}$, which is the predictive estimate obtained at the previous step by the other filter. Notice that after replacing $x_k^{(2)}$ with $\hat{x}_{k|k-1}^{(2)}$ in Equations (8) and (10), the first subsystem can be considered as a linear time varying system. Similar considerations hold true for the second subsystem.

The prediction step of each KF is characterized by the following equations:

$$\hat{x}_{k|k-1}^{(i)} = \tilde{A}_k^{(i)} \hat{x}_{k-1|k-1}^{(i)} + f^{(i)}(\hat{x}_{k-1|k-1}^{(j)}) \tag{10}$$

$$P_{k|k-1}^{(i)} = \tilde{A}_k^{(i)} P_{k-1|k-1}^{(i)} \tilde{A}_k^{(i)^T} + \tilde{Q}_k^{(i)} \tag{11}$$

with $(i = 1, \ j = 2)$ and $(i = 2, \ j = 1)$ for the first and second filters, respectively and

$$\tilde{A}_k^{(i)} = A_k^{(i)} + F^{(ij)}(x_{k-1|k-1}^{(j)}) \tag{12}$$

and

$$\tilde{Q}_k^{(i)} = Q_k^{(i)} + [J_{x,j}^{F,(ij)} + J_{x,j}^{f,(i)}]P_{k-1|k-1}^{(2)}[J_{x,j}^{F,(ij)} + J_{x,j}^{f,(i)}]^T \tag{13}$$

where $J_{x,j}^{F,(ij)}$ and $J_{x,j}^{f,(i)}$ are the Jacobian of $F_k^{(ij)}(\cdot)x_k^{(i)}$ and $f_k^{(i)}(\cdot)$ with respect to $x_k^{(j)}$.

The update equations can be put in the following form

$$\hat{x}_{k|k}^{(i)} = \hat{x}_{k|k-1}^{(i)} + K_k^{(i)}[z_k - C^{(i)}(\hat{x}_{k|k-1}^{(j)})\hat{x}_{k|k-1}^{(i)} + \\ - d^{(i)}(\hat{x}_{k|k-1}^{(j)})] \tag{14}$$

$$P_{k|k}^{(i)} = P_{k|k-1}^{(i)} - K_k^{(i)}C^{(i)}(\hat{x}_{k|k-1}^{(j)})P_{k|k-1}^{(i)} \tag{15}$$

where the Kalman gain is evaluated by applying the following Equation:

$$K_k^{(i)} = P_{k|k-1}^{(i)}C^{(i)}(\hat{x}_{k|k-1}^{(j)})^T \times \\ \times [C^{(i)}(\hat{x}_{k|k-1}^{(j)})P_{k|k-1}^{(i)}C^{(i)}(\hat{x}_{k|k-1}^{(j)})^T + R_k^{(i)}]^{-1} \tag{16}$$

in which

$$\tilde{R}_k^i = R_k + J_{x,j}^{d,i} P_{k|k-1}^{(j)} J_{x,j}^{d,iT}. \qquad (17)$$

From Equations (13) and (17) it can be noticed that positive semi-definite quantities conveniently increase both the process and measurement noise covariance matrices $Q_k^{(i)}$ and $R_k^{(i)}$, respectively. Such addition takes into account the error introduced by the decoupling operation, where modified noises are considered

$$\tilde{v}_k^{(i)} := v_k^{(i)} + J_{x,j}^{f,i} e_{k|k-1}^{(i)} + J_{x,j}^{F,ij} e_{k|k-1}^{(i)} \qquad (18)$$

$$\tilde{w}_k^{(i)} := w_k^{(i)} + J_{x,j}^{d,i}(\hat{x}_{k|k-1}^{(i)}) e_{k|k-1}^{(i)} \qquad (19)$$

where $e_{k|k-1}^{(i)} := x_k^{(i)} - \hat{x}_{k|k-1}^{(i)}$ is the estimation error.

As shown in Equations (8), (9), (10), and (11), the state transition and the observation models depend linearly (or affinely) on their arguments. When removing these assumptions, IKF can be applied on the linearized system. In this way, the obtained interlaced filter is no more linear and is known in the literature as Interlaced Extended Kalman Filter (Glielmo, Setola, & Vasca, 1999), which is largely applied to solve SLAM (Panzieri, Pascucci, & Setola, 2005).

5. DISTRIBUTED MULTI-ROBOT LOCALIZATION USING IEKF

This chapter deals with multi-robot localization using a probabilistic approach. In this framework, each robot takes advantage of the integration of measurements retrieved by teammates in a collaborative scheme. However, maintaining a single posterior pdf over all robot poses leads to an unpractical solution, due to the high computational load. To overcome such drawbacks and distribute the estimate computation over the entire team, the posterior pdf can be factorized as:

$$p(x_k \mid Z_k) = p(x_k^{(1)} \mid Z_k^{(1)}) \cdot p(x_k^{(2)} \mid Z_k^{(2)}) \cdot \ldots \cdot p(x_k^{(N)} \mid Z_k^{(N)}) \qquad (20)$$

where $x_k^{(i)}$ is the pose of the i-th robot and $Z_k^{(i)}$ the measurement history. Exploiting factorization, each robot maintains its own posterior pdf, able to model only its own uncertainty. Following particular events (i.e., when robots meet), probabilistic information can be shared. It is worth remembering that the factorial representation is only approximate; however, it shows the advantage of the estimation of the posteriors being carried out locally by each robot.

In this framework, the IEKF can be easily applied to solve multi-robot localization. The whole system is divided into N subsystems, each one devoted to estimate the pose of a single robot. In the prediction step, the kinematic model of the robot is used to produce an *a priori* estimate, without considering the interaction with the environment or the team. To integrate detection into posterior pdf, robots exchange information about their estimate. Specifically, when robot i is in the coverage area of robot j, they are able to exchange their estimated position and the related covariance matrix P, and Equations (16) and (17) are computed.

5.1 Complete System Model

The following considers a group of N independent robots, navigating in a partially known planar environment. Robots belonging to the team are supposed to have the same kinematics. In this case a unicycle forms the basis of kinematic model. Encoders represent the proprioceptive sensor, whereas exteroceptive sensors are exploited both to perceive the position and the bearing of landmarks in the environment, and to detect teammates. Information sharing in the team is achieved by limited-coverage antennas, available on each robot (see Figure 1). Note that when robots are close enough they can also compute

Figure 1. Robot 1 can communicate with robots 2 and 3 since they are closer than Rr (wireless connection range). Moreover, it is able to measure the relative position of robot 2 and of beacon A, which are closer than Rb (detection system range).

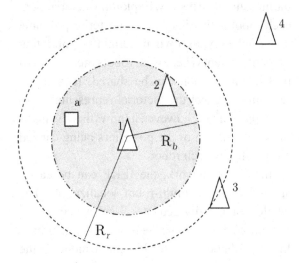

their relative distance, thus improving pose estimate.

The state x_k of the whole team at time step k is comprised of the pose of each the robot with respect to a global reference frame:

$$x_k^T = [x_k^{(1)T}, x_k^{(2)T}, \dots, x_k^{(N)T}] \tag{21}$$

Let us define the state vector for the robot i as its pose:

$$x_k^{(i)T} = [p_{x,k}^{(i)}, p_{y,k}^{(i)}, \phi_k^{(i)}]T \tag{22}$$

and the inputs as:

$$u_k^{(i)} = [\delta s_k^{(i)}, \delta\theta_k^{(i)}]^T \tag{23}$$

where $\delta s_k^{(i)}$ is the vehicle displacement and $\delta\theta_k^{(i)}$ rotation during the sampling time interval δt_k, both measured by proprioceptive sensors.

The state transition for a single robot is computed by unicycle equations:

$$x_k^{(i)} = f(x_{k-1}^{(i)}, u_{k-1}^{(i)}) = \tag{24}$$

$$= \begin{bmatrix} 1 & 0 & 0 \\ 0 & 1 & 0 \\ 0 & 0 & 1 \end{bmatrix} x_{k-1}^{(i)} + \begin{bmatrix} cos\tilde{\phi}_{k-1}^{(i)} & 0 \\ sin\tilde{\phi}_{k-1}^{(i)} & 0 \\ 0 & 1 \end{bmatrix} u_{k-1}^{(i)}$$

being $\tilde{\phi}_k^{(i)} = \phi_k^{(i)} + \delta\theta_k^{(i)}/2$ the average robot orientation during the time step interval δt_k.

Equation (24) can be rewritten in a compact form as:

$$x_k^{(i)} = x_{k-1}^{(i)} + G_{k-1}^{(i)} u_{k-1}^{(i)} \tag{25}$$

which is useful in writing the complete system model? In fact, assuming that robots move dependently, the *complete state transition model* of the multi-robot team can be formulated as:

$$\begin{bmatrix} x_k^{(1)} \\ x_k^{(2)} \\ \vdots \\ x_k^{(N)} \end{bmatrix} = \begin{bmatrix} x_{k-1}^{(1)} \\ x_{k-1}^{(2)} \\ \vdots \\ x_{k-1}^{(N)} \end{bmatrix} + \begin{bmatrix} G_{k-1}^{(1)} & 0 & \dots & 0 \\ 0 & G_{k-1}^{(2)} & \ddots & \vdots \\ \vdots & \ddots & \ddots & 0 \\ 0 & \dots & 0 & G_{k-1}^{(N)} \end{bmatrix} \begin{bmatrix} u_{k-1}^{(1)} \\ u_{k-1}^{(2)} \\ \vdots \\ u_{k-1}^{(N)} \end{bmatrix}. \tag{26}$$

The *observation model* describes both the relation between a robot pose and either a fixed landmark or another platform. Specifically, detection between robot i and robot j can be written as:

$$z_k^{(i,j)} = h^r(x_k^{(i)}, x_k^{(j)}) = \begin{bmatrix} \rho_k^{(i,j)} \\ \theta_k^{(i,j)} \end{bmatrix}$$

$$= \begin{bmatrix} \sqrt{(p_{x,k}^{(j)} - p_{x,k}^{(i)})^2 + (p_{y,k}^{(j)} - p_{y,k}^{(i)})^2} \\ atan2\left(\dfrac{p_{y,k}^{(j)} - p_{y,k}^{(i)}}{p_{x,k}^{(j)} - p_{x,k}^{(i)}}\right) - \phi_k^{(i)} \end{bmatrix} \tag{27}$$

being $\rho_k^{(i,j)}$ the distance between robot i and robot j, and $\theta_k^{(i,j)}$ the bearing of j-th robot in the reference of the i-th one.

The observation model of a landmark measurement is computed over a given map M:

$$
\begin{aligned}
z_k^{(i,j)} &= h^b(x_k^{(i)}, M) = \begin{bmatrix} \rho_k^{(i,j)} \\ \theta_k^{(i,j)} \end{bmatrix} = \\
&= \begin{bmatrix} \sqrt{(b_x^{(j)} - p_{x,k}^{(i)})^2 + (b_y^{(j)} - p_{y,k}^{(i)})^2} \\ atan2\left(\dfrac{b_y^{(j)} - p_{y,k}^{(i)}}{b_x^{(j)} - p_{x,k}^{(i)}}\right) - \phi_k^{(i)} \end{bmatrix}
\end{aligned} \tag{28}
$$

where landmarks are recorded in the map M as position tuple $(b_x^{(j)}, b_y^{(j)})$.

5.2 Single Robot Sub-System

A fully decentralized multi robot algorithm is designed exploiting IEKF. Each robot in the team runs its own localization procedure and exchanges pose information with the robots in the surrounding area. The state transition model for each sub-filter is given by:

$$
x_k^{(i)} = f(x_{k-1}^{(i)}, u_{k-1}^{(i)}) + w_k^{(i)} = x_{k-1}^{(i)} + G_k^{(i)} u_{k-1}^{(i)} + v_k^{(i)} \tag{29}
$$

being $v_k^{(i)} \in \mathcal{R}^3$ a zero mean white noise vector with covariance matrix $Q_k^{(i)}$.

Since $f(\cdot)$ is a nonlinear mapping with respect to the input, the Jacobian $J_{u,i}^f$ has to be computed to obtain the covariance $P_{k|k-1}^{(i)}$:

$$
P_{k|k-1}^{(i)} = P_{k-1|k-1}^{(i)} + J_{u,i}^f Q_k^{(u,i)} J_{u,i}^{fT} + Q_k^{(i)} \tag{30}
$$

where $Q_k^{(u,i)}$ represents the covariance matrix of the noise that affects the proprioceptive sensors. Note that the motion of a robot does not affect the motion of the other vehicles. Consequently, there is no need to compute $\tilde{Q}_k^{(i)}$.

The observations for a robot consist of m subvectors:

$$
z_k^{(i,j)} = h^b(x_k^{(i)}, \mathcal{M}) + w_k^{(i)} \tag{31}
$$

and l subvectors:

$$
z_k^{(i,j)} = h^r(x_k^{(i)}, x_k^{(j)}) + w_k^{(i)} \tag{32}
$$

where m and l are the number of landmarks and vehicles at the step k time in the range of the exteroceptive sensors of the i-th robot, respectively.

Due to the nonlinearity of the mapping, linearization is needed and the Jacobian of mapping $h^b(\cdot)$ and $h^r(\cdot)$ are used in Equations (14) and (16). When a robot detects another vehicle, the covariance update is calculated according to Equation (15), exploiting $\tilde{R}_k^{(i)}$.

Moreover, robots are able to use teammates as remote sensors. In this case, several data have to be exchanged to completely exploit the augmented perception in a probabilistic framework. Updated measurements are used to compute the innovation, while uncertainties are used to properly augment matrix $\tilde{R}_k^{(i)}$.

6. CASE STUDIES

Some simulations have been carried out using a MATLAB toolbox, including SimNav, developed by the Authors. The toolbox can simulate teams of robots navigating in an office-like environment. Some indoor environments are available in the package, however new environments can easily be added by users. SimNav is able to simulate robots having different kinematics and equipped with several sensor devices, including encoders, gyroscopes, GPS, laser and ultrasound rangefinders, and vision-based systems. It allows different path planning and navigation strategies and implements fuzzy obstacle avoidance (Balzarotti

& Ulivi, 1996). Hereafter, we present two case studies: the first one compares the behavior of the proposed procedure under different operating conditions in a free space, and the second one analyzes the performance of the proposed algorithm in an office-like environment where the number of robots in a team change and robots are used as remote sensors. In both case studies, robots working in the teams have the same kinematics and communication antennas, although they have different sensory abilities.

6.1 Free Space Navigation

In the first set of simulations consisted of an environment that is 12×12 meters in area (see Figure 5), which had 9 landmarks (small squares) and 10 robots. The sampling rate is set at 2 Hz for all robots in the team. The linear speed is 0.2 m/sec and the steering velocity is changes in a range of ± 0.2 rad/sec every 10 seconds so as to obtain random exploration (Indiveri, Aicardi, & Casalino, 2000). Gaussian white noises have been added over the inputs having a covariance of 0.1 m/s for the linear velocities and 0.1 rad/sec for the angular velocities. The initial pose of the robots is supposed to be known, so initial error is negligible. The communication range of the radio antenna is set to 5 m (R_r), while the laser rangefinders are able to compute distance within 4 m (R_b), as shown in Figure 1. These robots are equipped with Zigbee antennas because of their low transfer rate of the proposed localization protocol. During simulation, the communication channel is modeled as noisy, so data are sometimes not available, even if robots can perceive each other. In these cases, the localization filter is not able to perform the update step. We model the communication channel in accordance with the results reported in Gasparri, Panzieri, Pascucci, and Ulivi (2009).

The exteroceptive sensors measures are corrupted by white noise having a covariance of 0.22

m for the distances and 0.24 rad for the angles. The walls are not exploited for localization, although they represent constraints for the robot implementing a reactive navigation strategy. Five different operating conditions are tested:

- Single robot dead reckoning;
- Collaborative localization without beacons;
- Collaborative localization with heterogeneous team;
- Collaborative localization using robotic landmarks;
- Collaborative localization with homogenous team.

Results are reported in Figures 2, 3, and 4, where the position error (i.e., Euclidean distance) and the distribution of the bearing error are represented. The figures report the errors averaged for 10 robots; each iteration is obtained by filtering the trends with a center mean filter over 100 samples.

No cooperative localization was performed in the first trial because the exteroceptive sensors of the robots had been switched off and robots performed only dead reckoning to localize themselves. The upper curve in Figure 2 shows an error, which in 2000 steps reaches 2 meters. In Figure 4, the curve related to this simulation is the larger of the two.

The second set of simulations was performed considering only information sharing among robots. In this case, the sensing abilities were limited to distances and angular displacements of the robots within R_r. It is easy to argue that there is not any fixed reference frame; however, the drift for all estimated positions presents a slower trend than those obtained in the previous simulation.

The third simulation results were obtained by considering three robots that could sense beacons, thus creating a heterogeneous team. The presence of robots equipped with sensors able to detect

Figure 2. The free space environment including 10 robots and 9 landmarks

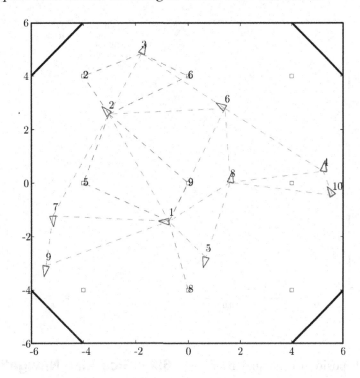

Figure 3. Trend of mean position errors (computed as a distance from the real robot) along simulation steps

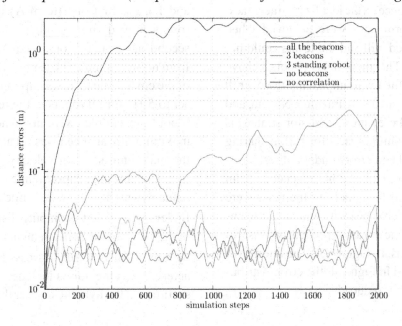

Figure 4. Distribution of angular errors. Numbers in the legend are covariances.

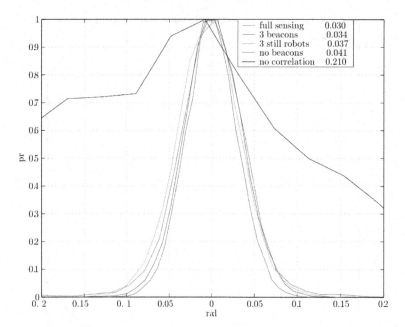

beacons reduces both position error and bearing error covariance. Indeed, the uncertainties linked to robot-beacon measurements are lower than robot-robot measurements, since, in the first case, no additional crossing terms are exchanged. Therefore, a robot that can perceive a beacon increases its confidence on the localization accuracy.

Similar performances were achieved when three robots stayed still and the teams could not detect beacons. The motionless robots behave like beacons for the team, improving the overall accuracy, as explained in Kurazume, Nagata, and Hirose (1994), where a move-and-stop strategy is adopted. In this simulation, information sharing allows robots to keep errors under a threshold.

In the last simulation, a homogeneous team was examined: all the robots in the team were able both to perceive beacons and detect teammates. The accuracy was improved because of beacon corrections. In Figure 2, the position error is under 2 cm and in Figure 4, the error distribution is considerably reduced.

6.2 Office-Like Navigation

In the second set of simulations, an office-like environment with 10 beacons (see Figure 5) was considered, along with five robots moving along the corridor. They are divided into two teams (A and B), the first one (team A) comprised of 3 robots moved from south to north, while the second one (team B), composed of 2 robots, moves from north to south. The teams meet in the center of the environment. The sampling frequency was set to 5 Hz for all robots in the teams; however, robot time samples were not synchronized. The maximum linear speed was set at 0.2 m/sec and the maximum steering velocity was ± 0.2 rad/sec. The communication range of the radio antenna was set at 3 m (R_r), while the laser rangefinders were able to compute distance within 2 m (R_b). All of the measurements were corrupted by noises. Such noises possessed the same characteristics as the noises included in the previous section and the hypothesis over the position error

Figure 5. The office like environment including 5 robots and 10 landmarks

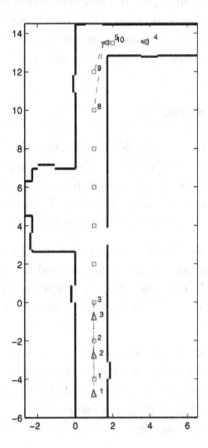

for single robots. During the simulations, only one robot in team A was able to perceive the beacons, while robots in team B were only able to detect each other's. As a consequence, at the beginning of the simulations (i.e., before rendezvous), robots in team B presented a larger value for error position in both trials. The errors were considerably reduced when the two teams meet.

In the first trial, information shared was limited to robot position, while in the second trial robots were used as virtual sensors, as proposed in Gasparri, Panzieri, and Pascucci (2008). In this case, the robot sensory system is augmented by the perceptual ability of its neighbors. As a consequence, the estimate error is considerably reduced, since more information is used to form the belief. A drawback of this approach is the communication cost since a large amount of information needs to be shared.

already formulated still held. During navigation, robots had to reach specific points, in accordance with the control algorithm proposed in Gasparri, Panzieri, and Pascucci (2008). Obstacles were avoided by means of local reactive behavior. Two different operating conditions were tested:

- Rendezvous without virtual sensors;
- Rendezvous using virtual sensors.

The results are reported in Tables 1 and 2. The tables point out the mean position errors (i.e., Euclidean distance) of the robots and their covariance before and after the rendezvous. The figures report data about the statistical information

Table 1. Rendezvous without virtual sensor

		Before Rendezvous		After Rendezvous	
		$e_r[m]$	$\sigma_e[m]$	$e_r[m]$	$\sigma_e[m]$
A	R1	0.23	0.7	0.17	0.5
	R2	0.07	0.3	0.6	0.3
	R3	0.21	0.6	0.19	0.6
B	R4	0.5	1.4	0.21	0.6
	R5	0.34	0.9	0.2	0.7

Table 2. Rendezvous using virtual sensor

		Before Rendezvous		After Rendezvous	
		$e_r[m]$	$\sigma_e[m]$	$e_r[m]$	$\sigma_e[m]$
A	R1	0.13	0.6	0.09	0.4
	R2	0.04	0.2	0.03	0.3
	R3	0.12	0.3	0.08	0.5
B	R4	0.34	0.9	0.14	0.8
	R5	0.21	0.7	0.12	0.7

7. CONCLUSION

This chapter deals with the design of a completely decentralized and distributed multi-robot localization algorithm. The localization procedure is based on nonlinear filtering. Specifically, it is an interlaced version of the EKF, able to produce accurate pose estimation for each robot of the team, thereby considerably reducing the computational load.

According to this approach, the localization problem is solved by means of a proper factorization of the posterior pdf. In this way, each robot runs single robot localization exploiting its own measurements. When robots meet, they improve the estimate accuracy by information sharing. Moreover, they consider robots in within line of sight as remote sensors. The information sharing is limited to few data so that the communication overhead can remain low. Specifically, robots exchange information about their states, their measurements, and the related uncertainties. There is no need for time synchronization since only the last pose estimate and its corresponding error covariance is shared.

Generally, this approach does not produce positioning estimates of uniform accuracy. However, it can be used even when some robots have lost their connection with the team.

In this work, we consider robots having the same kinematics and equipped with laser range-finders. The approach can easily be extended to teams of robots having different kinematics or sensory systems. Moreover, with a proper sensory system, this approach can also be applied outdoors. Simulation results exploiting different operating conditions for robots have been carried out. The absence of fixed landmarks in an environment was especially analyzed. In such a case, the algorithm can perform well regarding relative accuracy when some robots act as portable landmarks. Future work will be dedicated to further research this approach and apply the method we describe to multi-robot simultaneous localization and mapping. A strategy aimed at determining the initial pose of robots must also be considered.

REFERENCES

Antonelli, G., Arricchiello, F., Chiaverini, S., & Setola, R. (2006). Coordinated control of mobile antennas for ad hoc networks. *International Journal of Modelling. Identification and Control, 1*(1), 63–71. doi:10.1504/IJMIC.2006.008649

Balzarotti, M., & Ulivi, G. (1996). The fuzzy horizons obstacle avoidance method for mobile robots. In *Proceedings of the World Automation Conference, ISRAM 1996*. Montpellier, France: ISRAM.

Barnhard, D. H., Wimpey, B. J., & Potter, W. D. (2004). Odin and Hodur: Using Bluetooth communication for coordinated robotic search. In *Proceeding of the International Conference on Artificial Intelligence*. Las Vegas, NV: IEEE.

Dzung, D., Apneseth, C., Endersen, J., & Frey, J. E. (2005). Design and implementation of a real-time wireless sensor/actuator communication system. In *Proceedings of the 10th IEEE Conference on Emerging Technologies and Factory Automation*, (pp. 443-452). IEEE Press.

Fox, D., Burgard, W., Kruppa, H., & Thrun, S. (2000). A probabilistic approach to collaborative multi-robot localization. *Autonomous Robots, 8*(3), 325–344. doi:10.1023/A:1008937911390

Gasparri, A., Panzieri, S., & Pascucci, F. (2008). A fast conjunctive resampling particle filter for collaborative multi-robot localization. In *Proceedings of the Workshop on Formal Models and Methods for MultiRobot Systems*. AAMAS.

Gasparri, A., Panzieri, S., Pascucci, F., & Ulivi, G. (2009). An interlaced Kalman filter for sensors networks localization. *International Journal of Sensor Networks, 5*(3), 164–172. doi:10.1504/IJSNET.2009.026364

Gerkey, B. P., Vaughan, R. T., Stoy, K., Howard, A., Sukhatme, G. S., & Mataric, M. J. (2001). Most valuable player: A robot device server for distributed control. In *Proceedings of IEEE/RSJ IROS 2001,* (pp. 1226-1231). Wailea, Hawaii: IEEE Press.

Glielmo, L., Setola, R., & Vasca, F. (1999). An interlaced extended Kalman filter. *IEEE Transactions on Automatic Control, 44*(8), 1546–1549. doi:10.1109/9.780418

Indiveri, G., Aicardi, M., & Casalino, G. (2000). Robust global stabilization of an underactuated marine vehicle on a linear course by smooth time-invariant feedback. In *Proceedings of the Conference on Decision and Control, CDC 2000,* (pp. 2156-2161). Sydney, Australia: CDC.

Kahn, J. M., & Barry, J. R. (1997). Wireless infrared communication. *Proceedings of the IEEE, 85*(2). doi:10.1109/5.554222

Kurazume, R., Nagata, S., & Hirose, S. (1994). Cooperative positioning with multiple robots. [Los Alamitos, CA: IEEE Press.]. *Proceedings of IEEE ICRA, 1994,* 1250–1257.

Martinelli, A., Pont, F., & Siegwart, R. (2005). Multi-robot localization using relative observation. In *Proceedings of IEEE ICRA 2005.* Barcelona, Spain: IEEE Press.

Martinelli, A., & Siegwart, R. (2005). Observability analysis for mobile robot localization. In *Proceedings of the IEEE/RSJ IROS 2005,* (pp. 1264-1269). Edmonton, Canada: IEEE Press.

Nguyen, H. G., Pezeshkian, M. R., Raymond, M., Gupta, A., & Spector, J. M. (2003). Autonomous communication relays for tactical robots. In *Proceedings of the 11th International Conference on Advanced Robotics.* Coimbra, Portugal: IEEE.

Panzieri, S., Pascucci, F., & Setola, R. (2005). Simultaneous localization and map building algorithm for real-time applications. In *Proceedings of the 16th IFAC World Congress.* Praha, Czech Republic: IFAC.

Panzieri, S., Pascucci, F., & Setola, R. (2006). Multirobot localisation using interlaced extended Kalman filter. In *Proceedings of the IEEE/RSJ IROS 2006.* Beijing, China: IEEE Press.

Rekleitis, I. M., Dudek, G., & Milios, E. E. (2000). Multi-robot collaboration for robust exploration. [San Francisco, CA: IEEE Press.]. *Proceedings of the IEEE ICRA, 2000,* 3164–3169.

Rekleitis, I. M., Dudek, G., & Milios, E. E. (2002). Multi-robot cooperative localization: A study of trade-offs between efficiency and accuracy. In *Proceedings of the IEEE/RSJ IROS 2002,* (pp. 2690-2696). Lausanne, Switzerland: IEEE Press.

Rekleitis, I. M., Dudek, G., & Milios, E. E. (2003). Probabilistic cooperative localization and mapping in practice. [Taipei, Taiwan: IEEE Press.]. *Proceedings of the IEEE ICRA, 2003,* 1907–1912.

Ristic, B., Arulampalam, S., & Gordon, N. (2004). *Beyond the Kalman filter.* London, UK: Artech House.

Roumeliotis, S. I., & Bekey, G. A. (2000). Synergetic localization for groups of mobile robots. [Sidney, Australia: IEEE Press.]. *Proceedings of the IEEE CDC, 2000,* 3477–3482.

Roumeliotis, S. I., & Bekey, G. A. (2002). Distribuited multi-robot localization. *IEEE Transactions on Robotics and Automation, 18*(5), 781–795. doi:10.1109/TRA.2002.803461

Song, B., Tian, G., Li, G., Zhou, F., & Liu, D. (2011). *ZigBee based wireless sensor networks for service robot intelligent space.* Paper presented at the International Conference on Information Science and Technology. Nanjing, China.

Trawny, N., Roumeliotis, S. I., & Giannakis, G. B. (2005). Cooperative multi-robot localization under communication constraints. In *Proceedings of the IEEE ICRA 2005.* Kobe, Japan: IEEE Press.

Zhuang, Y., Gu, M. W., Wang, W., & Yu, H. Y. (2010). Multi-robot cooperative localization based on autonomous motion state estimation and laser data interaction. *Science China –. Information Sciences, 53*(11), 2240–2250.

Chapter 2
Indoor Surveillance Application using Wireless Robots and Sensor Networks:
Coordination and Path Planning

Anis Koubaa
Al-Imam Mohamed bin Saud University, Saudi Arabia
& Polytechnic Institute of Porto (ISEP/IPP), Portugal

Sahar Trigui
National School of Engineering, Tunisia

Imen Chaari
National School of Engineering, Tunisia

ABSTRACT

Mobile robots and Wireless Sensor Networks (WSNs) are enabling technologies of ubiquitous and pervasive applications. Surveillance is one typical example of such applications for which the literature proposes several solutions using mobile robots and/or WSNs. However, robotics and WSNs have mostly been considered as separate research fields, and little work has investigated the marriage of these two technologies. In this chapter, the authors propose an indoor surveillance application, SURV-TRACK, which controls a team of multiple cooperative robots supported by a WSN infrastructure. They propose a system model for SURV-TRACK to demonstrate how robots and WSNs can complement each other to efficiently accomplish the surveillance task in a distributed manner. Furthermore, the authors investigate two typical underlying problems: (1) Multi-Robot Task Allocation (MRTA) for target tracking and capturing and (2) robot path planning. The novelty of the solutions lies in incorporating a WSN in the problems' models. The authors believe that this work advances the literature by demonstrating a concrete ubiquitous application that couples robotic and WSNs and proposes new solutions for path planning and MRTA problems.

DOI: 10.4018/978-1-4666-2658-4.ch002

Copyright © 2013, IGI Global. Copying or distributing in print or electronic forms without written permission of IGI Global is prohibited.

1. INTRODUCTION

Robotics and Wireless Sensor Networks (WSNs) are two key technologies that integrate physical processes and phenomena with computation, and contribute to solving real-world problems in several areas such as industrial manufacturing, precision agriculture, smart homes and buildings, health care systems, intelligent transportation systems, military applications, and unknown environment exploration, etc. This technological development has naturally given birth to the paradigm of Cyber-Physical Systems where computation is no longer decoupled from its surrounding environment.

Although robotics and WSNs are typically designed to be deployed in similar contexts, little work has focused on the marriage of these two technologies to solve particular problems. In fact, robotic solutions basically rely on autonomous robots or a collaborative group of ad-hoc (i.e. decentralized) robots to accomplish certain tasks. In these typical robot applications, the intelligence, i.e. processing and data analysis, is embedded in the robot. Thus, the cooperative work between mobile robots in the field requires distributed and ah-hoc communications between the different mobile robots, which might increase the solution's complexity because of their high mobility. The cooperation becomes even more difficult if the real-time guarantee is a major requirement of the robotic distributed application.

Another alternative for mobile robotic applications consists in taking advantage from the Wireless Sensor Network paradigm as a complementary infrastructure allowing for acquiring additional knowledge about the environment. The idea is to feed the robots with additional cyber-physical information that can be carried out by the WSN, which provides substantial support to the robotic system to accomplish its missions. As a consequence, the performance will naturally improve. As an illustrative example, a WSN can afford localization services for robots to support its navigation in an indoor environment, in a similar fashion to what a GPS system can provide in an outdoor environment. Furthermore, a WSN deployed in the environment would effectively be able to inform the robotic system about crucial events and their locations so that they can act accordingly. In a nutshell, the pervasive and ubiquitous nature of the WSN deployment is of paramount importance to improve the quality of robotic system solutions.

In this chapter, we illustrate our vision by proposing a multi-robot indoor surveillance system, called SURV-TRACK, which takes advantage of WSN-based services to support the operation of a team of collaborative robots. SURV-TRACK is designed to track and capture intruders within a defined area. This application could be instantiated for the surveillance and real-time monitoring of large areas such as shopping centers, manufacturing plants, banks, and critical areas.

Surveillance, using a multi-robot team with the support of a WSN, encompasses several underlying challenges such as inter-robot and robot-WSN collaborations, path planning, target detection and tracking, navigation, mapping, etc. The key challenge is to devise solutions that take into account the efficient interaction between robots and the WSN infrastructure.

In this chapter, we first propose a model for the multi-robot indoor surveillance application supported by a wireless sensor network. Then, we tackle two different underlying issues in the SURV-TRACK application and propose solutions with respect to the following features: (1) multi robot task allocation for target capturing and (2) path planning.

The remainder of this chapter is organized as follows: Section 2 provides an overview of WSNs and mobile robot technologies. In Section 3, we describe the SURV-TRACK application and its network architecture model. In Section 4, we present the problem of multi-robot task allocation for target capturing in SURV-TRACK. Following this, we propose three solutions and evaluate their performance. Section 5, then, goes on to present the robot path planning problem using WSNs, show how to solve it with a new hybrid Ant Colony Optimization (ACO) and Genetic Algorithms (GA) approach and evaluate its performance. Finally, Section 6 concludes the chapter and presents insights into our future work.

2. BACKGROUND

In the following section, we present a brief overview of WSNs and mobile cooperative robots. The objective is to introduce these concepts and provide an adequate background on these technologies, without being comprehensive, because of the high density of research work in these two areas.

2.1 Wireless Sensor Networks

In the last decade, WSNs (Akyildiz, et al., 2002) have significantly contributed to the technological development of a vast array of ubiquitous and pervasive applications that were not possible before the birth of this technology. A sample of these applications includes health care monitoring (Jung, et al., 2009), precision agriculture (Prakashgoud, et al., 2011), military applications (Sang Hyuk, et al., 2009), structural health monitoring (Wijetunge, et al., 2010), home automation (Jian-She, et al., 2008), etc. WSNs represent an enabling technology for ambient intelligence and paved the way to revolutionize the design of emerging embedded systems leading to the broader concept of cyber-physical systems where computation and the surrounding environment are tightly coupled.

In general, a WSN consists of a set of sensor nodes deployed in a monitored environment, where they interact with the physical world. The role of sensor nodes is collecting specified sensory data required for the monitoring of the region of interest. Based on the collaborative work of the sensor nodes, data is delivered in a hop-by-hop fashion from the data source to the destination, which is typically a control station (also referred to as data sink). Roughly, sensor nodes can be seen as small computing systems that play two basic roles. First, a sensor node acts as a *data source* that captures physical data/phenomena (e.g. temperature, light, humidity, pressure) using its sensing capabilities, performs required buffering and processing, and transmits data packets through its low-power wireless interface. There are several types of advanced sensors such as seismic for earth motion, magnetic for acceleration and speed, thermal for temperature, acoustic for sound, medical sensors (EGC, heart rate, etc.), laser range sensors, etc. Second, it acts as a *data router* that is responsible for forwarding packets from neighbor-to-neighbor until they reach their destination (i.e. control station). The control station receives data, performs required processing, analyzes, and understands collected data to take/recommend adequate actions to end users.

WSNs encompass several challenges in terms of communication and coordination due to their inherent resource-constraints characteristics, including low power, restricted processing speed, small storage capacity, and low bandwidth (Akyildiz, et al., 2002). For instance, the TelosB mote (2011) is dotted with a random-access memory of 10 KB and a flash memory of 48 KB, a maximum data rate of 250 kbps, and a 16-bit RISC CPU for processing. For this reason, the design of sensor network systems is mainly driven by the major requirement of reducing energy consumption to the maximum to maintain network survivability as long as possible. Solutions proposing MAC protocols (Demirkol, et al., 2006), routing protocols (Akaya & Younis, 2005), topology control mechanisms (Jardosh & Ranjan, 2008), and others were fundamentally designed to achieve the energy-efficiency requirement.

Communication in WSNs is characterized by three features that make it different from communication in traditional wireless ad-hoc networks. First of all, they are *data-centric*. In data-centric networks, importance is given to data rather than to the devices where that data is produced. Data from multiple sources related to the same physical phenomenon need to be aggregated and sent to the control station. Secondly, they are *large-scale*. Typical WSN applications require the deployment of large numbers of nodes. Consequently, communication protocols need to be adequate for networks with large numbers of nodes and introduce a small communication overhead. Thirdly, *location-based routing* is used to forward packets. The identification of a node within a WSN should be based on its geographic position in the controlled area and not on a logical address. This kind of identification better fits the data-centric and large-scale properties of WSNs. In location-based routing, a WSN node is only required to know the position of its immediate neighbors, without having to maintain a large routing table based on logical addresses.

2.1.1 Medium Access Protocols

Communication protocols have received great attention in the literature, and a large number of solutions have been proposed to cope with WSN requirements, both at the levels of the MAC sub-layer and routing. With respect to MAC protocols (Demirkol, et al., 2006), they are roughly classified into (1) *scheduling-based* protocols, (2) *contention-based* protocols, and (3) *hybrid* approaches.

As for scheduling-based protocols, they rely on Time Division Multiple Access (TDMA) and avoid collisions by means of a centralized scheduling algorithm that determines the time at which a node can start its transmission. Examples of scheduling-based approaches include LEACH (Heinzelman, et al., 2000), one of the earliest MAC and routing protocols proposed for sensor networks, and Implicit-EDF (Crenshaw, et al., 2007) etc. The advantage of scheduling-based protocols is that they are energy efficient because they eliminate the collision problem and enable real-time Quality-of-Service (QoS). This comes, of course, at the expense of a greater complexity

because of the need for node synchronization, which limits the scalability of such techniques.

In the literature, most of the sensor network MAC solutions are contention-based, where nodes compete to gain access to the shared medium. Usually, collisions occur when multiple transmissions occur at the same time. These protocols rely on modified versions of the CSMA/CA mechanism to avoid collisions. To this end, a vast array of protocols has been proposed including B-MAC (Polastre, et al., 2004), S-MAC (Ye, et al., 2002, 2004), SIFT (Jamieson, et al., 2006), WiseMAC (El-Hoiydi & Decotignie, 2004), etc. Contention-based algorithms are preferred in WSNs because of their high flexibility and low communication complexity as compared to scheduling-based solutions. However, this comes at the cost of reduced throughput due to collisions and increased communication overhead due to collision avoidance mechanisms.

A third category of MAC protocols adopted is the hybrid approach, which combines both scheduling-based and contention-based mechanisms. Roughly, these protocols use TDMA to specify a superframe structure organized as time slots. Some time slots are exclusively allocated to nodes with real-time and QoS requirements, whereas other time slots are shared by several nodes using contention-based protocols. Apart from proprietary solutions such as RT-Link (Rowe, et al., 2006), TRAMA (Rajendran, et al., 2003), Adaptive Slot Assignment Protocol (ASAP) (Gobriel, et al., 2009), the most famous example of standardized solutions is the IEEE 802.15.4 standard protocol that defines both the physical layer and MAC sub-layer mechanisms from Low-Rate Wireless Local Area Networks (LR-WPANs). The IEEE 802.15.4 Medium Access Control (MAC) protocol provides very low duty cycles (from 100% to 0.1%), which is particularly useful for WSN applications, where energy consumption and network lifetime are main concerns. Additionally, the IEEE 802.15.4 protocol may also provide timeliness guarantees by using its Guaranteed-Time Slot (GTS) feature, which is quite attractive for time-sensitive WSNs. The IEEE 802.15.4 standard also provides a non-slotted version for purely contention-based medium access based on CSMA/CA. It results that IEEE 802.15.4 is highly flexible and has the ability to adapt to different applications requirements, turning it an interesting solution to WSNs. For that purpose, we also use the IEEE 802.15.4 protocol as the underlying MAC protocol for sensor networks in our surveillance application, as will be elaborated in Section 3. Interested readers can refer to Koubaa et al. (2007) for additional details about the IEEE 802.15.4 protocol.

2.1.2 Network Layer and Routing Protocols

As with the MAC protocol, network layer and routing protocols in WSNs have gained a lot of interest in the research community (Akaya & Younis, 2005). Initially, several proprietary attempts were proposed to specify routing protocols for WSNs while meeting their stringent energy-efficiency and resource scarcity requirements. In Akaya and Younis (2005), routing protocols were classified into four categories: (1) *data-centric* protocols, where the sink requests data based on certain attributes, known as attribute-based addressing. For example, a query may request all sensor nodes whose temperature is greater than a certain threshold. Several protocols were proposed, including Directed Diffusion (Intanagonwiwat, et al., 2000) based on flooding, Sensor Protocols for Information via Negotiation (SPIN) (Heinzelman, et al., 1999), energy-aware routing (Younis, et al., 2002), etc.; (2) *Hierarchical* routing, which consists of dividing the network into different clusters, where a central node (called cluster head) acts as a data aggregator then forwards fused data from one cluster-head to another until reaching the sink. Typical examples include the LEACH (Heinzelman, et al., 2000), TEEN (Manjeshwar & Agrawal, 2001), and APTEEN

protocols (Manjeshwar & Agrawal, 2002); (3) *location-based* routing, which relies on location information to forward packets to the destination such as Geographic Adaptive Fidelity (GAF) (Xu, et al., 2001) and Geographic and Energy-Aware Routing (GEAR) (Yu, et al., 2001); (4) *Network flow and QoS-aware* protocols, which consider QoS metrics, such as the delay, in setting the path from source to destination (e.g. SPEED [He, et al., 2003]).

In the recent years, routing protocol designs have not exclusively considered energy-efficiency as a main concern, but have increasingly studied interoperability with the Internet, thus opening the horizon towards the Internet-of-Things, where physical events and objects also form part of the Internet by means of a sensing infrastructure. This trend has given rise to 6LoWPAN (Montenegro, et al., 2007), which is a milestone protocol that bridges the gap between low-power devices and the IP world. It is an IP-based technology for Low-Power Wireless Personal Area Networks (LoWPANs), including WSNs, that combines the IEEE 802.15.4 and IPv6 protocols. This integration provides a new dimension in WSN design as it makes them fully interoperable with the Internet.

Since the specification of 6LoWPAN, routing in LoWPANs has been a key issue that has attracted significant effort. Indeed, much work towards specifying an efficient routing protocol for 6LoWPAN-compliant networks has been carried out, including Hydro (Tavakoli, 2009), Hilow (Kim, et al., 2005), and Dymo-low (Kim, et al., 2007). All of these proprietary solutions have not gained a lot of ground in the research community; as a consequence, the specification of standard routing protocols for LoWPANs becomes necessary. The IETF ROLL working group addressed this need and has proposed a routing protocol, referred to as RPL (Routing Protocol for Low power and lossy networks) (Winter & Thubert, 2010), which is currently the main candidate to act as the standard routing protocol for 6LoWPAN networks. A key feature of RPL is that it is designed for networks with lossy links, which are those exposed to high Packet Error Rate (PER) and link outages, common to WSNs. RPL is a Distance-Vector (DV) and a source routing protocol that is designed to operate on top of several link layer mechanisms including the IEEE 802.15.4 PHY and MAC layers. In particular, RPL considers two main features: (*1*) it has low data rate of less than 250 kbps, and (*2*) its communications are prone to high error rates, which results in low data throughput. For these reasons, the RPL protocol was designed to be highly adaptive to network conditions and to provide alternate routes, whenever default routes are inaccessible.

2.2 Mobile and Cooperative Robots

Robotics represent one of the oldest, yet still very active research areas in computer science and electrical engineering, continuously evolving with technology and emerging applications. Robots are currently becoming more and more tightly integrated in our lives and are enabling applications in several areas, including industrial manufacturing (Wu, et al., 2010; Sohn & Kim, 2008), medical applications (Feng, et al., 2011), search and rescue (Stopforth, et al., 2008), area exploration (Nagaoka, et al., 2009; Mita, et al., 2004), as well as others. With the increasing complexity of tasks that robots are supposed to perform, the standalone operation of one robot becomes insufficient. This raises the need to optimize the cooperation and coordination among numerous mobile robots to accomplish a common global mission. Cooperative robots, also referred to as networked robots, are typically equipped with wireless network interfaces to communicate with each other. Currently, several commercially available platforms such as Wifibot Lab robots (Wifibot, 2011), CoroWare CoroBot (CoroWare, 2011), Pioneer 3-DX (Mobile Robots, 2011), among others, possess IEEE 802.11 wireless capabilities, allowing them to communicate with each other and function in an interoperable

manner with the IP world. The main motivation of robot communications is to allow them to accomplish a common mission in a fully distributed and coordinated manner.

2.2.1 Mobile Robots' Communications

The communication architecture of mobile robots can be shaped into different forms: (1) a *star topology* is where, for example, robots must interact with each other by transmitting data through an access point. In this topology, robots cannot communicate directly from one to another. Instead, they must transmit their data through an access point to forward data among themselves. One limitation of this approach is that the zone coverage will be limited to the range of the access point. This problem can be bypassed by connecting several access points together with cables, thus extending the wireless coverage, (2) a *mesh topology* is where robots communicate (directly) with each other in an ad-hoc and distributed manner. One possible configuration is where all robots are fully meshed and have direct access to each other. In this case, no routing algorithm is needed as all robots are in the same radio range. This is the common architecture of swarm robotics (Yogeswaran & Ponnambalam, 2009). When robots are not in the same radio range, ad-hoc routing protocols must be used to ensure network self-organization, allowing it to maintain connectivity and communications among the robots.

Several routing protocols are proposed for ad-hoc networks which are commonly classified into two categories (Akkaya & Younis, 2005; Abolhasan, et al., 2004):

1. In *reactive* approaches, also called *on-demand* routing approaches, the routing protocol generates routes only to desired destinations in response to requests issued by nodes willing to transmit data to a certain destination. These protocols are adequate for networks with low to medium traffic

loads. Illustrative examples of on-demand protocols include Ad-Hoc On demand Distance Vector (AODV) (Perkins, et al., 2003), Dynamic Source Routing (DSR) (Johnson, et al., 2003), and Feasible Label Routing (FLR) (Rangarajan & Garcia, 2004). The AODV protocol is a distance-vector routing protocol. It is based on a Route Request mechanism that searches for a route towards the destination and a Route Reply mechanism to establish the discovered route between the source and the destination. It uses the destination sequence numbers to identify the most recent route. On the other hand, DSR is another simple, yet efficient, reactive protocol commonly used in Mobile Ad-Hoc Networks. It relies on the source routing paradigm rather than the hop-by-hop routing paradigm as with AODV. It was designed to reduce bandwidth waste due to control packet overhead in ad-hoc networks by avoiding the periodic update messages needed for updating routing tables, as in table-driven approaches. It is based on two processes: the route discovery process by which the source finds a path towards the destination, and route maintenance to maintain routes. The difference with other reactive protocols is that it does not send periodic beacons which results in reducing control packet overhead.

2. *Proactive* approaches, also called *table-driven approaches*, continuously evaluate routes to ensure the reachability of all nodes and maintain consistent and updated route states in the network. Proactive approaches present the advantage of building and maintaining routes before their use, unlike reactive approaches. The most commonly used proactive routing protocols are Optimized Link State Routing (OLSR) (Clausen & Jacquet, 2003) and Destination Sequenced Distance Vector Routing protocol (DSDV) (Perkins & Bhagwat, 1994). OLSR is a link state routing protocol that relies on the dissemination of the link state information between nodes in the network. OLSR reduces the overhead restricting the number of nodes that re-broadcast link state information. In OLSR, each node periodically sends Hello messages to discover 2-hop neighbor information and launches a distribution selection of relay points (called multipoint relays MPRs) that will form the path. On the other hand, DSDV is a table-driven protocol based on the traditional Bellman-Ford algorithm. Routing information consists of the next hop towards the destination, the cost metric of the path, and the destination sequence number, initiated by the destination. These sequence numbers allows for distinguishing active (fresh) routes from those that are outdated and stale, in addition to avoiding route loops.

Discussion: The design of ad-hoc mobile networks encompasses several challenges, mainly in terms of scalability and mobility. These two requirements directly impact the choice of routing protocols used in applications, as it is crucial to control routing protocol overhead in large and mobile deployments such those related to mobile and swarm robotics.

Proactive protocols need to store a routing table for each node in the network. With small size and networks that slowly change their network topology, this technique can be beneficial. Increasing the number of nodes leads to an increased overhead that can affect protocol scalability. In contrast, reactive protocols are meant to be more scalable for large-scale and quickly changing topologies as they do not need to maintain routing tables for long time (Wang, et al., 2005). In fact, it has been argued in Hong et al. (2002) that proactive and reactive protocols scale up to a certain degree. For proactive protocols, the sizes of routing tables grow more than linearly when network size increases, resulting in an overly congested channel and blocked data traffic. Reactive protocols, on the other hand, do not keep their routing tables updated. Instead, they manage to find a route only when a node wants to send a packet. The downside is that it induces a huge amount of flooding packets in large-scale networks, thus limiting its efficiency.

Mobility management is another critical requirement of mobile cooperative robots. It is meant to adapt network topology (and routes) to network changes. In Hong et al. (2002), the authors conclude that the performance of both reactive and proactive approaches is significantly impacted by frequent mobility, but reactive protocols are more efficient than proactive protocols in high mobility scenarios as they can still operate with lower performance, in contrast to proactive approaches which may cease working.

2.2.2 Challenges in Cooperative Robotics

The area of robotics is quite large and faces many challenges related to path planning, task allocation, localization and mapping (Carlone, et al., 2011), motion coordination (Flickinger, 2007), and perception (Grigorescu, et al., 2011), etc. In the following, we provide a brief introduction to a couple of problems.

Path planning has triggered a great deal of research in mobile robots (Jan, et al., 2008; Liu, et al., 2011). This area of research, is also known as robot navigation or motion planning from a point A to a point B. The literature presents many path-planning algorithms. Although the objective of these algorithms is to find the shortest path between two nodes A and B in a particular environment, there are several algorithms based on different approaches to find a solution to this complex problem. The complexity of algorithms depends on the underlying techniques, on other external parameters including the accuracy of the map, and the number of obstacles. A literature review of path planning approaches is presented in Section 5. Multi-Robot Task Allocation (MRTA)

represents another well-studied category of robotic problems. The problem roughly consists of finding an optimal allocation of tasks among several robots to reduce the mission cost to a minimum. In this chapter, this problem is investigated and a literature review of the main approaches is presented in Section 4.

Localization, mapping and exploration are other key issues that have attracted a lot of attention since 1986 (Durrant-Whyte & Bailey, 2006; Bailey & Durrant-Whyte, 2006). These problems are usually referred to as Simultaneous Localization and Mapping (SLAM), which is a mechanism that robots use to incrementally build up a consistent map of their environment without having a prior knowledge while simultaneously determining and keeping track of their location within this constructed map. Solutions to this problem represent important contributions to the robotic community as they make robots really autonomous. Solutions to SLAM are based on probabilistic methods including Extended Kalman Filter (EKF-SLAM) (Huang & Dissanayake, 2007) and Rao-Blackwellized Particle Filters (FastSLAM) (Grisetti, et al., 2006). The main idea is to collect environmental measurements affected by noise from special sensors including laser range finders, Light Detection And Ranging (LIDAR), sonar sensors, and video cameras. Filtering out noise is very important because reducing noise permits one to more accurately determine information about an environment's layout and location. Interested readers can read (Durrant-Whyte & Bailey, 2006; Bailey & Durrant-Whyte, 2006) for a tutorial on SLAM and its applications.

3. THE SURV-TRACK APPLICATION

3.1 System Model

This section presents an overview of the system model of the SURV-TRACK application and the underlying challenges to be addressed.

The objective of the SURV-RTRACK application is to provide an efficient monitoring service for the surveillance of a certain geographical region (office, building floor, industrial plants...) using a cooperative team of mobile robots and a fixed network infrastructure of sensor nodes.

Figure 1 illustrates the model of SURV-TRACK application, which consists of three main components:

- **Mobile Robots:** Also referred to as pursuers, which are responsible for following and catching intruders based on control commands sent by the central control station or from data provided by the WSN, depending on the collaboration strategy (refer to Section 4). Mobile robots are assumed to have wireless communication capabilities, such as WiFi to communicate with each other and with other components. Wifibot Lab and Wifibot 4G are two examples of wireless mobile robots (Wifibot, 2011).
- **A Static Wireless Sensor Network:** Deployed in the area in order to collaborate with the robots to accomplish the whole mission. The sensor nodes are responsible for detecting intruders, locating mobile robots, and supporting their missions.
- **A Central Control Station:** Represents a data collection center used to remotely monitor and control the area of interest. It typically plays two roles: First, it collects data from the mobile robots and the WSN to obtain an updated system status (robot locations, sensor events, etc.). Second, it remotely controls and commands the robots and the WSN, based on the available information.

The interaction between these three components depends on the coordination strategy. However, in our model, the WSN should be active in the sense that it contributes to the intelligence of the entire system. Unlike the case of autonomous robotic systems, where intelligence is embedded in the robots, one particular goal of SURV-TRACK is to migrate the intelligence—or part of it—to the WSN to help make the correct decision and build efficient plans. This issue will be presented in detail in Section 4 and Section 5.

Figure 1. The SURV-TRACK application model

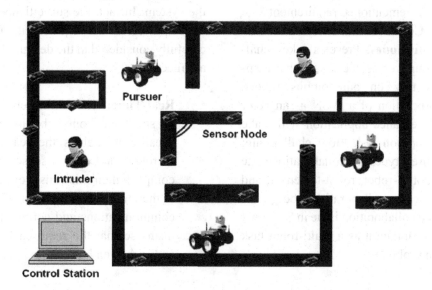

Roughly, once an intruder is detected by a subset of wireless sensor nodes, an alert will be sent to trigger the tracking and capturing actions of the intruder.

The design of the SURV-TRACK application raises several challenges. There are typically three underlying questions that should be answered in the SURV-TRACK framework: (1) How should the system components effectively communicate with each other? (2) How can the tasks be efficiently subdivided among the different actors (mobile robots and sensor nodes) of the system? (3) How can a mobile robot plan its path to accomplish its mission?

These three questions can be then summarized in the following three issues to be addressed:

- **Network Architecture:** Central to devising robust, reliable, and Quality-of-Service (QoS) aware communication services that ensure efficient interaction among the different system components, namely, the sensor nodes, mobile robots, and the control station. The network architecture must consider the heterogeneity of resources of the different actors since sensor nodes are more constrained than mobile robots and the control station. In the next subsection, we propose a network architecture that satisfies the aforementioned requirements.

- **Effective Collaboration between Robots and Sensor Nodes:** Presents a key challenge in our work since most robotic applications rely on autonomous robots. The incorporation of a WSN as an actor in the surveillance application will make the collaboration task more challenging; indeed, three types of collaboration take place; robot-to-robot, robot-to-sensor, and sensor-to-sensor collaboration models. We address the collaboration issue in Section 4 and we instantiate it as a multi-robot task allocation problem.

- **Robot Path Planning using a WSN:** In surveillance applications, a robot needs to know how to plan its path to a target location based on the available information. In contrast to autonomous robot systems, where the robot relies on an a priori knowledge of the environment (e.g. grid map), in SURV-TRACK, this restrictive assumption is not admitted. Instead, the location-aware wireless sensor nodes are considered as landmarks that feed the mobile robot with required information to find its path to the target. This issue is presented in details in Section 5.

3.2 Design Considerations of the System Architecture

In this section, we present the design considerations to be taken into account in the SURV-TRACK surveillance application. We do not describe a concrete prototype of the application, but we highlight the high-level features and requirements needed to design an efficient architectural design for SURV-TRACK in terms of networking and application issues, and we provide insights on possible solutions.

The network architecture of the SURV-TRACK model plays a central role in the performance of the system. In fact, the surveillance application encompasses crucial requirements that must be carefully considered in the design of the network architecture:

- **Real-Time:** The end-to-end delay must be short and bounded. This is important because it enables the real-time control of robots and allows sensor nodes to accomplish their missions more efficiently. It is therefore essential to provide real-time communications and coordination mechanisms so that the required levels of QoS and end-to-end delays are met.

- **Energy Efficiency:** This requirement is particularly crucial for battery-powered sensor nodes, although it is also necessary for mobile robots. In fact, communications are known to be the main cause of energy dissipation for sensor networks.

- **Reliability:** It is central to provide a reliable service for critical messages such as alarms when an intruder is detected. Failing to do this, the surveillance application will be exposed to the non- detection of intruders, which compromises system performance. The reliability issue becomes even more challenging with wireless communications, which are known to be very unreliable and prone to errors (Baccour, et al., 2012).

- **Scalability:** The surveillance application may comprise a large number of sensor nodes and mobile robots, and the complexity greatly increases with the network size and the extent of the monitored area. Therefore, the network architecture must

be scalable and able to meet real-time, the energy, and reliability requirements inherent in large-scale deployments.

- **Self-Organization:** Mobile robots and sensor nodes must be self-organizing to cope with the dynamicity of the system as the robots move within the monitored area. Each mobile robot should be able to automatically discover other peers, identify their services and capabilities, and announce its own services and capabilities to others.

In what follows, we describe the main characteristics of the network model of the SURV-TRACK application and we discuss the potential standardized technologies that fit the application. To meet the aforementioned requirements, we propose a two-tier architecture for the SURV-TRACK application, as depicted in Figure 2.

The Tier-1 Network: Represents a basic WSN, which ensures the communication between the sensor nodes themselves, and between sensor

Figure 2. SURV-TRACK two-tier network architecture

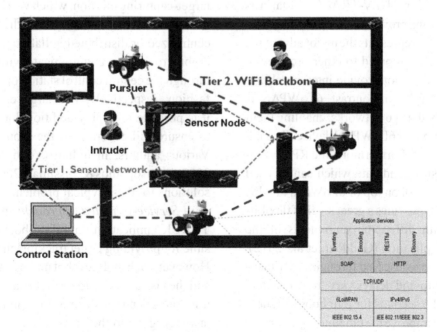

nodes, mobile robots and the control station. To cope with the energy efficiency issue, the choice of adequate communication technologies is of paramount importance. Specifically, the IEEE 802.15.4/6LoWPAN is a promising protocol stack candidate for the WSN tier characterized by its low-power radio features and short transmission range (10-30 m) that is suitable for indoor environments. This protocol presents enough flexibility to meet different requirements of the SURV-TRACK application. Indeed, for small-scale networks, the beacon-enabled mode of IEEE 802.15.4 can be used effectively. In fact, it offers a means to provide real-time QoS through its GTS mechanism, which is critical for sending out alarms (intruder detection) in a timely fashion. In large-scale networks, synchronization is harder to deploy. Consequently, the non-beacon enabled mode can be used to allow a mesh self-organizing network. QoS can be achieved by adopting priority levels to differentiate traffic in the CSMA/CA protocol. On the other hand, the recently emerging 6LoWPAN protocol represents a de-facto standard for the Network Layer of LR-WPANs because it is designed to operate on top of the IEEE 802.15.4 MAC protocol. The advantage of using 6LoWPAN in SURV-TRACK is that it is energy efficient and provides interoperability. The latter requirement represents the major advantage of 6LoWPAN as compared to other proprietary solutions, which do not provide interoperability with other networks. In contrast, 6LoWPAN is interoperable with IPv6 networks, enabling them to control and monitor 6LoWPAN devices through the legacy Internet. Furthermore, the RPL protocol is a promising candidate which can be used as a routing protocol on top of 6LoWPAN in this tier. In fact, RPL provides a great level of flexibility to deal with different requirements of the surveillance application (Gaddour, et al., 2012). However, we have observed through experimentation that end-to-end delays vary from 1 second per hop up to 2.5 seconds for four hops. As such, there is room to improve the timeliness guarantees

of this protocol to cope with the requirements of real-time applications such as SURV-TRACK. In our architecture, we specify a tier-2 backbone network that provides more powerful capabilities in terms of QoS, scalability, and reliability, which is presented next.

The Tier 2 Network: A wireless network that can potentially act as a backbone for the underlying WSN. This tier provides another alternative for mobile robots to communicate among each other or with the control station through a more powerful backbone. This is important because it contributes to extending the scalability of the network and consolidating its reliability. Even though the architecture is independent from the communication technology, the IEEE 802.11b standard is highly suitable in this tier for several reasons: (*1*) it is a widely adopted and cost effective technology, (*2*) it offers a high bandwidth that reaches 11 Mbps, and has a longer range of up to 100 meters, (*3*) it is integrated in typical commercially-available wireless robots (e.g. Wifibot Lab [Wifibot, 2011]) and computers. It should be noted that the communication paradigm between mobile robots and the control station depends on the collaborative strategy adopted for the target-capturing mission, which will be discussed in Section 4. Depending on whether to adopt a centralized or distributed collaboration strategy, mobile robots can communicate with each other through the central control station or in an ad-hoc fashion, respectively. For large environments, it is possible to deploy additional access points to ensure full coverage of the monitored areas. Various authors, including (Youngseok, et al., 2002; Koutsopoulos & Tassiulas, 2007), propose solutions for these typical problems.

A Service-Oriented Approach: In the SURV-TRACK application, mobile robots are responsible for providing target tracking and capturing. However, each mobile robot makes part of a team, and the task assigned to a robot in a collaborative mission depends on the mission nature and on the tasks assigned to the other robots teammates. It is therefore important for active robots exchange information about the services they can provide as well as continuously expressing their capabilities to perform any specific mission. In other words, it is vital to ensure a scalable and seamless service deployment in the context of surveillance applications.

A Service-Oriented Architecture (SOA) is a commonly used approach to achieve this objective. In fact, it represents a core mechanism for enabling service deployment on the Internet, which has been subsequently adapted to WSNs and robotics (Rouached, 2011) for improved service deployment. The typical "Find-Bind-Execute" paradigm can be used in the context of SURV-TRACK. Mobile robots and wireless sensor nodes represent *service providers* that register their services in a public registry. Service providers must report their identities, resources and capabilities (memory, battery level, bandwidth, etc.), as well as the list of services they can provide. This registry is used by a consumer, namely the control station, to obtain information about available services in order to plan for missions, including assigning tasks adequately to active robots. The mobile robots and sensor nodes might also play the role of consumers as they may want to acquire information about other components of the system in case they need to make local decisions like, for instance, in a distributed approach (see Section 4).

There are roughly two possible approaches to implementing SOA in SURV-TRACK: (1) the Web Services approach, (2) the Representational State Transfer (REST) approach (Fielding, 2000). In what follows, we give a brief overview of these two approaches and discuss their advantages and drawbacks.

Web Services technology is the leading implementation of SOA on the Web and can be used to specify service description, publication, and discovery. In the literature, several research papers propose using Web Services in robotic applications (Kim, et al., 2005; Levine & Vickers, 2001). In the context of SURV-TRACK, a mobile robot defines

and exposes its services as Web Services using Simple Object Access Protocol (SOAP) to encode exchanged messages in a common XML format and the Web Services Description Language (WSDL) for service description. Consumers, i.e. the control station, use the Universal Description Discovery and Integration (UDDI) protocol as a service discovery protocol to look for the services provided by mobile robots. The target capturing mission can then be considered as a Web service composition and the coordination between robots can be seen as composition of composite Web services (Mokarizadeh, et al., 2009). While this approach can be supported by typical mobile robots, it faces serious limitations with respect to its implementation over WSNs, mainly due to the high cost of using SOAP for message flow and the heavyweight nature of the XML data representation. The Device Profile for Web Service (DPWS) (Moritz, et al., 2010) is another alternative that makes it possible to implement Web Services over resource-constrained devices, e.g. a sensor node, and can be applied to reduce the overhead of native Web Services protocols. This allows not only mobile robots but also resource-constrained sensor nodes to offer, discover, and use services in a lightweight manner.

The REST-based approach is another paradigm of SOA implementation, known as Resource Orientation Architecture (ROA), which follows the classical client/server model. It considers a service as a *resource* that can be accessed through the Web using traditional HTTP requests. These resources can be manipulated using the CRUD (Create, Read, Update, Delete) style via the typical HTTP methods, namely GET to request/retrieve a resource, POST to modify/update a resource, PUT to create a resource and DELETE to remove a resource. In the context of SURV-TRACK, the service provider or simply the *server* (i.e. a mobile robot or a sensor node) creates a resource for each service they provide and makes it accessible to service consumers or simply *clients* (i.e. control station) through a Uniform Resource Identifier

(URI). The server describes its functionality with a certain service description language such as the Web Application Description Language (WDAL), which permits simple and easily scalable service description (Moritz, et al., 2010). Further optimizations are possible by relying on lightweight representation formats such as JSON, instead of XML (Baccar, 2011).

In Section 3, we presented the main networking and application layer protocols and system requirements of the SURV-TRACK application. These provide useful guidelines for any real-world implementation of an indoor surveillance application encompassing mobile robots and a sensor network. The aforementioned network and application layers architecture offers a means to the different components of the surveillance system to communicate; however, it is necessary to devise intelligent coordination mechanisms between mobile robots, sensor nodes, and the control station to effectively accomplish the target capturing missions. This is the purpose of the next section.

4. MULTI-ROBOT TASK ALLOCATION FOR TARGET CAPTURING

4.1 Problem Statement

In this section, we investigate the problem of multiple robots coordination with the support of a WSN for the SURV-TRACK application. This problem can be seen as an instantiation of the Multi-Robot Task Allocation (MRTA) protocol for target capturing. We consider the same system model presented in Figure 1. In SURV-TRACK, it is essential to devise effective coordination mechanisms between mobile robots with the support of the WSN, for accomplishing the mission of target tracking and capturing by a robot team.

The problem can roughly be formulated as follows:

Given a set of n mobile robots and a set of m intruders and a sensor network distributed in the environment, how can tasks be efficiently allocated to mobile robots to capture the intruder(s) with a minimum cost, using a WSN.

The cost can take several forms including mission duration, consumed energy, path traveled, etc. The intruder is considered captured if the distance between the intruder and the robot is less than a specified threshold.

Several researchers (Pizzocaro & Preece, 2009; Gerkey & Mataric, 2004) have conducted studies to answer the following question: Which robot should execute which task? Dias *et al.* (2006) provide a good survey on multi-robot coordination (based on market-based approach) in which they propose the following formal definition of the Multi-Robot Task Allocation Problem:

Given a set of tasks T, a set of robots R, and a cost function for each subset of robots $r \in R$ specifying the code for completing each subset of tasks, $c_r : 2^T \rightarrow IR^+ \cup \{\infty\}$, find the allocation of $A^* \in R^T$ that minimizes a global objective function $C : R^T \rightarrow IR^+ \cup \{\infty\}$

It just happens that many multi-robot task allocation problems are NP-Hard problems; these problems have no exact solutions in a polynomial time, therefore heuristics and approximate algorithms are typically used to estimate solutions. The purpose of this chapter is to propose solutions to this problem and apply them to the SURV-TRACK application.

In the literature, there are three typical approaches devised for ensuring multi-robot coordination (Dias, et al., 2006), namely: (*1*) *centralized approach*: it assumes the knowledge of global information by a central agent (e.g. control station), which calculates an optimal (or near-optimal) solution to the allocation problem, (*2*) *distributed approach*: it makes decisions (or local solutions) based on local information for each agent performing a task (e.g. robot), (*3*) *market-based approach*: it assumes that solutions are built based on a bidding-auctioning procedure between the working agent (e.g. robot) and a coordinator that allocates tasks for low-cost bidders.

In the following section, we present solutions to the MRTA problem for SURV-TRACK based on the three aforementioned approaches.

4.2 Literature Review of the MRTA Problem

In Gerkey and Mataric (2004), a taxonomy for some variants of the MRTA problems is introduced and proposes the following:

- **Single-Task Robots (ST) vs. Multi-Task Robots (MT):** In the ST approach, robots can execute one task at most. However, in the MT approach, robots can simultaneously execute multiple tasks.
- **Single-Robot Tasks (SR) vs. Multi-Robot Tasks (MR):** Using the SR approach, each robot requires exactly one robot to complete it. However, in the MR approach, some tasks need multiple robots.
- **Instantaneous Assignment (IA) vs. Time-Extended Assignment (TA):** In the IA approach, robots are only concerned with the task they are handling at the moment and they cannot plan for future allocations. However, in the TA approach, robots have more global information and they are able to build a more efficient plan for future allocations by defining a schedule.

In the literature, several research works propose different solutions to the MRTA problem applied to different areas. Elmogy *et al.* (2009) propose a solution for the Multi Robot Task Allocation

problem in mobile surveillance systems (Elmogy, et al., 2009; Khamis, et al., 2011). Their problem consists of allocating a set of tasks to a group of robots, in addition to providing coordination among robots. To solve their problem, the authors propose a solution based on the market mechanism. They propose centralized and hierarchical, dynamic and fixed tree task allocation approaches. To validate their solution, the authors calculate the average cost as a function of the number of deployed robots. They conclude that the hierarchical dynamic tree allocation approach outperforms all other approaches.

Also, Viguria *et al.* (2008) propose a market-based approach called S+T (Services and Tasks) algorithm to solve the MRTA problem. They introduce the concept of services, which means that a robot provides services to another robot to execute its task. The proposed approach is based on two roles: bidder and auctioneer. In the bidding process, if a task requires a service to be executed, it is necessary for a robot willing to perform the task to send an initial bid to the auctioneer, which must select the best bid with the lowest cost. To demonstrate the efficiency of their solution, the authors considered a surveillance mission. As a result, they conclude that the S+T algorithm can help complete tasks that need more than one robot to be executed.

To solve the multi-robot task allocation in a distributed manner, (Werger & Mataric, 2000) present the Broadcast of Local Eligibility (BLE) technique to ensure fully distributed robot coordination. Interestingly, the authors consider solving the Cooperative Multi robot Observation of Multiple Moving Targets (CMOMMT) problem. BLE involves the comparison of locally determined eligibility for a given task in a robot with the best eligibility calculated by other robots. If a robot has the best eligibility for a certain task, it exclusively executes that task and prevents the other robots from performing it. If a robot fails to execute the required task, one of the other robots will immediately take over the task.

In addition, Parker (1998) designed a software architecture called "ALLIANCE" based on a fully distributed behavior. The ALLIANCE architecture allows mobile robots to individually determine the appropriate actions to perform. To accomplish this, the ALLIANCE architecture uses two mathematically modeled motivations: impatience and acquiescence motivations. Using the impatience characteristic, some robots can examine situations when other robots fail to accomplish their tasks. The acquiescence characteristic lets the robot examine the situation when it fails to accomplish its own task. In addition, the ALLIANCE architecture allows robots to handle abrupt environmental changes robustly, reliably, coherently and flexibly, including the addition of new mobile robots or the failure of others.

On the other hand, a few researchers have incorporated Wireless Sensor Networks (WSNs) to support robotic applications, unlike previous research dealing with autonomous robots. WSNs have been shown to considerably improve the performance of applications. In Batalin and Sukhatme (2004), the problem of online task allocation for a team of robots using a static WSN is tackled. To accomplish this, they introduce two approaches: DINTA (Distributed In-network Task Allocation) and DINTA-MF (Multi Field Distributed In-network Task Allocation). Moreover, Zheng et al. (2008) tackles the problem of task allocation and localization using a WSN. The WSN is responsible for allocating tasks to the robot. It also provides information to help robots' localization. They prove that the robots have successfully finished their tasks with the support of the WSN.

4.3 Solution Design

In all approaches, we assume that the sensor nodes at the border of the monitored region are responsible for detecting the intruder(s) when entering that region. Then, subsequent actions depend on the chosen collaborative strategy. As stated above,

three approaches (centralized, distributed, and market-based) can be applied to solve the task allocation problem.

4.3.1. The Centralized Strategy

The centralized strategy consists in making the decision in the central control station that is responsible for ensuring coordination among mobile robots, assuming that it possesses global knowledge of the environment of interest (i.e. locations of robots, sensor nodes' locations, and locations of intruders). To accomplish this, the control station is assumed to receive continuous updates of the system states (via the sensor nodes or WiFi interface). The centralized approach is illustrated in Figure 3.

When a sensor node detects an intruder, it estimates its location (e.g. camera or ultrasound sensors) and sends an alarm to the control station through the WSN interface. Because the central agent already possesses up-to-date information about its environment, it uses its global knowledge to find an (near) optimal solution to follow and capture the intruder(s) by the mobile robots. This mechanism can be seen as an example of the Travel Salesman Problem (TSP) that is NP-Hard problem (Khamis, et al., 2011), whose complexity depends on the number of acting agents (i.e. pursuers and intruders). In order to find the optimal assignment considering the positions of the deployed robots and the intruder position, the central control station calculates the cost that each robot would pay to reach the intruder. The cost is evaluated as the length of the path that separates the current intruder's position from the robot, which is calculated using a path-planning algorithm. The best assignment is the one with the minimum cost, i.e. shortest path.

After evaluating the necessary cost for all the robots, the central unit sends a command to the winner robot (the robot that has the lowest cost) either directly (through WiFi connection) or via

Figure 3. Centralized strategy for SUVR-TRACK multi-robot coordination

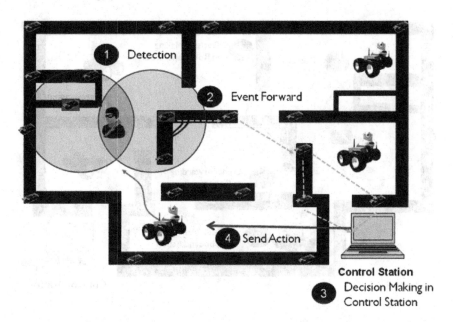

the wireless sensor network to start tracking the intruder until capturing it. During the mission, the WSN keeps updating the control station periodically with the new intruder's location and feeds it with other potential changes of the environment.

The centralized approach suffers from poor tractability in large-scale systems where information gathering will be much harder than in small-scale systems. In addition, this strategy is demanding in terms of communication overhead and is not reactive to fast changes. This approach also suffers from the typical single-point-of-failure problem as it heavily relies on the central control station. However, the advantage of this approach is that it can provide very accurate solutions because it is based on global knowledge of the environment. These issues will be investigated in the performance evaluation study presented in Section 4.4.

4.3.2. The Distributed Strategy

In the distributed strategy, the control station does not necessarily have recent state information on the network status (i.e. locations of robots, sensor

nodes' locations, and locations of intruders). Even if a recent state is available, the decision is made locally by each robot based on the knowledge it has, in contrast to the centralized strategy where the decision is made by the control station. The distributed strategy is illustrated in Figure 4.

Unlike the centralized strategy, once an intruder is detected by sensor nodes at the border, an alarm will be sent directly, not only to the control station, but also to each pursuer robot throughout the WSN infrastructure. A robot that receives the intrusion detection alarm will decide whether (or not) to follow the intruder, based on the local information it has collected from the environment (its location, location of the intruder, etc.). One limitation of this strategy is that, in some cases, a subset of mobile robots may follow the same intruder because each robot acts autonomously and makes its decision independently of the others, since it does not have a global knowledge about the other robots deployed in the same environment. A major challenge is to specify intelligent heuristics to optimize robot-initiated decisions based on local information. In fact, it is unwise for all robots receiving an alarm

Figure 4. Distributed strategy for SUVR-TRACK multi-robot coordination

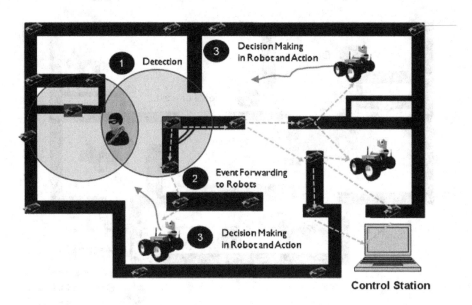

to go into action independently of their locations and that of the intruder(s); only those that are most likely to capture the intruder(s) should accept the mission. The distances and paths between robots and intruders play an important role in decision-making. The kinematics of robots, the number of robots in the team, the speed of intruders and their number are also parameters that affect the decision making process.

The WSN continues to update the robots with their locations relative to the intruder(s) so that they may quickly adapt to changes occurring in the system.

Sensor nodes bring additional intelligence to the application by providing a continuous information update of changes in the environment to help robots improve their visibility of the environment and its relative changes. This significantly improves the system performance as compared to an autonomous robotic team with no WSN support since robots cannot communicate all the times due to their restricted communication range. In contrast, with the support of the WSN, communication is always possible between robots throughout the WSN, making the coordination more effective. The control station will mostly have a passive role in the distributed approach; however, it can still supply the robots with useful information it may receive to optimize their plans.

Of course, solutions of the distributed strategy are expected to be less optimal than those in the centralized strategy due to the lack of global information as robots only maintain local knowledge. The distributed strategy usually produces sub-optimal solutions, since good local solutions may not necessarily produce a good global solution. However, the distributed approach is typically reactive to environmental changes, highly scalable and fault-tolerant, as it does not rely on a single agent for solution calculation. The aforementioned characteristics of the distributed approach will be discussed in the performance evaluation section. The advantage of this strategy is fault-tolerance. In fact, if a robot fails to accomplish its specific task, the whole system continues operating, which is especially true when using a homogenous robot team.

4.3.3. The Market-Based Strategy

The market-based approach provides a good trade-off between centralized and distributed strategies. It eliminates the need for global information maintenance at the control station, while it provides more efficient solutions as compared to the distributed approach (Dias, et al., 2006). The market-based strategy is illustrated in Figure 5.

Once an intruder is detected, sensor nodes notify the control station with an alarm. In contrast to the centralized approach, which requires maintenance of an updated status of the system in the control station, in the market-based approach, the control station will not search for a solution based on its local information, but will act as an auctioneer asking for the best price to accomplish the intruder tracking and capturing task. In this strategy, the control station will initiate the announcement phase in which it sends an auction to all available robots, informing them about the list of intruder(s) and its (their) location(s). Each robot receiving the auction will make an offer for these intruders by submitting a bid for each intruder to the control station. When the control station receives all the bids or a predefined time has passed, the auction then passes to the winner determination phase and the control station decides which intruder(s) to allocate to which robot(s). The robot with the lowest bid for tracking a particular intruder is selected. The properties of the three multi-robot coordination strategies are summarized in Table 1.

4.4 Performance Evaluation

4.4.1 Simulation Model

This section presents the performance evaluation study of the three proposed coordination strategies to (1) validate and analyze their behaviors and (2) compare their performance under different conditions. The simulation model is implemented using the well-known Player/Stage robotic simulator under the Ubuntu operating system. Specifically, Stage v2.1.1 and Player v2.1.2 were used.

Figure 5. Market-based strategy for SUVR-TRACK multi-robot coordination

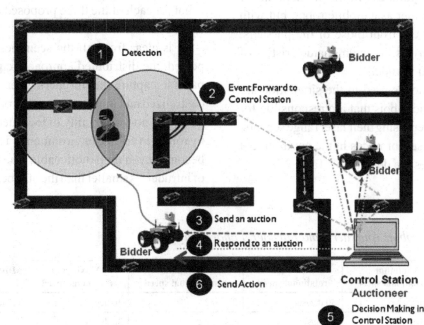

Table 1. Properties of multi-robot coordination strategies

	Centralized	**Distributed**	**Market-based**
Optimality	Optimal Solution	Sub-optimal solution	Near optimal solution
Complexity	Hardl tractable for large teams	Easily tractable for large teams	Tractable
Scalability	Non-scalable	Well-scalable	Scalable
Real-Time	Not easily reactive to changes	Reactive to change	Medium reactivity to changes

Each simulation scenario was repeated 30 times and the result of each scenario represents the average of the results obtained in the 30 iterations. The locations of the robots (intruders and pursuers) were also changed between iterations to randomize the initial system state.

We simulated the same environment depicted in Figure 1, which represents a house plan of 20m by 13m. A number of mobile pursuers were deployed in the area in different locations. For each scenario, we changed the number and the locations of the pursuers. In the simulation, we modeled an intruder as a mobile robot, but with properties different from those of the pursuers. Table 2 summarizes the main characteristics of the pursuers and intruders.

The sensor nodes used in the simulation are modeled as static robots that are responsible for detecting intruders using their laser range sensors. We deployed 8 sensor nodes in the environment whose properties are depicted in Table 3.

To evaluate the proposed strategies under different settings, we designed several scenarios in which we varied the number of pursuers and the number of intruders. Our objective was to assess the proposed strategies in three different cases: (1) the number of intruders is equal to the number of pursuers, (2) the number of intruders is smaller than the number of pursuers, and (3) the number of intruders is greater to the number of pursuers. We studied the impact of the variation of the numbers of intruders and pursuers on the average traveled distance per robot, and the mission time as metrics of interest. In each simulation run (iteration), the average traveled distance of pursuers is calculated. As shown in the x-axis of Figure 6, simulation scenarios are indexed so that *scenario$_{nk}$* means that there are *n* intruders and *k* pursuers. We alert the reader that simulation models and results do not consider communication delays and provide insights only on the pure behavior of the coordination strategies, without considering the communication impact.

4.4.2 Simulation Results

Figure 6 shows the average distance traveled per robot for each of the three proposed coordination strategies.

It is clear from all the scenarios that, as expected, the distributed approach required longer paths to capture the intruders when compared to centralized and market-based approaches. This is due to the non-optimality of local decisions taken by robots in ad-hoc environments. The difference becomes even more noticeable when the number of intruders is smaller than the number of pursuers.

Table 2. Mobile robots (pursuers and intruders) characteristics

	Minimum translational speed	**Minimum rotational speed**	**Maximum translational speed**	**Maximum rotational speed**	**Distance to avoid obstacles**
Pursuers	0.05 m/sec	30 deg/sec	3 m/sec	30 deg/sec	0.5 meters
Intruders	0.0 m/sec	0 deg/sec	0.4 m/sec	30 deg/sec	1 meter

Table 3. Sensors characteristics

	Minimum range	Maximum range	Field of view
Sensors	0 m	8 m	180 deg

This particular case seems to be less complex than the two other cases (i.e., case [1] and case [3]), and the centralized and market-based approaches quickly find efficient near optimal solutions to the problem in case (2). The difference between centralized and market-based approaches is not significant, although the centralized approach provides the lowest cost in most scenarios.

Figure 7 shows the average mission time for the three distributed approaches. The mission time for each strategy was calculated based on the average mission times of all scenarios. We observe that the centralized approach exhibits the smallest mission time with the smallest variation, which is natural as it calculates the most optimal solution for target capturing. On the other hand, the distributed approach not only has the greatest mission time, but it is extremely variable. This means that it varies a lot from one scenario to another, in contrast to centralized and market-based scenarios, whose mission times seem to be less dependent on the scenario.

5. ROBOT PATH PLANNING USING WIRELESS SENSOR NETWORKS

5.1 Problem Statement

In this section, we investigate the problem of robot path planning with the support of a WSN in the SURV-TRACK application. The objective of path planning is to establish a collision-free path for a robot starting from the initial (or current) location to the target location; the robot must avoid a variety of obstacles scattered in a workspace, based on its knowledge about the environment. Designing an efficient path-planning algorithm is an essential objective since path quality influences the efficiency of the entire application. The constructed path must satisfy a set of optimization criteria including traveled distance, processing time, and energy consumption. The traveled distance represents a typical metric of interest since it has a direct impact on time and energy.

The path-planning problem can be done in both *static* and *dynamic* environments: a static environment is unchanging, the start and goal positions are fixed, and obstacles do not change locations over time. However, in a dynamic environment, mobile robots are exposed to unexpected situations, including the locations of obstacles and the target, which may change over time.

Figure 6. Average traveled distance per

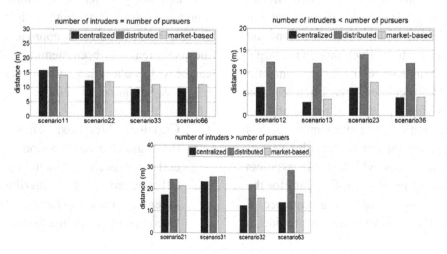

Figure 7. Mission time

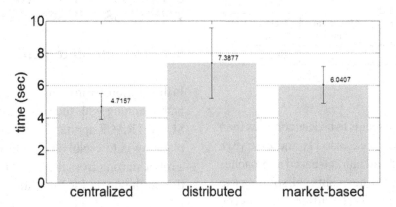

According to the robot's knowledge of the environment, path planning can be divided into two classes: In the first class, the robot has a prior knowledge of the environment modeled as a map. As such, the path can be planned offline based on the available map. This category of path planning is known as *global path planning*. The second class of path planning assumes that the robot does not have a prior knowledge of its environment. Consequently, it has to sense the locations of the obstacles and construct an estimated map of the environment in real-time during the search process to avoid obstacles and acquire a suitable path toward the goal state. This type of path planning is known as *local path planning* or *reactive navigation*.

In SURV-TRACK, we consider the path-planning problem in a static environment with an a priori knowledge. The environment is static in the sense that obstacles have fixed locations and do not change over time. Even though the target location changes as it moves in the environment, a new path must be calculated for each new location of the intruder after a predefined period of time. The difference with other environment models is that we represent the environment by a set of connected sensor nodes, which act as waypoints that help construct an obstacle-free path for the mobile robot. A path is seen as a sequence of sensor nodes. The following section provides a review of current literature regarding robot path planning research. Following this, we propose a new hybrid path planning method based on Ant Colony Optimization and Genetic Algorithms.

5.2 Literature Review of the Path-Planning Problem

Path planning is a computationally complex problem that has been proven to be PSPACE-hard (Latombe, 1991; Reif, 1979). This means that its complexity grows exponentially with the space's dimension. This problem is generally intractable and its complexity triggers the need for a powerful and efficient path-planning algorithm.

Research on the path-planning problem started in late 60s. Afterwards, several research initiatives, although many different solutions have since then been proposed for both static and dynamic environments. Numerous approaches to design these solutions have been attempted, although they can be widely classified into two main categories (Ellips & Sedighizadeh, 2007):

- **Classical Methods:** These are variations and/or a combination of a few general approaches, including Roadmap (Bhattacharya & Gavrilova, 2008), Potential Field (Warren, 1989), Cell Decomposition (Lingelbach, 2004), and

Mathematical programming. These methods dominated this field during the first 20 years. However, they were deemed to have some deficiencies regarding their global optimization, time complexity, robustness, etc.

- **Heuristic Methods:** These methods were designed to overcome the aforementioned limits of classical methods. Several techniques to solve the path planning problem have emerged including Genetic Algorithms (Tang, et al., 1996), Neural Networks (Simon & Meiig, 1999), Tabu Search (Masehian & Naseri, 2006), Ant Colony Optimization (Dorigo & Stützle, 2004), Particle Swarm Optimization (PSO) (Eberhart & Shi, 2001), Fuzzy Logic (Vachtsevanos & Hexmoor, 1986), and Simulated Annealing (Hui & Yu-Chu, 2008), just to mention the most important techniques.

Among all these techniques, ACO and GA are the most widely used heuristics to solve the path planning problem (Fan, et al., 2003; Porta Garcia, et al., 2009; Nagib & Gharieb, 2004; Zhao, et al., 2009; Joon-Woo & Ju-Jang, 2010). Some other research efforts have proposed hybrid solutions combining ACO and GA. The difference with our work is that the aforementioned works use different models of the environment.

For instance, in Tewolde and Weihua (2008), two solutions are proposed for path planning; the first is based on GA and the second is based on ACO. A comparison study of the techniques was performed in a real-world deployment of multiple robotic manipulators using specific spraying tools in an industrial environment. Results of this study show that both solutions provide very comparable results for small problem sizes, but when the size and the complexity of the problem increase, the ACO-based algorithm achieves a better quality of solution, but with a higher execution time, as compared to the GA-based algorithm.

In Buniyamin *et al.* (2011), the authors present an intensified ACO algorithm for the global path planning problem and compare the performance of their proposal with GA. They were able to prove that both solutions can find the optimal path, although the ACO algorithm proved to be more robust and effective in finding the optimal path.

In Sariff and Buniyamin (2009), the authors perform a comparative performance study of the two aforementioned approaches. The algorithms were tested in three workspaces with different levels of complexity. They demonstrated that the ACO algorithm was more effective and outperformed GA in terms of time complexity and number of iterations.

Although ACO and GA are effective in resolving the path-planning problem, these two techniques suffer from some significant limitations. Of the two, ACO possesses a stronger local search capability and faster convergence speed, but the algorithm can easily sink into a local optimum if the size of the problem is large. On the other hand, GA belongs to the class of random optimization processes, so the local convergence problem does not appear; however, this makes its convergence speed slower.

Therefore, we believe that a hybrid ACO and GA approach could be a promising alternative, which is the focus of our proposal.

Current literature proposes solutions based that combine the ACO and GA algorithms. For instance, in Ma and Hou (2010), the authors present a path planning method based on a hierarchical hybrid algorithm whose goal is to find a path in a road traffic network. The network is divided into several sub-networks and ACO is applied to each sub-network; the optimal paths generated by ant colony optimization algorithms represent the initial population of genetic algorithms. Simulations were conducted and showed the effectiveness of the hybrid algorithm.

The authors in Gao et al. (2008), introduce an approach which combines the GA and ACO algorithms to solve the robot navigation problem.

They propose a special function to improve the path quality generated by the ants at the beginning of the algorithm. Crossover and mutation operators are applied to improve the quality of the solution generated by the ACO algorithm.

This chapter proposes a new solution, which combines ACO and GA. Our solution fundamentally differs from others as it adopts a different environment map model based on WSNs. The optimal paths generated at the end of each iteration of the ACO algorithm are taken as an initial population of a crossover operator in the GA. The crossover is a kind of post-optimization or local search that avoids getting trapped in a local optimum. Our solution is presented in the next section.

5.3 Hybrid ACO/GA for Robot Path-Planning using WSN-Based Map

In this section, we present a new hybrid approach to solve path-planning problem of a mobile robot. The hybrid algorithm idea integrates ACO and GA to overcome their weaknesses and to exploit their advantages. The proposed path-planning algorithm can be divided into two phases: the first phase consists in generating a set of optimal paths by an intensified ACO algorithm. The second phase consists in using a modified crossover operator between the optimal paths generated. The algorithm is illustrated in Figure 8.

Figure 8. Flowchart of ACO-GA algorithm

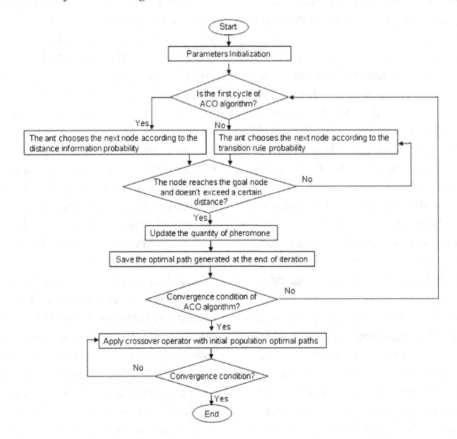

5.3.1 Environment Model

The mobile robot workspace is described by a 2-D map. The map represents the starting point, the goal point, and differentiates the areas that contain obstacles and free spaces.

In free space, a collection of connected sensor nodes are arranged in random locations with each node representing a location in the environment which is characterized by an identifier, (X,Y) coordinates and a set of adjacent nodes, as depicted in Figure 9.

The key idea of our model is to use wireless sensor nodes as:

- Signposts to locate the mobile robot, and
- Waypoints to guide the robot towards the desired destination. In fact, sensor nodes can collaborate and help the mobile robot to navigate in its surroundings. The robot must choose the best series of nodes to follow in order to find the best path, which is encoded as a sequence of sensor node IDs; for example {0-4-5-10-9-15} is a path with a start node 0 and goal node 15.

This reduces the complexity of environment's representation as compared to a fine-grain map model like a grid map. The configuration space of the mobile robot can be abstracted to a graph $G(V, E)$, where V denotes the set of wireless sensor nodes and E is the set of links between the nodes.

5.3.2 The Path-Planning Hybrid Algorithm

The flowchart of our ACO-GA algorithm for path planning is presented in Figure 8. The hybrid algorithm idea integrates ACO and GA approaches to overcome their weaknesses and to exploit their advantages. The algorithm comprises two phases; the first phase applies an ACO algorithm whose output is processed later by a GA via a crossover operator.

Figure 9. Simulation environments. The dashed line shows the path generated by the ACO-GA algorithm.

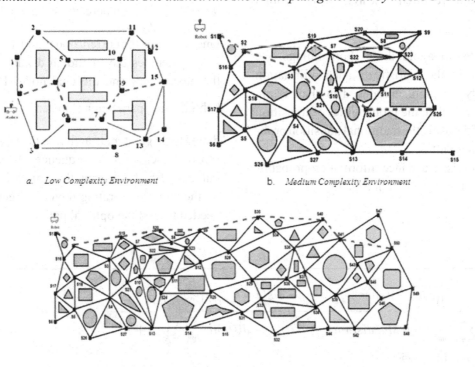

a. *Low Complexity Environment*   b. *Medium Complexity Environment*

c. *High Complexity Environment*

The ACO Phase: The ACO phase generates a set of optimal paths by employing an intensified ACO algorithm. The model of the workspace is abstracted to a graph and the ants must find the shortest start/end positions in the graph. Each ant has a current position in the graph and it can move from the current sensor node to another as it makes a decision about its new position. We propose two new functions to optimize the decision of an ant in the path planning process:

- A heuristic distance information probability is used in the beginning of the algorithm: indeed, in the conventional ACO algorithm, the ant's decision is made according to the transition rule probability function, which depends on the quantity of pheromone. However, at the beginning of the algorithm, particularly in the initial path-building phase, the quantity of pheromone is not important and does not have a great impact on the construction of the solution; this makes the choice of the ant not obvious and random, thus increasing the time required to search for the optimal path.

In order to avoid these shortcomings, we propose using the heuristic distance information probability function, introduced in (Fan et al., 2003), to calculate the probabilities of transition instead of the using the traditional transition probability function.

The heuristic distance information probability function is expressed in Box 1.

Where p_{ij}^k is the probability of transition of the kth ant from sensor node i to sensor node j, *allowed (1)* is the set of sensor nodes in the neighborhood of sensor node i, which the kth ant has not yet visited, $distance_{i,target}$ is the Euclidian distance from the sensor node j to the destination, $Maxdistance_{allowed(i),target}$ is the maximum of all $distance_{i,target}$ and ω, μ and λ are three constant calibration parameters as proposed in Fan *et al.* (2003).

A modified transition rule probability for the rest of the algorithm. This function is defined by the following formula:

$$p_{ij}^k = \frac{\tau_{ij}^{\alpha} * D^{\beta}}{\sum_{j \in allowed(i)} \tau_{ij}^{\alpha} * D^{\beta}}$$

where τ_{ij} denotes the quantity of pheromone between node i and node j, *allowed (1)* is the set of neighboring nodes of node i which the kth ant has not visited yet, α and β are two parameters to weight the significance of pheromone and distance in the selection of the next node, and $D = \frac{1}{d_{ij} + d_{jgoal}}$, where d_{ij} is the Euclidian distance between the current node i and the next node j. $d_{i\,goal}$ denotes the Euclidian distance between the next node j and the goal node. The reason behind using parameter D instead of $n_{ij} = \frac{1}{d_{ij}}$ is that D has a greater attraction to the goal wireless sensor node. As a consequence, it reduces the number of bad solutions that might be selected by the ants, accelerating the convergence speed needed to find the optimal path.

Box 1.

$$p_{ij}^k = \begin{cases} \dfrac{\left(\left(Maxdistance_{allowed(i),target} - distance_{j,target}\right) \cdot \omega + \mu\right)^{\lambda}}{\sum_{j \in allowed(i)} \left(\left(Maxdistance_{allowed(i),target} - distance_{j,target}\right) \cdot \omega + \mu\right)^{\lambda}} & \text{if } j \in allowed(i) \\ 0 & \text{otherwise} \end{cases}$$

It has to be noted that α and β are two important parameters of the transition rule probability. These parameters indicate the importance of the remaining pheromone on each node and the importance of the heuristic information, respectively. In the conventional ACO algorithm, α and β are numerical constants and cannot be changed during the execution of the algorithm; this, however, induces a negative impact on the performance of the algorithm. Thus, we propose varying the values of α and β as follows. In the beginning of the algorithm, the impact of the distance on the transition probability is more significant than the impact of the pheromone, so we must consider values so that $\alpha < \beta$. After a period of time the influence of the pheromone becomes more important as several valid paths appear; thus, we consider values so that $\alpha > \beta$.

Ants are monitored during path construction, meaning that if an ant walks more than a certain threshold distance, it will be discarded, as it will certainly not produce the optimal path. This helps reduce the search space and the execution time by quickly eliminating bad solutions.

A near optimal path is generated after each iteration of the ACO algorithm. Then, a mutation operator is applied on this path in search of a better solution. The main idea of the mutation operator is to check all the nodes, except the start and the goal nodes, and try to change one or more nodes in the path if the length of the resulting new path is shorter than the length of the initial path. At the end of this step, the quantity of pheromone is updated using the pheromone update rule expressed by the following formula:

$$\tau_{ij}\left(t+1\right) = \rho * \tau_{ij}\left(t\right) + \sum_{k=1}\Delta_{ij}^{k}$$

where τ_{ij} is the pheromone value associated to the edge (i,j) at iteration t, $0 \le \rho \le 1$ is a coefficient such that $\left(1-\rho\right)$ represents the evaporation of the trail between iterations t and $t+n$ and

$$\sum_{k}^{m}\Delta_{ij}^{k} = \begin{cases} \dfrac{Q}{L^{K}\left(t\right)} \\ 0 \end{cases}$$

where Q is a constant and $L^{k}\left(t\right)$ is the tour length of the kth ant at time t.

The GA Phase: it consists in applying a modified crossover operator on the set of optimal paths generated by the ACO algorithm after N iterations in order to improve the quality of solution found by the ants. The key idea of our proposed crossover operator consists in selecting two common nodes $N1$ and $N2$ from two parent paths $P1$ and $P2$ and comparing the three sub-parts of $P1$ and $P2$ existing (*1*) before $N1$, (*2*) between the two selected nodes $N1$ and $N2$, and (*3*) after the second node $N2$. The best parts are selected to form the new path.

Considering the following example, we have two paths:

P1: S0 S4 S6 S7 S9 S10 S11 S12 S15
P2: S0 S3 S6 S4 S5 S10 S9 S15 S15

The two common nodes selected are S4 and S10. We compare the different sub-parts ("S0" and "S0-S3-S6"), ("S4-S5-S10" and "S4-S6-S7-S9-S10"), and ("S11-S12-S15" and "S9-S15-S15") by calculating the Euclidian distance between the different sensor nodes. The new crossover generates only one path formed by the shortest parts from the two parents P1 and P2:

New Path: S0 S4 S5 S10 S9 S15 S15 S15 S15

5.4 Performance Evaluation

5.4.1 Simulation Model

In this section, we present a performance evaluation study of the proposed ACO-GA path planning solution applied to three environments with different complexities illustrated in Figure 9. We show how the proposed algorithm improves on the native ACO and GA path-planning algorithms. We used MATLAB for our simulation and all of simulations were implemented on a PC with an Intel Core $i3$ CPU @ 2.27GHz and 4GB of RAM under Windows 7.

- **Environment 1**: (Figure 9(a)) this environment is the simplest one because it has the smallest number of obstacles and the smallest number of nodes (16 sensor nodes). The mobile robot has to find the shortest collision-free path from sensor node 0 to sensor node 15.

- **Environment 2**: (Figure 9(b)) this environment is of medium complexity and consists of 27 sensor nodes. The robot has to reach sensor node 25 from starting sensor node 1.

- **Environment 3**: (Figure 9(c)) this environment is of high complexity as it contains the largest number of sensor nodes and obstacles, with 50 sensor nodes. The start position of the mobile robot is sensor node 1 and the goal position is sensor node 50.

The parameters of the ACO algorithm are presented in Table 4.

5.4.2 Simulation Results

To evaluate the efficiency of our path-planning algorithm, two performance metrics were assessed: (1) *the path length*: the length of the shortest path found by the algorithm, (2) *the execution time*: the time required by the algorithm to find the optimal solution. For each environment, we performed 30 different runs of the ACO-GA algorithm and we recorded the length of the shortest path generated by the algorithm and its execution time. An average value was taken for each environment. The generated paths highlighted in dashed lines are depicted in Figure 9.

To demonstrate the added value brought by our hybrid ACO-GA algorithm, we compared its performance against those of pure ACO and GA algorithms. The results of simulations are presented in Table 5, Table 6, and Figure 10.

Table 5 and Figure 10(a) depict the quality of solution found by the three algorithms in different environments. We have verified that the ACO-GA algorithm generates the same optimal paths generated by the Bellman-Ford shortest path exact algorithm. Table 6 and Figure 10(b) represent the

Table 4. ACO parameter specifications

Parameters		Value
m	Number of ants	Varies, depend on cases
α	Pheromone trail coefficient: determines the relative importance of the pheromone value	0.5 and 1
β	Heuristic coefficient: controls the relative importance of the distance (visibility)	0.5 and 1
ρ	Evaporation rate: controls how fast the old path evaporates.	0.99
Q	Constant	100
τ(0)	The initial pheromone value	0.05
NC_ACO	Convergence condition (fixed number of iterations)	NC_MAX=30
w	Calibration parameter of the heuristic distance information probability	10
μ	Calibration parameter of the heuristic distance information probability	2
λ	Calibration parameter of the heuristic distance information probability	2

time that each algorithm needed to find the optimal path.

We observe that the ACO-GA algorithm provides the optimal paths for all the three environments. In contrast, the pure ACO and pure GA algorithms both failed to find the optimal solution in medium- and large-scale environments. However, for a small problem size (small number of sensor nodes as in environment 1), the quality of solution generated by the three algorithms after 30 runs is the same, expect for native ACO, which finds sub-optimal paths (0-4-6-7-13- 15), (0-3-8-13-15) most of the time (27 out of 30 runs), as well as more quickly. However, ACO-GA finds the optimal solution in all iterations in a slightly greater time than the time required by the native ACO to find sub-optimal paths, as illustrated in Table 6 and Figure 10(b).

On the other hand, for a large-scale problem, the hybrid ACO-GA algorithm clearly outperforms the ACO algorithm and the GA algorithm in terms of solution quality, i.e. shortest path.

Figure 10(b) and Table 6 show that the ACO algorithm finds its best solution, which is not optimal, in a shorter time than ACO-GA, which finds a better and more optimal solution in a slightly greater time. This does not mean that pure the ACO algorithm is faster than ACO-GA because the former fails to find the optimal solution after

the 30 iterations of each of the 30 runs. The GA algorithm always exhibits the lowest quality and longest execution time, particularly for large problems. The results clearly show the benefit of using a hybrid approach for the path-planning problem.

The good performance of the ACO-GA algorithm in terms of quality of solution is due to the marriage of these two approaches. Indeed, the GA algorithm is a kind of post optimization or local search that improves the quality of solution found by the ACO algorithm, which can easily sink into a local optimum if the size of the problem is large.

Furthermore, Figure 10(b) shows that the hybrid algorithm converges faster after 30 generations than the GA algorithm. This is because the hybrid algorithm begins the search with the ACO algorithm which is characterized by a faster convergence speed as it determines the path accurately, based on the heuristic distance information probability and the modified transition rule probability. However, this is different with the GA algorithm that is based on a random process, which contributes to an increase in the computational time.

We are presently working to reduce the execution time of the ACO-GA algorithm and comparing the ACO-GA approach with other approaches used to solve the path-planning problem.

Table 5. Generated paths of ACO, GA, and ACO-GA algorithms in different complexity of environment

	ACO-GA Path	Path length (ACO-GA)	ACO Path	Path length (ACO)	GA Path	Path length (GA)	Bellman Ford Exact Method	Path length (Bellman Ford)
Environment 1nm	0-4-6-7-9-15	17.97	0-4-6-7-9-15 0-4-6-	17.97	0-4-6-7-9-15	17.97	0-4-6-7-9-15	17.97
Environment 2	1-2-3-21-10-24-25	19.52	1-2-19-20-8-23-11-24-25	25.61	1-16-2- 19-20-9 -23-12-25	27.17	1-2-3-21-10-24-25	19.52
Environment 3	1-2-19-20-9-30-40-41-50	35.22	1-16-2-19-20-9-28-29-36-41-50	40.99	1-16-2-19-7-8-23-9-30-28-34-36-35-37-39-45-43-50	59.52	1-2-19-20-9-30-40-41-50	35.22

Figure 10. Performance evaluation results of the ACO/GA path-planning algorithm

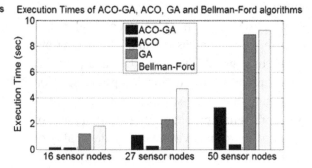

a. *Generated Paths*   b. *Execution Times*

6. CONCLUSION AND FUTURE WORK

Mobile robots and wireless sensor networks are evolving at a dramatic pace and will have a growing impact on our daily lives. These technologies have been deployed in several common application areas, but have been considered somewhat separately in the literature. This chapter illustrated the benefits of coupling these two technologies through a typical indoor surveillance application where a team of robots is responsible for target capture with the support of a WSN infrastructure. First, we presented a system model of a multi-robot indoor surveillance application using WSNs and

Table 6. Computation time (in seconds) of ACO, GA, and ACO-GA algorithms in different environments

	ACO-GA	ACO	GA	Bellman Ford Exact Method
Environment 1	0,1685	0,12844	1,2157	1,8088
Environment 2	1,11436	0,2574	2,3264	4,7244
Environment 3	3,2447	0,39	8,9092	9,2454

we went on to discuss the design considerations of the network and application layers that are necessary to provide efficient communications between the robots and the sensor network. It is clear that the use of the WSN provides several advantages as it: (*1*) increases the availability of the system as sensor nodes are pervasively embedded in the environment, (*2*) fosters the interaction between the robots' team, and (*3*) reduces the complexity of the environment representation by modeling the environment map as a set of sensor nodes considered as waypoints.

Second, we tackled the problem of multi-robot task allocation for target capturing and we proposed three coordination strategies: centralized, distributed and market-based. Centralized and market-based approaches were shown to provide better solution qualities in terms of traveled distance per robot and mission time as compared to the distributed strategy. This comes at the expense of the need for global knowledge of the system status either periodically in case of centralized approach, or on demand in case of market-based approach. These results did not consider communication delay and overhead, which should impact the performance of these three strategies. We plan to study this issue in future work, as it is important to quantify the impact of communications on the system performance, mainly on mission time.

This chapter also presented a third contribution about robot path planning using WSNs. We modeled the environment as a collection of connected sensor nodes and devised a hybrid ACO-GA path planning strategy to discover the path between two sensor nodes (waypoints). We demonstrated the effectiveness of using a hybrid ACO-GA approach in two fronts: improving the solution quality (i.e. shortest path) and reducing the execution time to find the solution. The advantage of the hybrid path planning approach is that it takes advantage of the fast convergence speed of the ACO algorithm and avoids falling into a local optimum using the GA technique. We are currently planning to devise other path planning algorithms for the same environment model using other heuristics to compare their performance against our hybrid ACO-GA algorithm and indentify new ways to solve the problem more efficiently and in the shortest time possible.

We are currently working towards implementing a real-world prototype of the SURV-TRACK application using a team of Wifibot Lab robots supported by a TelosB sensor network. This will help us assess the real behavior of the coordination between robots and sensor nodes and provide an experimental framework to design, implement and validate solutions to both MRTA and path planning problems.

We believe that this chapter provided a comprehensive technical insight on the design and challenges of multi-robot surveillance applications using WSNs that will pave the way toward other innovative ideas mixing different ubiquitous and pervasive technologies.

REFERENCES

Abolhasan, M., Wysocki, T., & Dutkiewicz, E. (2004). A review of routing protocols for mobile ad hoc networks. *Ad Hoc Networks*, *2*(1), 1–22. doi:10.1016/S1570-8705(03)00043-X

Akaya, K., & Younis, M. F. (2005). A survey on routing protocols for wireless sensor networks. *Ad Hoc Networks*, *3*(3), 325–349. doi:10.1016/j.adhoc.2003.09.010

Akyldiz, I. F., Su, W., Sankarasubramaniam, Y., & Cayirci, E. (2002). Wireless sensor network: A survey. *Computer Networks*, *38*, 393–422. doi:10.1016/S1389-1286(01)00302-4

Baccar, S. (2011). *Service-oriented architecture for wireless sensor networks: A RESTful approach.* (Unpublished Master Thesis Dissertation). National School of Engineering. Sfax, Tunisia.

Baccour, N., Koubâa, A., Mottola, L., Zuniga, M., Youssef, H., Boano, C., & Alves, M. (2012). Radio link quality estimation in wireless sensor networks: A survey. *ACM Transactions on Sensor Networks, 8*(4).

Bailey, T., & Durrant-Whyte, H. (2006). Simultaneous localisation and mapping (SLAM): Part II state of the art. *Robotics and Automation Magazine*. Retrieved from http://www.cs.berkeley.edu/~pabbeel/cs287-fa09/readings/Bailey_Durrant-Whyte_SLAM-tutorial-II.pdf

Batalin, M. A., & Sukhatme, G. S. (2004). Using a sensor network for distributed multi-robot task allocation. In *Proceedings of the IEEE International Conference on Robotics Automation (ICRA)*, (pp. 158-164). IEEE Press.

Bhattacharya, P., & Gavrilova, M. L. (2008). Roadmap-based path planning - Using the Voronoi diagram for a clearance-based shortest path. *IEEE Robotics & Automation Magazine*, *15*, 58–66. doi:10.1109/MRA.2008.921540

Buniyamin, N., Sariff, N., Wan Ngah, W. A. J., & Mohamad, Z. (2011). Robot global path planning overview and a variation of ant colony system algorithm. *International Journal of Mathematics and Computers in Simulation, 5,* 9–16.

Carlone, L., Aragues, R., Castellanos, J. A., & Bona, B. (2011). A first-order solution to simultaneous localization and mapping with graphical models. In *Proceedings of the 2011 IEEE International Conference on Robotics and Automation (ICRA),* (pp. 1764-1771). IEEE Press.

Clausen, T., & Jacquet, P. (2003). Optimized link state routing protocol (OLSR). *Network Working Group, 3626.*

CoroWare Inc. (2011). *Website.* Retrieved from http://www.CoroWare.com

Crenshaw, T. L., Hoke, S., Tirumala, A., & Caccamo, M. (2007). Robust implicit EDF: A wireless MAC protocol for collaborative real-time systems. *ACM Transactions on Embedded Computing Systems, 6*(4).

Demirkol, I., Ersoy, C., & Alagoz, F. (2006). MAC protocols for wireless sensor networks: A survey. *IEEE Communications Magazine, 44*(4), 115–121. doi:10.1109/MCOM.2006.1632658

Dias, B. M., Zlot, R., Kalra, N., & Stentz, A. (2006). Market-based multirobot coordination: A survey and analysis. *Proceedings of the IEEE, 94,* 1257–1270. doi:10.1109/JPROC.2006.876939

Dorigo, M., & Stützle, T. (2004). *Ant colony optimization.* Cambridge, MA: The MIT Press. doi:10.1007/b99492

Durrant-Whyte, H., & Bailey, T. (2006). Simultaneous localisation and mapping (SLAM): Part I: The essential algorithms. *Robotics and Automation Magazine.* Retrieved from http://www.cs.berkeley.edu/~pabbeel/cs287-fa09/readings/Durrant-Whyte_Bailey_SLAM-tutorial-I.pdf

Eberhart, Y., & Shi, Y. (2001). Particle swarm optimization: Developments, applications and resources in evolutionary computation. In *Proceedings of the 2001 Congress on Evolutionary Computation,* (pp. 81-86). IEEE.

El-Hoiydi, A., & Decotignie, J. D. (2004). WiseMAC: An ultra low power mac protocol for the downlink of infrastructure wireless sensor networks. In *Proceedings of the Ninth IEEE Symposium on Communication, ISCC 2004,* (pp. 244-251). IEEE Press.

Ellips, M., & Sedighizadeh, D. (2007). Classic and heuristic approaches in robot motion planning – A chronological review. *Proceedings of World Academy of Science. Engineering and Technology, 23,* 101–106.

Elmogy, A. M., Khamis, A. M., & Karray, F. O. (2009). Dynamic complex task allocation in multisensor surveillance systems. In *Proceedings of the International Conference on Signals, Circuits and Systems (SCS),* (pp. 1-6). SCS.

Fan, X., Luo, X., Yi, S., Yang, S., & Zhang, H. (2003). Optimal path planning for mobile robot based on intensified ant colony optimization algorithm. In *Proceedings of the IEEE International Conference on Robotics, Intelligent Systems and Signal Processing,* (Vol. 1, pp. 131-136). IEEE Press.

Feng, M., Fu, Y., Pan, B., & Liu, C. (2011). A medical robot system for celiac minimally invasive surgery. In *Proceedings of the 2011 IEEE International Conference on Information and Automation (ICIA),* (pp. 33-38). IEEE Press.

Fielding, R. T. (2000). *Architectural styles and the design of network-based software architectures.* (Doctoral Dissertation). University of California. Irvine, CA.

Flickinger, M. D. (2007). *Motion planning and coordination of mobile robot behavior for medium scale distributed wireless network experiments.* (Unpublished Master Dissertation). University of Utah. Salt Lake City, UT.

Gaddour, O., Koubaa, A., Chaudhry, S., Tezeghdanti, M., Chaari, R., & Abid, M. (2012). Simulation and performance evaluation of DAG construction with RPL. In *Proceedings of the Third International Conference on Communication and Networking, COMNET.* COMNET.

Gao, M., Xu, J., & Tian, J. (2008). Mobile robot path planning based on improved augment ant colony algorithm. In *Proceedings of the Second International Conference on Genetic and Evolutionary Computing,* (pp. 273-276). IEEE.

Gerkey, B., & Mataric, M. (2004). A formal analysis and taxonomy of task allocation in multi-robot systems. *The International Journal of Robotics Research, 23*(9), 939–954. doi:10.1177/0278364904045564

Gobriel, S., Mosse, D., & Cleric, R. (2009). TDMA-ASAP: Sensor network TDMA scheduling with adaptive slot-stealing and parallelism. In *Proceedings of the 29th IEEE International Conference on Distributed Computing Systems, ICDCS 2009,* (pp. 458-465). IEEE Press.

Grigorescu, S. M., Macesanu, G., Cocias, T. T., & Moldoveanu, F. (2011). On the real-time modelling of a robotic scene perception and estimation system. In *Proceedings of the 15th International Conference on System Theory, Control, and Computing (ICSTCC),* (pp 1-4). ICSTCC.

Grisetti, G., Diego, G., Stachniss, C., Burgard, W., & Nardi, D. (2006). *Fast and accurate SLAM with Rao-Blackwellized particle filters.* Retrieved from http://www.informatik.uni-freiburg.de/~grisetti/pdf/grisetti06jras.pdf

He, T., Stankovic, J. A., Lu, C., & Abdelzaher, T. (2003). SPEED: A stateless protocol for real-time communication in sensor networks. In *Proceedings of International Conference on Distributed Computing Systems.* Providence, RI: IEEE.

Heinzelman, W., Chandrakasan, A., & Balakrishnan, H. (2000). Energy-efficient communication protocol for wireless sensor networks. In *Proceedings of the Hawaii International Conference System Sciences.* IEEE.

Heinzelman, W., Kulik, J., & Balakrishnan, H. (1999). Adaptive protocols for information dissemination in wireless sensor networks. In *Proceedings of the 5th Annual ACM/IEEE International Conference on Mobile Computing and Networking (MobiCom 1999).* Seattle, WA: ACM/IEEE Press.

Hong, X., Xu, K., & Gerla, M. (2002). Scalable routing protocols for mobile ad hoc networks. *IEEE Network, 16,* 11–21. doi:10.1109/MNET.2002.1020231

Huang, S., & Dissanayake, G. (2007). Convergence and consistency analysis for extended Kalman filter based SLAM. *IEEE Transactions on Robotics, 23,* 1036–1049. doi:10.1109/TRO.2007.903811

Hui, J. W., & Culler, D. E. (2008). Extending IP to low-power, wireless personal area networks. *IEEE Internet Computing, 12*(4), 37–45. doi:10.1109/MIC.2008.79

Hui, M., & Yu-Chu, T. (2008). Robot path planning in dynamic environments using a simulated annealing based approach. In *Proceedings of the 10th International Conference on Control, Automation, Robotics and Vision, ICARCV 2008,* (pp. 1253-1258). ICARCV.

Intanagonwiwat, C., Govindan, R., & Estrin, D. (2000). Directed diffusion: A scalable and robust communication paradigm for sensor networks. In *Proceedings of the 6th Annual ACM/IEEE International Conference on Mobile Computing and Networking (MobiCom 2000).* Boston, MA: ACM/IEEE Press.

Jamieson, K., Balakrishnan, H., & Tay, Y. C. (2006). sift: a mac protocol for event-driven wireless sensor networks. In *Proceedings of the Third European Workshop on Wireless Sensor Networks (EWSN),* (pp. 260-275). EWSN.

Jan, G. E., Ki, Y. C., & Parberry, I. (2008). Optimal path planning for mobile robot navigation. *IEEE/ASME Transactions on Mechatronics, 13,* 451–460. doi:10.1109/TMECH.2008.2000822

Jardosh, S., & Ranjan, P. (2008). A survey: Topology control for wireless sensor networks. In *Proceedings of the International Conference on Signal Processing, Communications and Networking,* (pp. 422-427). IEEE.

Jian-She, J., Jing, J., Yong-Hui, W., Ke, Z., & Jia-Jun, H. (2008). Development of remote-controlled home automation system with wireless sensor network. In *Proceedings of the Fifth IEEE International Symposium on Embedded Computing,* (pp 169-173). IEEE Press.

Johnson, D. B., Maltz, D. A., & Hu, Y.-C. (2003). *The dynamic source routing protocol for mobile ad hoc networks (DSR).* Unpublished.

Joon-Woo, L., & Ju-Jang, L. (2010). Novel ant colony optimization algorithm with path crossover and heterogeneous ants for path planning. In *Proceedings of the IEEE International Conference on Industrial Technology,* (pp. 559-564). IEEE Press.

Jung, S. J., Kwon, T. H., & Chung, W. Y. (2009). A new approach to design ambient sensor network for real time healthcare monitoring system. [IEEE Press.]. *Proceedings of IEEE Sensors, 2009,* 576–580.

Khamis, A. M., Elmogy, A. M., & Karray, F. O. (2011). Complex task allocation in mobile surveillance systems. *Journal of Intelligent & Robotic Systems, 64*(1), 33–55. doi:10.1007/s10846-010-9536-2

Kim, B. K., Miyazaki, M., Ohba, K., Hirai, S., & Tanie, K. (2005). Web services based robot control platform for ubiquitous functions. In *Proceedings of the International Conference on Robotics and Automation,* (pp. 690-696). IEEE.

Kim, K., Park, S., Chakeres, I., & Perkins, C. (2007). *Dynamic MANET on-demand for 6LoWPAN (DYMO-low) routing.* Retrieved from http://tools.ietf.org/html/draft-montenegro-6lowpan-dymo-low-routing-03

Kim, K., Yoo, S., Park, J., Park, S. D., & Lee, J. (2005). *Hierarchical routing over 6LoWPAN (HiLow).* Retrieved from http://tools.ietf.org/html/draft-daniel-6lowpan-hilow-hierarchical-routing-01

Koubaa, A., Alves, M., & Tovar, E. (2007). IEEE 802.15.4: A federating communication protocol for time-sensitive wireless sensor networks. In *Sensor Networks and Configurations: Fundamentals, Techniques, Platforms, and Experiments* (pp. 19–49). Berlin, Germany: Springer-Verlag.

Koutsopoulos, I., & Tassiulas, L. (2007). Joint optimal access point selection and channel assignment in wireless networks. *IEEE/ACM Transactions on Networking, 15*(3), 521–532. doi:10.1109/TNET.2007.893237

Latombe, J. C. (1991). *Robot motion planning.* Norwell, MA: Kluwer Academic Publishers. doi:10.1007/978-1-4615-4022-9

Levine, J., & Vickers, L. (2001). Robots controlled through web services. *Technogenesis Research Project.* Retrieved from http://reference.kfupm.edu.sa/content/l/e/learning_robots_____91__a_class_63381.pdf

Lingelbach, F. (2004). Path planning for mobile manipulation using probabilistic cell decomposition. In *Proceedings of the IEEE/RSJ International Conference on Intelligent Robots and Systems,* (Vol. 3, pp. 2807-2812). Stockholm, Sweden: Centre for Autonomous Systems.

Liu, S., Sun, D., & Zhu, C. (2011). Coordinated motion planning for multiple mobile robots along designed paths with formation requirement. *IEEE/ASME Transactions on Mechatronics, 16,* 1021–1031. doi:10.1109/TMECH.2010.2070843

Ma, Y. J., & Hou, W. J. (2010). Path planning method based on hierarchical hybrid algorithm. In *Proceedings of the International Conference on Computer, Mechatronics, Control and Electronic Engineering,* (Vol. 1, pp. 74-77). IEEE.

Manjeshwar, A., & Agrawal, D. P. (2001). TEEN: A protocol for enhanced efficiency in wireless sensor networks. In *Proceedings of the 1st International Workshop on Parallel and Distributed Computing Issues in Wireless Networks and Mobile Computing.* San Francisco, CA: IEEE.

Manjeshwar, A., & Agrawal, D. P. (2002). AP-TEEN: A hybrid protocol for efficient routing and comprehensive information retrieval in wireless sensor networks. In *Proceedings of the 2nd International Workshop on Parallel and Distributed Computing Issues in Wireless Networks and Mobile Computing.* Ft. Lauderdale, FL: IEEE.

Masehian, E., & Amin-Naseri, M. R. (2006). A tabu search-based approach for online motion planning. In *Proceedings of the IEEE International Conference on Industrial Technology,* (pp. 2756-2761). Tehran, Iran: IEEE Press.

Mita, M., Mizuno, T., Ataka, M., & Toshiyoshi, H. (2004). A 2-axis MEMS scanner for the landing laser radar of the space explorer. In *Proceedings of the 30th Annual Conference of IEEE,* (pp. 2497-2501). IEEE Press.

Mobile Robots, Inc. (2011). *Pioneer 3D datasheet.* Retrieved from http://www.mobilerobots.com/ResearchRobots/PioneerP3DX.aspx

Mokarizadeh, S., Grosso, A., Matskin, M., Kungas, P., & Haseeb, A. (2009). Applying semantic web service composition for action planning in multi-robot systems. In *Proceedings of the Fourth International Conference on Internet and Web Applications and Services, 2009, ICIW 2009,* (pp. 370-376). ICIW.

Montenegro, G., Kushalnagar, N., Hui, J., & Culler, D. (2007). *Transmission of IPv6 packets over IEEE 802.15.4 networks.* Retrieved from http://tools.ietf.org/html/draft-montenegro-lowpan-ipv6-over-802.15.4-02

Moritz, G., Zeeb, E., Pruter, S., Golatowski, F., Timmermann, D., & Stoll, R. (2010). Devices profile for web services and the REST. In *Proceedings of the 2010 8th IEEE International Conference on Industrial Informatics (INDIN),* (pp. 584-591). IEEE Press.

Nagaoka, K., Kubota, T., Otsuki, M., & Tanaka, S. (2009). Robotic screw explorer for lunar subsurface investigation: Dynamics modelling and experimental validation. In *Proceedings of the International Conference on Advanced Robotics,* (pp. 1-6). IEEE.

Nagib, G., & Gharieb, W. (2004). Path planning for a mobile robot using genetic algorithms. *IEEE International Conference on Electrical, Electronic and Computer Engineering*, (pp. 185-189). IEEE Press.

Parker, L. E. (1998). ALLIANCE: An architecture for fault-tolerant multi-robot cooperation. [IEEE Press.]. *IEEE Transactions on Robotics and Automation*, (n.d.), 220–240. doi:10.1109/70.681242

Perkins, C. E., & Bhagwat, P. (1994). Highly dynamic destination-sequenced distance-vector routing (DSDV) for mobile computers. In *Proceedings of the ACM Special Interest Group on Data Communications (SIGCOMM)*, (pp. 234-244). ACM Press.

Perkins, C. E., Royer, E. M., & Das, S. R. (2003). *Ad hoc on-demand distance vector (AODV) routing*. Retrieved from http://www.ietf.org/rfc/rfc3561.txt

Pizzocaro, D., & Preece, A. (2009). Towards a taxonomy of task allocation in sensor networks. In *Proceedings of the IEEE International Conference on Computer Communications Workshops*, (pp. 413-414). IEEE Press.

Polastre, J., Hill, J., & Culler, D. (2004). Versatile low power media access for wireless sensor networks. [ACM Press.]. *Proceedings of the SenSys, 2004*, 3–5.

Porta Garcia, M. A., Montiel, O., Castillo, O., Sepulveda, R., & Melin, P. (2009). Path planning for autonomous mobile robot navigation with ant colony optimization and fuzzy cost function evaluation. *Journal of Applied Soft Computing*, 9(3), 1102–1110. doi:10.1016/j.asoc.2009.02.014

Prakashgoud, P., Vidya, H., Shreedevi, P., & Umakant, K. (2011). Wireless sensor network for precision agriculture. In *Proceedings of the 2011 International Conference on Computational Intelligence and Communication Networks (CICN)*, (pp. 763-766). CICN.

Rajendran, V., Obraczka, K., & Garcia Luna Aceves, J. J. (2003). Energy efficient, collision free medium access control for wireless sensor networks. In *Proceedings of SenSys 2003*. ACM Press. doi:10.1145/958491.958513

Rangarajan, H., & Garcia-Luna-Aceves, J. J. (2004). Using labeled paths for loop-free on-demand routing in ad hoc networks. In *Proceedings of the 5th ACM MOBIHOC*, (pp. 43-54). ACM Press.

Reif, J. H. (1979). Complexity of the mover's problem and generalizations. In *Proceedings of the 20th Annual Symposium on Foundations of Computer Science*, (pp 421-427). ACM.

Rouached, M., Chaudhry, S., & Koubaa, A. (2011). Service-oriented architecture meets LowPANs: A survey. *The International Journal of Ubiquitous Systems and Pervasive Networks*, 1(1).

Rowe, A., Mangharam, R., & Rajkumar, R. (2006). RT-link: A time-synchronized link protocol for energy-constrained multi-hop wireless networks. In *Proceedings of the 3rd Annual IEEE Communications Society on Sensor and Ad Hoc Communications and Networks, 2006*, (pp. 402-411). IEEE Press.

Sang Hyuk, L., Soobin, L., Heecheol, S., & Hwang Soo, L. (2009). Wireless sensor network design for tactical military applications: Remote large-scale environments. In *Proceedings of the Military Communications Conference, 2009*, (pp. 1-7). IEEE Press.

Sariff, N. B., & Buniyamin, N. (2009). Comparative study of genetic algorithm and ant colony optimization algorithm performances for robot path planning in global static environments of different complexities. In *Proceedings of the IEEE International Symposium on Computational Intelligence in Robotics and Automation (CIRA)*, (pp.132-137). Mara, Malaysia: IEEE Press.

Simon, X. Y., & Meiig, M. (1999). Real-time collision-free path planning of robot manipulators using neural network approaches. In *Proceedings of the IEEE International Symposium on Computational Intelligence in Robotics and Automation,* (pp. 47-52). Guelph, Canada: IEEE Press.

Sohn, S. Y., & Kim, M. J. (2008). Innovative strategies for intelligent robot industry in Korea. In *Proceedings of the IEEE International Conference on Industrial Engineering and Engineering Management,* (pp. 101-106). IEEE Press.

Stopforth, R., Holtzhausen, S., Bright, G., Tlale, N. S., & Kumile, C. M. (2008). Robots for search and rescue purposes in urban and underwater environment- A survey and comparison. [IEEE.]. *Proceedings of the Mechatronics and Machine Vision in Practice, 2008,* 476–480.

Tang, K. S., Man, K. F., & Kwong, S. (1996). Genetic algorithms and their applications. *IEEE Signal Processing Magazine, 13,* 22–37. doi:10.1109/79.543973

Tavakoli, A. (2009). *HYDRO: A hybrid routing protocol for Lossy and low power networks.* Retrieved from http://tools.ietf.org/html/draft-tavakoli-hydro-01

TelosB mote, Inc. (2011). *TelosB mote datasheet.* Retrieved from http://www.willow.co.uk/TelosB_Datasheet.pdf

Tewolde, G. S., & Weihua, S. (2008). Robot path integration in manufacturing processes: Genetic algorithm versus ant colony optimization. *IEEE Transactions on Systems, Man, and Cybernetics. Part A, Systems and Humans, 38,* 278–287. doi:10.1109/TSMCA.2007.914769

Vachtsevanos, G., & Hexmoor, H. (1986). A fuzzy logic approach to robotic path planning with obstacle avoidance. In *Proceedings of the 25th IEEE Conference on Decision and Control,* (Vol. 25, pp. 1262-1264). IEEE Press.

Viguria, A., Maza, I., & Ollero, A. (2008). S+T: An algorithm for distributed multirobot task allocation based on services for improving robot cooperation. In *Proceedings of the IEEE International Conference on Robotics and Automation (ICRA 2008),* (pp. 3163-3168). IEEE Press.

Wang, Z., Liu, L., & Zhou, M. (2005). Protocols and applications of ad-hoc robot wireless communication networks: An overview. *International Journal of Intelligent Control and Systems, 10,* 296–303.

Warren, C. W. (1989). Global path planning using artificial potential fields. In *Proceedings of the IEEE International Conference on Robotics and Automation,* (Vol. 1, pp. 316-321). IEEE Press.

Werger, B. B., & Mataric, M. J. (2000). Broadcast of local eligibility for multi-target observation. *Distributed Autonomous Robotic Systems, 4,* 347–356. doi:10.1007/978-4-431-67919-6_33

WifiBot Inc. (2011). *Website.* Retrieved from http://www.wifibot.com

Wijetunge, S., Gunawardana, U., & Liyanapathirana, R. (2010). Wireless sensor networks for structural health monitoring: Considerations for communication protocol design. In *Proceedings of the 2010 IEEE 17th International Conference on Telecommunications (ICT),* (pp. 694-699). IEEE Press.

Winter, T., & Thubert, P. (2010). *RPL: IPv6 routing protocol for low power and Lossy networks.* Retrieved from http://tools.ietf.org/html/draft-ietf-roll-rpl-19

Wu, T., Duan, Z. H., & Wang, J. (2010). The design of industry mobile robot based on LL WIN function blocks language and embedded system. In *Proceedings of the 2nd International Conference on Computer Engineering and Technology,* (pp. 622-625). IEEE.

Xu, Y., Heidemann, J., & Estrin, D. (2001). Geography-informed energy conservation for ad hoc routing. In *Proceedings of the 7th Annual ACM/IEEE International Conference on Mobile Computing and Networking (MobiCom 2001)*. Rome, Italy: ACM/IEEE Press.

Ye, W., Heidemann, J., & Estrin, D. (2002). An energy-efficient MAC protocol for wireless sensor networks. In *Proceedings of the IEEE INFOCOM*, (pp. 1567-1576). IEEE Press.

Ye, W., Heidemann, J., & Estrin, D. (2004). Medium access control with coordinated, adaptive sleeping for wireless sensor networks. *IEEE/ACM Transactions on Networking, 12*(3), 493–506. doi:10.1109/TNET.2004.828953

Yogeswaran, M., & Ponnambalam, S. G. (2009). An extensive review of research in swarm robotics. In *Proceedings of the World Congress on Nature & Biologically Inspired Computing, 2009*, (pp. 140-145). IEEE.

Youngseok, L., Kyoungae, K., & Yanghee, C. (2002). Optimization of AP placement and channel assignment in wireless LANs. In *Proceedings of the 27th Annual IEEE Conference on Local Computer Networks, 2002*, (pp. 831-836). IEEE Press.

Younis, M., Youssef, M., & Arisha, K. (2002). Energy-aware routing in cluster-based sensor networks. In *Proceedings of the 10th IEEE/ACM International Symposium on Modeling, Analysis and Simulation of Computer and Telecommunication Systems (MASCOTS 2002)*. Fort Worth, TX: IEEE/ACM Press.

Yu, Y., Estrin, D., & Govindan, R. (2001). *Geographical and energy aware routing: A recursive data dissemination protocol for wireless sensor networks. UCLA Computer Science Department Technical Report, UCLA-CSD TR-01-0023*. Los Angeles, CA: UCLA.

Zhao, J., Zhu, L., Liu, G., & Han, Z. (2009). A modified genetic algorithm for global path planning of searching robot in mine disasters. In *Proceedings of the IEEE International Conference on Mechatronics and Automation*, (pp. 4936-4940). IEEE Press.

Zheng, T., Li, R., Guo, W., & Yang, L. (2008). Multi-robot cooperative task processing in great environment. In *Proceedings of the IEEE Conference on Robotics, Automation and Mechatronics*, (pp. 1113-1117). IEEE Press.

KEY TERMS AND DEFINITIONS

Ant Colony Optimization (ACO): It is a metaheuristic, which is an instance of the swarm intelligence; it is inspired from the real life behavior of a colony of ants, when searching for food, to find approximate solutions to difficult optimization problems.

Cooperative Mobile Robots Team: It amplifies the capabilities of a single mobile robot in order to perform complex tasks that require additional skills.

Genetic Algorithm (GA): It is a stochastic method of optimization used to solve optimization and search problems. It is based on the laws of natural selection and genetics. The notions of chromosome, gene and population constitute the base of GA.

Multi Robot Task Allocation: It consists in finding the efficient allocation mechanism in order to assign different tasks to the set of available robots. It aims at ensuring an efficient execution of tasks under consideration and thus minimizing the overall system cost.

Path Planning: It aims at the construction of an obstacles-free path for the mobile robot from a starting state to a goal state, through several obstacles scattered in an environment.

Wireless Sensor Network: Is an instance of wireless networks consisting of a set of small and battery powered devices equipped with sensing and communication capabilities. These devices are intended to collect measurements from their environment, and then to report them over radio links to a base station.

Chapter 3
Local Path–Tracking Strategies for Mobile Robots Using PID Controllers

Lluis Pacheco
University of Girona, Spain

Ningsu Luo
University of Girona, Spain

ABSTRACT

Accurate path following is an important mobile robot research topic. In many cases, radio controlled robots are not able to work properly due to the lack of a good communication system. This problem can cause many difficulties when robot positioning is regarded. In this context, gaining automatic abilities becomes essential to achieving a major number of mission successes. This chapter presents a suitable control methodology used to achieve accurate path following and positioning of nonholonomic robots by using PID controllers. An important goal is to present the obtained experimental results by using the available mobile robot platform that consists of a differential driven one.

PATH FOLLOWING AND MOTION-CONTROL BACKGROUND

In mathematics and physics, a nonholonomic system is a system in which a return to the original internal configuration does not guarantee return to the original system position (Bloch, 2003). In other words, unlike with a holonomic system, the outcome of a nonholonomic system is path-dependent, and the number of generalized coordinates required to represent a system completely is more than its control degrees of freedom. In this case the WMR has two control inputs, which correspond to the voltages applied to each motor wheel and three degrees of freedom for generalized coordinates (position coordinates x, y, and heading orientation,θ). Thus, return to the internal configuration does not guarantee return to the original system position. Cars and bicycles are other examples of nonholonomic systems.

DOI: 10.4018/978-1-4666-2658-4.ch003

Copyright © 2013, IGI Global. Copying or distributing in print or electronic forms without written permission of IGI Global is prohibited.

From a control science point of view, accuracy and performance of WMR trajectory tracking are subject to nonholonomic constraints, and consequently it is usually difficult to achieve stabilized tracking of trajectory points using linear feedback laws (Brockett, 1983). In the research, results presented by Hauser and Hindman (1995), demonstrate through Lyapunov stability theory that asymptotic stability exists in the control system with respect to the desired trajectory. In the case considered in Hauser and Hindman (1995), where trajectory tracking is performed using different linear models, the Euclidean distance between the robot and the desired trajectory can be used as a potential function. Such functions are CLF (Control Lyapunov Function), and consequently asymptotic stable with respect to the desired trajectory can be achieved. Path-following stabilization of nonholonomic systems can be achieved using time varying, discontinuous, or hybrid feedback laws.

The time-varying smooth state feedback asymptotically stabilizes a mobile robot to a point (Coron, 1991). Coron showed that it is possible to generate time-periodic feedbacks, which assure the finite-time convergence. Feedback stabilization by using homogeneous structures has been proposed in Dayawansa *et al.* (1995). The homogeneous control for autonomous systems has depicted homogeneous Lyapunov function results (Rosier, 1992; Aicardi, et al., 1995). Homogeneous time periodic control laws allow, in nonholonomic WMR, use feedback that gives the finite time convergence (M'Closkey & Murray, 1997). An alternative to time-dependent smooth controllers are discontinuous or piecewise smooth controllers (Bloch, et al., 1990). The stabilization about an arbitrary point in the state space is presented in Canudas de Wit and Sørdalen (1991). The use of hybrid techniques considering discontinuous and time variant control laws has been the focus of efforts by many researchers. These controllers operate by switching at discrete time instants between various time-periodic feedback functions

(Kolmanovky & McClamroch, 1995). Hence, the use of discontinuous control law with time variant sinusoids is proposed in Kolmanovky and McClamroch (1996). The use of force and torque as useful inputs for changing the control law is proposed in Aguiar and Pascoal (2000). The placement of integrators relative to the nonholonomic integrator has shown leading to a class of second order systems which can be distinguished by their controllability properties. Furthermore, studies of cascade dynamic system to the nonholonomic integrator lead to optimal solutions by using calculus of variations (Struemper & Krishnaprasad, 1997).

The use of piecewise analytic feedback laws in discontinuous stabilization approach to an arbitrary point is reported by Sørdalen and Canudas de Wit (1993). The approach is used to track a sequence of fixed points consisting of positions and orientations. Thus, the desired path to be followed can be composed of a sequence of straight lines and circle segments (Reeds & Shepp, 1990). Therefore, instead of regulating the robot about one point, the sequence of points to track is solved moving the WMR from point to point in the state space by using feedback combined with some path planning (Astolfi, 1996). The importance of these results is in the fact that linear control is suitable by using motion discontinuities as a trajectory-tracking function. Within this scope, PID solutions are widely adopted as low-level DC (direct current) motor speed controllers (Ogata, 2009). Path-following achievement is accomplished by developing high-level strategies for motion planning. For instance, the dynamic window approach is based on the dynamic constraints of the WMR and uses the available robot speeds to plan goal achievement while collision avoidance and safety stops are considered (Ögren & Leonard, 2005). Other suitable control approaches have proposed the evolution of a neural controller as a function of the sensor inputs to switch the WMR behavior to perform different tasks (Capi & Kaneko, 2009). Capi and Kaneko point out the importance of using real robot systems to achieve effective learning by neural controllers. The empirical results are used as an important data source to achieve proper WMR navigation. Cheng (Cheng, et al., 2009) reports successful path following by mobile robots using fuzzy control systems. The use of adaptive control capabilities that include robot-model uncertainties was also proposed in Pourboughrat and Karlsson (2002).

Some path following strategies presented in the previous subsections are comparatively tested in WMR (De Luca, et al., 2001). The acquired experience can be summarized in general observations that can be useful guidelines for implementation of the same control strategies for other vehicles. Therefore, good motion performances have been achieved considering some important aspects reported by researchers:

- The idea of using sate-space transformation that is singular at the goal configuration has been proposed (Astolfi, 1996).
- The use of appropriate coordinates involving Lyapunov functions is suggested by reported bibliography (Aicardi, et al., 1995).
- The performance of posture stabilization is increased by using dynamic feedback linearization methods (Struemper & Krishnaprasad, 1997).

INTRODUCTION

Control science is based on the model of the system knowledge. Usually, linear control assumes linear models and linear feedback control laws. Nowadays, the majority of control laws are computed by digital systems and hence discrete system control should be considered. Therefore, the classic continuous linear control theory should be considered as an important source of information. In this sense, it becomes an important engineering objective to obtain accurate dynamic models suitable for controlling the system. System

identification is usually used due to the complexity of the mathematical modelling in many cases. However, some *a priori* knowledge of the physical laws is useful for planning the experiment design as well as choosing the identification methodology.

This chapter, first of all, presents the basic system identification methodology. The parameter identification process is based on black-box models. The system dealt in this work is considered initially to be a MIMO (Multiple-Input Multiple-Output) system. This MIMO system is composed of a set of SISO (Single-Input Single-Output) subsystems with coupled connections. The parameter estimation is performed using a PRBS (pseudo-random binary signal) such as an excitation input signal. It guarantees the correct excitation of all dynamic sensible modes of the system along the whole spectral range, resulting in accurate high-precision parameter estimation. The experiments to be performed consist of exciting the two DC (direct current) motors in various (low, medium, and high) speed ranges. An ARX (autoregressive with external input) structure was used to identify the system parameters. The problem consists of finding a model that minimizes the error between the real and estimated.

Once the model of the WMR (wheeled mobile robot) is obtained, the problem is focused on the speed regulation by using linear control laws. The above nonlinear system can be tackled using different intervals, in which the transfer function is considered as linear. The control of a linear system can be done by using classic feedback controllers using linear design such as PID (proportional-integral-derivative). The controller design can be done using classic control methods like pole placement or frequency design methods; even heuristic methods such as "Ziegler Nichols" can be used. The performance of the controller can be easily simulated using software tools like SIMU-LINK. This work focuses on root locus methods.

Once the control of speed is introduced, the issues concerning to the trajectory and path-following control are presented. Path following stabilization of differential driven WMR nonholonomic is reached using discontinuous feedback laws. In this work, linear control is implemented by considering control law discontinuities referred to the path planning. Such techniques have been developed in the WMR by using heuristic concepts attaining the experimental knowledge of the system. Therefore, when orientation error of path distance of the robot is less than a heuristic threshold, the control law can minimize the distance to the path by commanding the same speed to both wheels. Otherwise, the angular error correction control laws are used by performing a turning action and consequently are commanded different speeds to each wheel. Thus, the desired path to be followed can be composed of a sequence of straight lines and circle segments.

IDENTIFICATION, PID CONTROLLER DESIGN, AND PATH FOLLOWING

This section presents the available WMR platform that can be used for testing the research developed in this work. It is a differential driven WMR with a free rotating wheel. This section also deals with the problem of modeling the dynamics of the WMR system. The aim of this work is to obtain a set of dynamic models for high, medium, and slow velocities, in order to control the WMR speed. Once dynamic models are obtained, this section presents the design of low-level PID speed controllers. High-level motion planning is also described in order to plan accurate and safe navigation while goal achievement is considered. Moreover, factor tuning of control law parameters is also analyzed by regarding the kind of path to be followed. Finally, experimental results are analyzed by using a representative set of trajectories.

Hardware and Architecture of the Robot System

The robot structure is made from aluminum (Pacheco, et al., 2011). Figure 1 shows the different levels where different parts are placed. The first level has two differential driven wheels, controlled by two DC motors and an omni-directional wheel that is the third contact point with the floor. The second level contains an embedded PC computer and the third level has specific hardware and the sonar sensors. A fourth level can be used because of the flexibility of the system to place a machine vision system and/or a multimedia set up, depending on the platform application.

The basic electrical hardware is also depicted in this subsection. The system can be powered by 12V DC batteries or by an external AC power source through a voltage of 220V, and a switch selects either mode of operation. The battery system is actually composed of a set of 4 12V batteries with 7Ah each which provides between 3 and 4 hours of autonomy. The robot is equipped with two DC motors that act over the two independent traction wheels. The WMR has the following sensorial system:

- Two encoders connected to the rotation axis of each DC motor.
- An array of sonar composed of 8 ultrasounds sensors.

Figure 1. The different platform hardware levels

- A machine vision system composed of a monocular camera.

The meaningful hardware consists of the following electronic boards:

- The DC motor power drivers are based on a MOSFET bridge that controls the energy supplied to the actuators.
- A set of PCB (printed circuits boards) based on PLD (programmable logic devices) act as the interface between the embedded PC system, the encoders, and the DC motors. The interface between the PLD boards and the PC is carried out by the parallel port.
- A μc processor board controls the sonar sensors. Communication between this board and the embedded PC is made through a serial port. This board is also in charge of a radio control module that enables the teleoperation of the robot.
- The embedded PC is the core of the basic system and it is where the high-level decisions are taken.

The PLD boards generate 23 khz PWM (pulse width modulation) signals for each motor and the consequent timing protection during the command changes. This protection system provides a delay during the power connection and at the change of the rotation motor sense. A hardware ramp is also implemented in order to give a better transition between command changes. The speed is commanded by a byte. In this way, it can generate from 0 to 127 advancing or reverse speed commands through the use of the parallel port as interface between the PC and the PLD's boards. The PLD boards also measure the pulses provided by the encoders during an adjustable period of time, giving the PC the speed of each wheel at every 25ms. The absolute position of each encoder is also measured by two absolute counters used to measure the position and orientation of the robot by odometer. The shaft encoders provide 500 counts/rev since encoders are placed at the motor axes; this means that the encoders provide 43,000 counts for each turn of the wheel. Moreover, the μc has control of the sonar sensors, so for each sensor a distance measure is obtained. The ultrasound sensor range is between 3cm and 5m. The data provided by these boards is gathered through the serial port in the central computer. The rate of communication with these boards is 9,600 b/s. The flexibility of the system allows for different hardware configurations as a function of the desired application and consequently the ability to run different programs on the μc or PLD boards. However, the platform is actually being tested under the configuration explained previously. The open platform philosophy is reinforced by the use of similar μc and PLD boards to those used as teaching tools at our school.

The main decision system arises from the embedded PC that controls the hardware shown before. The data gathering and control by digital computer is set to 100ms. The software of the PC is implemented in C language and runs under the LINUX operating system. However, this platform acts as an open system that allows the connection of other PCs through a LAN. These possibilities cover two complementary aspects of the system, which are the multimedia point of information and the machine vision system, as an advanced sensor system. Figure 2 shows the system architecture. The multimedia system is composed of a PC with a tactile screen that allows interaction with people. The implementation of this application as a multimedia information point means that this computer should be configured with the software information that the users applications demand. A wireless Internet connection allows communication with the whole world, and a set of multiple possibilities. The machine vision system is composed of the following components: a remote camera with motorized focus, overture, and zoom control by a serial port, two steep motors that control the pan and tilt position of the camera, and specific hardware boards running on a PC, exclusively used by the machine vision system. The system is connected to the main control system through a LAN. At present, the vision system is being studied for many applications but its integration in the context of a mobile multimedia information point has not yet been done. Hence, several studies should be carried out in the near future with the goal of obtaining sensor integration.

Kinematics of the Robot System

The WMR is a rigid body so non-deforming wheels are considered. It is assumed that the vehicle moves without slipping on a plane, so there is pure rolling contact between the wheels and the ground. Denoting (x, y, θ) as the coordinates of position and orientation, and $u = [v, w]^T$ as the velocity vector; where v and w respectively denote the tangential and angular velocities. Then, the kinematic model of the WMR is given by:

$$dx = v \cos \theta$$
$$dy = v \sin \theta \qquad (1)$$
$$d\theta = \omega$$

By using the discrete time representation (with T being the sampling period and k the time instant) and Euler's approximation, the following discrete time model is obtained for the robot dynamics:

$$x(k+1) = x(k) + v(k) \cos \theta(k) T$$
$$y(k+1) = y(k) + v(k) \sin \theta(k) T \qquad (2)$$
$$\theta(k+1) = \theta(k) + \omega(k) T$$

The WMR platform uses incremental encoders to obtain the position and orientation coordinates. Figure 3 describes the positioning of robot as a function of the radius of left and right wheels (R_e, R_d), and the angular incremental positioning (θ_e, θ_d), with E being the distance between two wheels and dS the incremental displacement of the robot. The position and angular incremental displacements are expressed as:

$$dS = \frac{R_d d\theta_d + R_e d\theta_e}{2} \qquad (3)$$

The coordinates (x, y, θ) can be expressed:

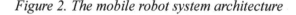

Figure 2. The mobile robot system architecture

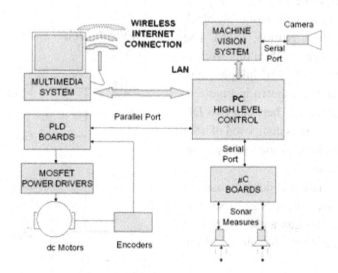

Figure 3. The WMR position as function of the angular displacement of each wheel

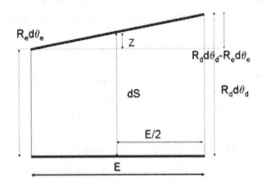

$$x(k+1) = x(k) + dS \cos\big(\theta(k) + d\theta\big)$$
$$y(k+1) = y(k) + dS \sin\big(\theta(k) + d\theta\big) \qquad (4)$$
$$\theta(k+1) = \theta(k) + d\theta$$

Thus, the incremental position of the robot can be obtained by the odometer system through the available encoder information from (3) and (4).

System Identification: The WMR Dynamic Models

The model is obtained through the approach of a set of linear transfer functions that includes the nonlinearities of the whole system. The parametric identification process is based on black box models (Ljung, 1987; Norton, 1986). The nonholonomic system dealt with in this work is considered initially to be a MIMO (Multiple-Input Multiple-Output) system, as shown in Figure 4, due to the dynamic influence between two DC motors. This MIMO system is composed of a set of SISO (single input single output) subsystems with coupled connections.

The parameter estimation is done by using a PRBS (Pseudo Random Binary Signal) such as excitation input signal. This guarantees the correct excitation of all dynamic sensible modes of the system along the whole spectral range and thus results in an accurate precision of parameter estimation. The experiments to be realized consist in exciting the two DC motors at different (low, medium, and high) speed ranges. The ARX (Auto-Regressive with External input) structure has been used to identify the parameters of the system. The problem consists in finding a model that minimizes the error between the real and estimated data. By expressing the ARX equation as a linear regression, the estimated output can be written as:

$$\hat{y} = \lambda \phi \qquad (5)$$

with $\hat{y}$ being the estimated output vector, λ the vector of estimated parameters, and φ the vector of measured input variables. By using the coupled system structure, the transfer function of the robot can be expressed as follows:

$$\begin{pmatrix} Y_D \\ Y_E \end{pmatrix} = \begin{pmatrix} G_{DD} & G_{ED} \\ G_{DE} & G_{EE} \end{pmatrix} \begin{pmatrix} U_D \\ U_E \end{pmatrix} \qquad (6)$$

where Y_D, and Y_E represent the speeds of right and left wheels, and U_D and U_E the corresponding speed commands, respectively. In order to have information about the coupled system, the matrix

Figure 4. MIMO system structure

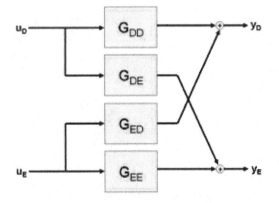

of transfer function should be identified. Figure 5 shows the speed responses of the system corresponding to the PRBS input signals. The filtered data, which represent the average value of five different experiments with the same input signal, are used for identification. The system is identified by using the identification toolbox "ident" of Matlab for the second order models. Table 1 shows the continuous transfer functions obtained for the three different linear speed models.

The coupling effects are studied as a way of obtaining a reduced-order dynamic model. It can be seen from Table 1 that the dynamics of two DC motors are different and the steady gains of coupling terms are relatively small (less than 20% of the gains of main diagonal terms). Thus, it is reasonable to neglect the coupling dynamics so as to obtain a simplified model. In order to verify the above facts from real results, a set of experiments have been done by sending a zero speed command to one motor and different non-zero speed commands to the other motor. In Figure 6, a response obtained on the left wheel is shown, when a medium speed command is sent to the right wheel.

The experimental result confirms that the coupled dynamics can be neglected. The existence of different gains in steady state is also verified experimentally. As shown in Figure 7, the gain of the left DC motor is greater than that of the right motor at low speed ranges. Finally, the order reduction of the system model is carried out through the analysis of pole positions by using the root locus method. It reveals the existence of a dominant pole and consequently the model order can be reduced from second order to first order.

Table 2 shows the first order transfer functions obtained. Afterwards, as shown in Figure 8, the system models are validated through the experimental data by using the PRBS input signal.

Design of PID Controllers and Motion Planning

PID control design is a well know issue that is solved by considering a closed loop system that includes the plant and the controller (Ogata, 2009). The problem is firstly tackled in continuous space domain, considering the first order models shown in Table 2 and PI control systems, by using the root locus methodology. A second order transfer-

Figure 5. The speed system outputs corresponding to the PRBS input signals

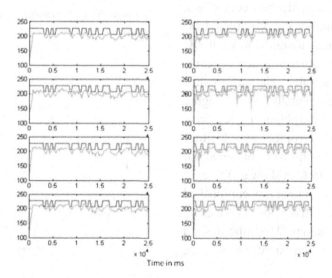

Table 1. The second order WMR model

Linear Transfer Function	High velocities	Medium velocities	Low velocities
G_{DD}	$\dfrac{0.20s^2 - 3.15s + 9.42}{s^2 + 6.55s + 9.88}$	$\dfrac{0.20s^2 + 3.10s + 8.44}{s^2 + 6.17s + 9.14}$	$\dfrac{0.16s^2 + 2.26s + 5.42}{s^2 + 5.21s + 6.57}$
G_{ED}	$\dfrac{-0.04s^2 - 0.60s - 0.32}{s^2 + 6.55s + 9.88}$	$\dfrac{-0.02s^2 - 0.31s - 0.03}{s^2 + 6.17s + 9.14}$	$\dfrac{-0.02s^2 - 0.20s + 0.41}{s^2 + 5.21s + 6.57}$
G_{DE}	$\dfrac{-0.01s^2 - 0.08s - 0.36}{s^2 + 6.55s + 9.88}$	$\dfrac{0.01s^2 + 0.13s + 0.20}{s^2 + 6.17s + 9.14}$	$\dfrac{-0.01s^2 - 0.08s - 0.17}{s^2 + 5.21s + 6.57}$
G_{EE}	$\dfrac{0.31s^2 + 4.47s + 8.97}{s^2 + 6.55s + 9.88}$	$\dfrac{0.29s^2 + 4.11s + 8.40}{s^2 + 6.17s + 9.14}$	$\dfrac{0.25s^2 + 3.50s + 6.31}{s^2 + 5.21s + 6.57}$

function is obtained where PI parameters can be adjusted by selecting an appropriate damping ratio, and considering other design parameters such as overshoot ratio or settling time. In this research, the damping ratio δ is set to 0.7, the overshoot ratio to 10% and the settling time to 2 s. After computing PI time space parameters, appropriate computer control factors are obtained by computing discrete space transforms. The step response of the system is used to compute an adequate sampling time (Astrom & Wittenmark, 1997). In this work the sampling time is set to 100 ms. Table 3 shows the obtained continuous and discrete controllers, where k denotes the sampling period of time, $U(k)$ denote the power signals assigned to each DC motor and $e(k)$ the difference between the commanded and obtained velocities. The performance of PI controllers is experimentally verified. Figure 9 shows the commanded and obtained speed, error and first derivative of the error for low velocities, respectively. Similar results were obtained for the complete set of controllers show at Table 3.

Path-following achievement is accomplished by tracking a sequence of fixed points consisting in positions and orientations. Thus, the desired path to be followed can be composed of a sequence of straight and turn motion planning actions. To perform a straight-trajectory tracking action, both wheels must have same velocity; meaning the robot must keep a constant straight heading. Right and left turn actions are performed by the following algorithm:

$$U_R = U_R + K_R \left(\theta_d - \theta \right)$$
$$U_L = U_L + K_L \left(\theta - \theta_d \right)$$
(7)

Figure 6. Experimental coupling effects on the left wheel

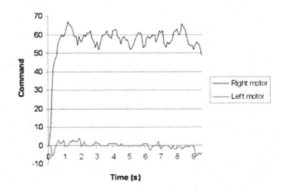

Figure 7. Experimental test of the gains for small velocities

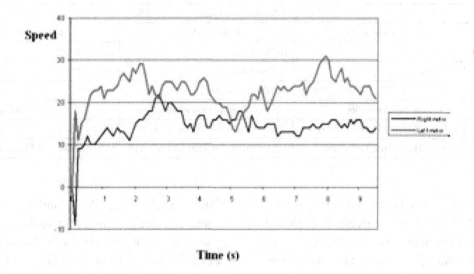

Table 2. The reduced WMR model

Linear Transfer Function	High velocities	Medium velocities	Low velocities
G_{DD}	$\dfrac{0.95}{0.42s+1}$	$\dfrac{0.92}{0.41s+1}$	$\dfrac{0.82}{0.46s+1}$
G_{EE}	$\dfrac{0.91}{0.24s+1}$	$\dfrac{0.92}{0.27s+1}$	$\dfrac{0.96}{0.33s+1}$

Figure 8. System model validation through the comparison of the acquired experimental data

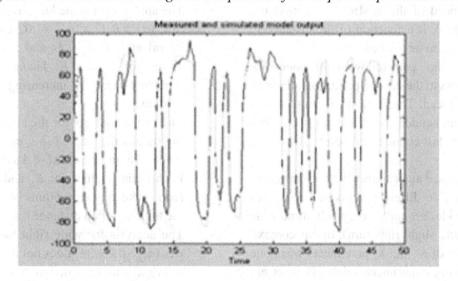

Table 3. Continuous and discrete PI controllers

Continuous space			
	High velocities	**Medium velocities**	**Low velocities**
Right wheel	(0.59s+3.36)/s	(0.70s+3.65)/s	(1.02s+4.59)/s
Left wheel	(-0.04s+2.16)/s	(0.33s+2.40)/s	(0.33s+2.81)/s
Discrete space			
	High velocities	**Medium velocities**	**Low velocities**
Right wheel	U(k) = U(k-1) + 0.59e(k)-0.254e(k-1)	U(k) = U(k-1) + 0.7e(k)-0.54e(k-1)	U(k) = U(k-1) + e(k)-0.54e(k-1)
Left wheel	U(k) = U(k-1) -0.04e(k)+0.54e(k-1)	U(k) = U(k-1) + 0.33e(k)-0.09e(k-1)	U(k) = U(k-1) + 0.33e(k)-0.049e(k-1)

U_R and U_L denote the commanded speed for right and left wheels respectively. θ_d depicts the desired orientation and θ the WMR heading orientation. K_R and K_L are turning factors for wheels right and left. It means that when a straight trajectory is commanded, the difference between θ_d and θ is zero and straight path following is done by commanding the same velocity to each wheel. When a left turn has to be done, θ_d is larger than θ and computation of (7) increases U_R while U_L is decreased. Otherwise, when right turns have to be done θ is larger than θ_d and computation of (7) increases U_L while U_R is decreased. Moreover, high level motion planner considers the following aspects:

- Acceleration ramps have been implemented.
- The speed of the WMR, (slow, medium, and high), is a function of the distance to the point to be tracked.
- When the path-followed distance approaches to the desired point, the velocities are reduced. Therefore, if no more points are commanded the WMR must stop before the last commanded point.

The proposed algorithms have been tested by considering five different kinds of trajectories (straight, wide left turn, slight left turn, wide right turn, and slight right turn). In this context, factor tuning of K_R and K_L parameters was done by considering experimental data (Box, et al.,

2005). Tables 4, 5, 6, 7, and 8, respectively show measured statistics for each trajectory for parameters such as Time (T), Trajectory Error (TE), Travelled Distance (TD), and Averaged Speed (AS). The results were tested for straight trajectories, wide and slight left turns, and wide and slight right turns. From computed means and standard deviations, main and lateral effects of factors K_R and K_L are obtained. Main and lateral effects are presented in Table 9. The importance of the factor effects can be determined using a rough rule that considers the effects when mean value differences are close to or greater than two or three times their standard deviations.

Main and lateral effects are detected for wide left turning, slight left turning, and wide right turning:

- The analysis for wide left turning trajectories, ($\sigma_T = 0.17$ s, $\sigma_{AS} = 1.02$ cm/s) shows lateral effects of -0.4s and 2.4 cm. High values for K_R and K_L factors reduce the trajectory time by increasing the WMR velocity.
- The factor analysis for the slight left turning trajectories ($\sigma_T = 0.2$ s, $\sigma_{TD} = 0.51$ cm) reveals lateral effects of -0.4 s and -1.2 cm. High tuning values for K_R and K_L factors reduce the trajectory time by decreasing the total travelled distance.
- The analysis for wide right turn trajectories ($\sigma_{TE} = 0.26$ cm) does not provide much relevant information, but trajectory error

Figure 9. (a) Comanded signal and mesured speed of left wheel for low velocities; (b) measured error; (c) first derivative of the error

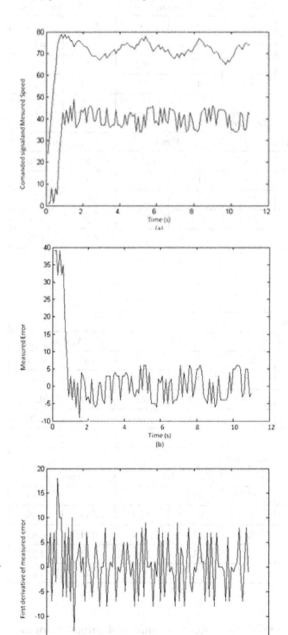

improvement seems to occur as a lateral effect.

Finally, it is noted that lateral effects are not very significant. Therefore, computed values are always less than three times their standard deviations.

Path-Following Experimental Results with PID Controllers

In this subsection previous obtained results of PID are analyzed. The research previously achieved is used to perform function factor adjustments with the aim of obtaining better control laws. PID controllers are tested by using larger trajectories that include a set of the basic trajectories. Figure 10 shows PID results when tracking of the different five trajectories is commanded. Turning asymmetries can be caused because of differences between the right and left velocity models.

Quantitative results were computed in Table 9. The analysis is done by considering the different trajectories:

- There are no significant effects of K_R and K_L factors when straight-line trajectories are commanded.
- Analysis of wide left turning depicts time and averaged-speed lateral effects. Therefore, when high values of K_R and K_L factors are both selected, trajectory-time is decreased while averaged-speed is increased.
- Concerning slight left turns, statistical analysis reveals that K_L factor effect reduces the travelled distance. Therefore, when high values of K_R and K_L factors are selected, a reduction of the travelled-distance and time is accomplished at the same time.

Table 4. Straight trajectory (x, y) in cm: (0, 0) to (0, 120)

Time (s) and Trajectory Error (cm)									
$K_R K_L$	Run 1		Run 2		Run 3		Mean		Variance	
L L (0)	5.1	3.5	4.5	2.7	4.6	2.8	4.7	3	0.07	0.13
L H (1)	4.6	2.1	4.4	2.5	4.7	4.1	4.6	2.9	0.01	0.80
H L (2)	5.6	5.0	4.9	4.5	4.6	3.0	5	4.2	0.18	0.72
H H (3)	5.3	4.7	4.8	3.6	4.7	3.7	4.9	4.0	0.07	0.25
Average of estimated variances (σ_{ID}=0.28s) (σ_{IE}=0.069cm)									0.08	0.47
Travelled distance (cm) and Averaged speed (cm/s)										
$K_R K_L$	Run 1		Run 2		Run 3		Mean		Variance	
L L (0)	117.5	23	117.2	26	114.5	25	116.4	24.7	1.82	1.56
L H (1)	117.3	25.5	116.5	26.5	117.5	25	117.1	25.7	0.19	0.39
H L (2)	118.1	21.1	118.3	24.1	118	25.7	118.1	23.6	0.01	3.63
H H (3)	117.2	22.1	118.3	24.6	116.9	24.9	117.5	23.9	0.36	1.58
Average of estimated variances (σ_{ID}=0.77 cm) (σ_{AS}=1.34 cm/s)									0.59	1.79

Table 5. Wide left turn (x, y) in cm: (0, 0) to (-40, 60)

Time (s) and Trajectory Error (cm)									
$K_R K_L$	Run 1		Run 2		Run 3		Mean		Variance	
L L (0)	3.5	4.0	3.5	3.7	3.9	3.5	3.7	3.7	0.03	0.04
L H (1)	3.7	3.8	3.7	4.1	3.3	4.2	3.4	4.0	0.05	0.03
H L (2)	3.4	2.9	3.4	3.7	3.2	3.9	3.3	3.5	0.01	0.19
H H (3)	3.3	3.9	3.3	4.0	3.5	3.0	3.3	3.6	0.03	0.20
Average of estimated variances (σ_I=0.17 s) (σ_{IE}=0.34 cm)									0.03	0.12
Travelled distance (cm) and Averaged speed (cm/s)										
$K_R K_L$	Run 1		Run 2		Run 3		Mean		Variance	
L L (0)	65.8	18.8	65.9	17.3	64.7	16.6	65.5	17.6	0.30	0.84
L H (1)	64.7	17.5	65.4	20.4	65.6	19.9	65.2	19.3	0.15	1.60
H L (2)	65.1	19.1	65.5	19.8	66.2	20.7	65.6	19.9	0.21	0.43
H H (3)	65.2	19.8	66.5	21.5	65.3	18.7	65.7	20.0	0.15	1.33
Average of estimated variances (σ_{ID}=0.45 cm) (σ_{AS}=1.02 cm/s)									0.20	1.05

- Analysis of wide right turns show a trajectory-error reduction when K_R and K_L factors are set to high values.
- Slight right turn analysis from control law statistics has no relevant effects.

Considering qualitative results, we suggest using a fixed control law for PID controllers when less trajectory-time and larger velocities are pursued. In this way, high values for K_R and K_L factors are proposed for the complete set of trajectories.

Previously obtained control law results are tested by commanding two different trajectories. These trajectories have to include the different five basic trajectories studied before. In order to do

Table 6. Slight left turning (x, y) in cm: (0, 0) to (-20, 80)

	Time (s) and Trajectory Error (cm)									
$K_R K_L$	Run 1		Run 2		Run 3		Mean		Variance	
L L (0)	3.5	3.7	4.2	4.7	3.4	3.3	3.7	3.9	0.13	0.34
L H (1)	3.3	3.4	3.4	3.2	3.2	3.2	3.3	3.3	0.01	0.01
H L (2)	3.4	3.0	3.5	3.3	3.3	3.1	3.4	3.1	0.01	0.02
H H (3)	3.4	3.7	3.3	3.4	3.3	3.3	3.3	3.5	0.01	0.03
Average of estimated variances (σ_I=0.17 s) (σ_{IE}=0.34 cm)									0.04	0.1
	Travelled distance (cm) and Averaged speed (cm/s)									
$K_R K_L$	Run 1		Run 2		Run 3		Mean		Variance	
L L (0)	80.2	22.9	81.0	19.3	80.5	23.7	80.6	22.0	0.12	3.66
L H (1)	78.5	23.8	78.8	23.2	79.5	24.9	78.9	24.0	0.13	0.50
H L (2)	81.1	23.8	80.6	23.0	79.1	24.0	80.3	23.6	0.72	0.23
H H (3)	79.3	23.3	79.7	24.2	79.1	24.0	79.4	23.8	0.06	0.15
Average of estimated variances (σ_{ID}=0.51 cm) (σ_{AS}=1.07 cm/s)									0.26	1.14

Table 7. Wide right turn (x, y) in cm: (0, 0) to (40, 60)

	Time (s) and Trajectory Error (cm)									
$K_R K_L$	Run 1		Run 2		Run 3		Mean		Variance	
L L (0)	3.3	2.7	3.0	1.7	3.2	1.8	3.2	2.1	0.01	0.20
L H (1)	3.0	1.8	3.0	1.4	3.1	1.7	3.0	1.6	0.003	0.03
H L (2)	3.0	2.1	3.1	2.2	3.0	2.2	3.0	2.2	0.003	0.003
H H (3)	3.2	1.4	3.3	1.6	3.2	1.2	3.2	1.4	0.003	0.03
Average of estimated variances (σ_I=0.07 s) (σ_{TE}=0.26 cm)									0.005	0.07
	Travelled distance (cm) and Averaged speed (cm/s)									
$K_R K_L$	Run 1		Run 2		Run 3		Mean		Variance	
L L (0)	65.3	19.8	66.1	22.0	66.3	20.7	65.9	20.8	0.19	0.82
L H (1)	65.2	21.7	65.1	21.7	67.2	21.7	65.8	21.7	0.94	0
H L (2)	65.7	21.9	66.3	21.4	65.3	21.8	65.8	21.7	0.17	0.05
H H (3)	65.8	20.6	66.2	20.1	66.7	20.8	66.2	20.5	0.14	0.09
Average of estimated variances (σ_{ID}=0.6 cm) (σ_{AS}=0.49 cm/s)									0.36	0.24

this, a set of twelve points with a shape similar to a dodecagon is proposed. The trajectory tracking is tested in clockwise and anticlockwise sense for testing both right and left turns. In order to analyze trajectory accuracy the sequence of points is commanded each time that the previous point is reached. It is pointed that this produces a reduction of speed each time that the WMR is close to the final commanded point. However, a better trajectory-tracking accuracy is accomplished. As in previous sections, three different runs were tried for each control law and trajectory. In this way, the averaged value of the three runs makes possible a statistical analysis for each control law and trajectory. The measured parameters are also Time (T), Trajectory Error (TE), Travelled Distance (TD),

Table 8. Slight right turning (x, y) in cm: (0, 0) to (20, 80)

$K_R K_L$	Time (s) and Trajectory Error (cm)									
	Run 1		Run 2		Run 3		Mean		Variance	
L L (0)	3.3	1.0	3.3	0.4	3.4	1.0	3.3	0.8	0.003	0.08
L H (1)	3.5	1.6	3.3	0.8	3.3	0.9	3.4	1.1	0.01	0.13
H L (2)	3.4	1.2	3.4	2.2	3.4	1.3	3.4	1.6	0	0.20
H H (3)	3.4	0.6	3.6	1.1	3.3	0.5	3.4	0.7	0.016	0.07
Average of estimated variances (σ_I=0.09 s) (σ_{TE}=0.35 cm)									0.007	0.12
$K_R K_L$	Travelled distance (cm) and Averaged speed (cm/s)									
	Run 1		Run 2		Run 3		Mean		Variance	
L L (0)	80.6	24.4	79.3	24.0	80.5	23.7	80.1	24.0	0.41	0.08
L H (1)	80.6	23.0	80.0	24.2	79.6	24.1	80.1	23.8	0.17	0.56
H L (2)	79.8	23.5	80.2	23.6	80.0	23.5	80.0	23.5	0.01	0
H H (3)	80.1	23.6	80.3	22.3	79.5	24.1	80.0	23.3	0.12	0.58
Average of estimated variances (σ_{ID}=0.45 cm) (σ_{AS}=0.55 cm/s)									0.18	0.31

and Averaged Speed (AS). Table 10 depicts the statistical results obtained for the clockwise and anticlockwise dodecagon trajectories.

Figures 11 and 12 show the path following obtained results for clockwise and anticlockwise trajectories.

Comparing clockwise and anticlockwise trajectories with PID control laws shows that negative time, error, and travelled distance effects are obtained while positive averaged speed are also detected. In this context, using the available WMR platform, right-turning trajectories can be performed with less time and error than left turning trajectories. Moreover, right turning trajectories are performed with less travelled distance and with more averaged speed.

CONCLUSION AND FUTURE RESEARCH DIRECTIONS

The strategies presented in this chapter are simple and effective in order to perform path-following strategies. However, some aspects should be mentioned and can be improved. Reported studies show that right turning trajectories are imple-

mented with better performance than left turning trajectories. The reason of such asymmetries can be found in Tables 1 and 2, due to the different models obtained for left and right DC motors. In this way, future work should be oriented to test other motion planning strategies of PID controllers that reduce the total travelled distance, error, and time during WMR navigation. Therefore, including other parameters that consider the error deviation of the WMR by formulating new control laws will address our future research efforts. In spite of expending time performing larger trajectories, the goal is to reduce the trajectory error as a means to decrease additional path so that the task can be performed with less time and greater accuracy. Moreover, studies of strategies that do not stop at the different desired points will also focus future work. To compare PID controllers with other control methods is another interesting objective.

ACKNOWLEDGMENT

This work has been partially funded by the Ministry of Science and Innovation of Spain through

Table 9. Main and lateral effects

Parameter Performance	Main Effect K_R factor	Main Effect K_L factor	Lateral Effect K_R & K_L factors
Straight line trajectory			
Time (σ_T = 0.28s)	0.3	-0.1	0.2
Trajectory Error (σ_{TE} = 0.69cm)	1.2	-0.1	1.0
Travelled distance (σ_{TD} = 0.77 cm)	1.1	0	1.1
Averaged speed (σ_{AS} = 1.34 cm/s)	-1.45	0.7	-0.8
Wide left turn trajectory			
Time (σ_T = 0.17 s)	-0.25	-0.15	-0.4
Trajectory Error (σ_{TE} = 0.34 cm)	-0.3	-0.2	-0.1
Travelled distance (σ_{TD} = 0.45 cm)	0.3	-0.1	0.2
Averaged speed (σ_{AS} = 1.02 cm/s)	1.5	1.4	2.4
Slight left turn trajectory			
Time (σ_T = 0.2 s)	-0.15	-0.25	-0.4
Trajectory Error (σ_{TE} = 0.32 cm)	-0.3	-0.1	-0.4
Travelled distance (σ_{TD} = 0.51 cm)	0.1	-1.3	-1.2
Averaged speed (σ_{AS} = 1.07 cm/s)	0.7	1.1	1.8
Wide right turn trajectory			
Time (σ_T = 0.07 s)	0	0	0
Trajectory Error (σ_{TE} = 0.26 cm)	-0.05	-0.65	-0.7
Travelled distance (σ_{TD} = 0.6 cm)	0.15	0.15	0.3
Averaged speed (σ_{AS} = 0.49 cm/s)	-0.15	-0.15	-0.3
Slight right turn trajectory			
Time (σ_T = 0.09 s)	0.05	0.05	0.1
Trajectory Error (σ_{TE} = 0.35 cm)	0.2	-0.3	-0.1
Travelled distance (σ_{TD} = 0.45 cm)	-0.1	0	-0.1
Averaged speed (σ_{AS} = 0.55 cm/s)	-0.5	-0.2	-0.7

Figure 10. Trajectory-tracking results for PID controllers

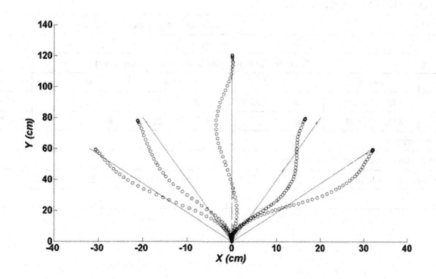

Figure 11. Clockwise dodecagon trajectory tracking results for PID. PID is drawn using dots and desired trajectory with solid lines

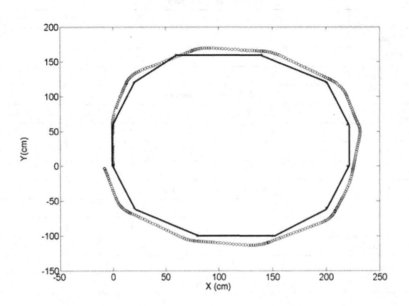

Table 10. Clockwise and anticlockwise decagon analysis for PID control laws

	Time (s) & Trajectory Error (cm)									
Trajectory	**Run1**		**Run2**		**Run3**		**Mean**		**Variance**	
Clockwise	35.8	5.7	36	5.3	36.9	4.9	36.2	5.3	0.23	0.11
Anticlockwise	41.5	6.8	38.3	6.4	38.6	6.5	39.5	6.6	2.08	0.03
Average of estimated variances									1.16	0.07
Difference between clockwise and anticlockwise means										
Time effect (σ_T=1.1s)				Trajectory error effect (σ_{TE}=0.26cm)						
-3.3s				-1.3cm						
Travelled distance (cm) & Averaged speed (cm/s)										
Trajectory	**Run 1**		**Run 2**		**Run 3**		**Mean**		**Variance**	
Clockwise	832	23.2	826	22.9	819	22.2	826	22.8	28.3	0.18
Anticlockwise	851	20.5	846	22.1	847	21.9	848	21.5	4.7	0.51
Average of estimated variances									16.5	0.35
Difference between clockwise and anticlockwise means										
Travelled distance effect (σ_{TD}=4.1cm)				Averaged speed effect (σ_{AS}=0.59cm/s)						
-22cm				1.3cm/s						

Figure 12. Anticlockwise dodecagon trajectory tracking results for PID. PID is drawn using dots and desired trajectory with solid lines

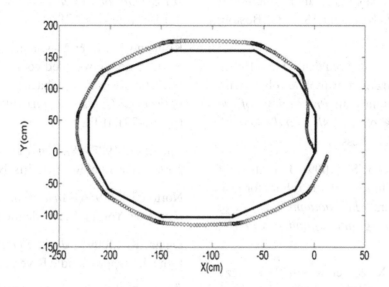

the project DPI2011-27567-C02-01. The authors wish to thank the reviewers for their comments that improved the work quality.

REFERENCES

Aguiar, A. P., & Pascoal, A. (2000). Stabilization of the extended nonholonomic double integrator via logical based hybrid control: An application to point stabilization of mobile robots. In *Proceedings of the 6th International IFAC Symposium on Robot Control*. Vienna, Austria: IFAC.

Aicardi, M., Casalino, G., Bicchi, A., & Balestrino, A. (1995). Closed loop steering of unicycle-vehicles via Lyapunov techniques. *IEEE Robotics & Automation Magazine*, 2(1), 27–35. doi:10.1109/100.388294

Astolfi, A. (1996). Discontinuous control of non-holonomic systems. *Systems & Control Letters*, 27, 37–45. doi:10.1016/0167-6911(95)00041-0

Astrom, K. J., & Wittenmark, B. (1997). *Computer-controlled systems: Theory and design* (3rd ed.). Upper Saddle River, NJ: Prentice Hall.

Bloch, A. M. (Ed.). (2003). *Nonholonomic mechanics and control*. New York, NY: Springer. doi:10.1007/b97376

Bloch, A. M., McClamroch, N. H., & Reyhanoglu, M. (1990). Controllability and stabilizability properties of a nonholonomic control system. In *Proceedings of the IEEE Conference on Decision and Control*. Honolulu, HI: IEEE.

Box, G. E. P., Hunter, J. S., & Hunter, W. G. (2005). *Statistics for experimenters*. Hoboken, NJ: Wiley.

Brockett, R. W. (Ed.). (1983). Asymtotic stability and feedback stabilization. In R. W. Brockett, R. S. Millman, & H. S. Sussman (Eds.), *Differential Geometric Control Theory*, (pp. 181-191). Boston, MA: Birkhäuser.

Canudas de Wit, C., & Sørdalen, O. J. (1991). Exponential stabilization of mobile robots with nonholomic constraints. In *Proceedings of the IEEE Conference on Decision and Control.* Brighton, UK: IEEE Press.

Capi, G., & Kaneko, S. (2009). Evolution of neural controllers in real mobile robots for task switching behaviors. *International Journal of Innovative Computing, Information, & Control, 5*(11), 4017–4024.

Cheng, Y., Wang, X., & Lei, R. (2009). A fuzzy control system for path following of mobile robots. *ICIC Express Letters, 3*(3), 403–408.

Coron, J. M. (1991). Global asymtotic stabilization for controllable systems without drift. *Mathematical Constrained System Design, 15*, 295–312.

Dayawansa, W. P., Martin, C. F., & Samuelson, S. (1995). Asymtotic stabilization of a generic class of 3-dimensional homogeneus quadratic systems. *Systems & Control Letters, 24*(2), 115–123. doi:10.1016/0167-6911(94)00040-3

De Luca, A., Oriolo, G., & Vendittelli, M. (2001). Control of wheeled mobile robots: An experimental overview. In *RAMSETE: Articulated and Mobile Robots for Service and Technology.* Berlin, Germany: Springer.

Hauser, J., & Hindman, R. (1995). Maneuver regulation from trajectory tracking: Feedback linearizable systems. In *Proceeding World Congress of International Federation Automatic Control, Symposium of Nonlinear Control System Design,* (pp. 638-643). IEEE.

Kolmanovky, I., & McClamroch, N. H. (1995). Developments in nonholomic control problems. *IEEE Control Systems Magazine, 15*, 20–36. doi:10.1109/37.476384

Kolmanovky, I., & McClamroch, N. H. (1996). Stabilization of wheeled vehicles by hybrid non-linear time-varying feedback laws. In *Proceedings of the IEEE Conference on Control Applications,* (pp. 66-72). IEEE Press.

Ljung, L. (1987). *System identification: Theory for the user.* Englewood Cliffs, NJ: Prentice Hall.

Norton, J. P. (1986). *An introduction to identification.* New York, NY: Academic Press.

Ogata, K. (2009). *Modern control engineering* (5th ed.). Upper Saddle River, NJ: Prentice Hall.

Ögren, P., & Leonard, N. (2005). A convergent dynamic window approach to obstacle avoidance. *IEEE Transactions on Robotics, 21*(2), 188–195. doi:10.1109/TRO.2004.838008

Pacheco, L., Luo, N., Ferrer, I., Cufí, X., & Arbusé, R. (2012). Gaining control knowledge through an applied mobile robotics course. In *Mobile Robots Current Trends* (pp. 69–86). Rijeka, Croatia: InTech. doi:10.5772/27994

Pourboughrat, F., & Karlsson, M. P. (2002). Adaptive control of dynamic mobile robots with nonholomic constraints. *Computers & Electrical Engineering, 28*, 241–253. doi:10.1016/S0045-7906(00)00053-7

Reeds, J. A., & Shepp, L. A. (1990). Optimal paths for a car that goes forwards and backwards. *Pacific Journal of Mathematics, 145*.

Rosier, L. (1992). Homogeneous Lyapunov functions for homogeneous continuous vector field. *Systems & Control Letters, 19*(6), 467–473. doi:10.1016/0167-6911(92)90078-7

Siciliano, B., Bicchi, A., & Valigi, P. (Eds.). (2001). Control of wheeled mobile robots: An experimental overview. *Lecture Notes in Computer Information Science, 270*, 181-226.

Sørdalen, O. J., & Canudas de Wit, C. (1993). Exponential control law for a mobile robot: Extension to path following. *IEEE Transactions on Robotics and Automation, 9*(6), 837–842. doi:10.1109/70.265927

Struemper, H., & Krishnaprasad, P. S. (1997). On approximate tracking for systems on three-dimensional matrix lie groups via feedback nilpotentization. In *Proceedings of the International Federation Automatic Control Symposium on Robot Control*. IEEE.

ADDITIONAL READING

Butler, Z. (2006). Corridor planning for natural agents. In *Proceedings of the IEEE International Conference Robotics and Automation*, (pp. 499-504). IEEE Press.

Camacho, E. F., & Bordons, C. (2002). *Model predictive control*. Berlin, Germany: Springer-Verlag.

Campbell, J., Sukthankar, R., & Nourbakhsh, I. (2004). Techniques for evaluating optical flow in extreme terrain. In *Proceedings of IEEE International Conference of Robots and Systems*. IEEE Press.

DeSouza, G. N., & Kak, A. C. (2002). Vision for mobile robot navigation: A survey. *IEEE Transactions on Pattern Analysis and Machine Intelligence, 24*(2). doi:10.1109/34.982903

Elfes, A. (1989). Using occupancy grids for mobile robot perception and navigation. *IEEE Computer, 22*(6), 46–57. doi:10.1109/2.30720

Fernandez, J., & Casals, A. (1997). Autonomous navigation in ill-structured outdoor environments. In *Proceedings of the IEEE International Conference on Intelligent Robots and Systems*, (pp. 395-400). IEEE Press.

Fox, D., Burgard, W., & Thun, S. (1997). The dynamic window approach to collision avoidance. *IEEE Transactions on Robotics and Automation, 4*, 23–33. doi:10.1109/100.580977

Gaussier, P., Joulain, C., & Zrehen, R. A. (1997). Visual navigation in a open environment without map. In *Proceedings of the IEEE International Conference on Intelligent Robots and Systems*, (pp. 545-550). IEEE Press.

Gluer, D., & Schmidt, G. (2000). A new approach for context based exception handling in autonomous mobile service robots. In *Proceedings of IEEE International Conference on Robotics and Automation*, (vol. 4, pp. 3272-3277). IEEE Press.

Gonzalez, R. C., & Woods, R. E. (2002). *Digital image processing* (2nd ed.). Englewood Cliffs, NJ: Prentice Hall.

Hashima, M., Hasegawa, F., Kanda, S., Maruyama, T., & Uchiyama, T. (1997). Localization and obstacle detection for a robot carrying food trays. In *Proceedings of the IEEE International Conference on Intelligent Robots and Systems*, (pp. 345-351). IEEE Press.

Horn, B. K. P. (1998). *Robot vision*. New York, NY: McGraw-Hill.

Jones, S. D., Andresen, C., & Crowley, J. L. (1997). Appearance based processes for visual navigation. In *Proceedings of the IEEE International Conference on Intelligent Robots and Systems*, (pp. 551-557). IEEE Press.

Kim, D., & Nevatia, R. (1998). Recognition and localization of generic objects for indoor navigation using functionality. *Image and Vision Computing, 16*(11), 729–743. doi:10.1016/S0262-8856(98)00067-5

Klancar, G., & Skrjanc, I. (2007). Tracking-error model-based predictive control for mobile robots in real time. *Robotics and Autonomous Systems, 55*, 460–469. doi:10.1016/j.robot.2007.01.002

Lozano-Perez, T. (1983). Spatial planning: A configuration space approach. *IEEE Transactions on Computers, 32*, 108–120. doi:10.1109/TC.1983.1676196

Maciejowski, J. M. (2002). *Predictive control with constraints*. Upper Saddle River, NJ: Prentice Hall.

Michalska, H., & Mayne, D. Q. R. (1993). Robust receding horizon control of constrained nonlinear systems. *IEEE Transactions on Automatic Control, 38*, 1623–1633. doi:10.1109/9.262032

Moravec, H. P. (1988). Sensor fusion in certainty grids for mobile robots. *AI Magazine, 9*(2), 61–74.

Murray, R. M., Aström, K. J., Boyd, S. P., Brockett, R. W., & Stein, G. (2003). Future directions in control in an information-rich world. *IEEE Control Systems Magazine*. Retrieved from http://www.stanford.edu/~boyd/papers/pdf/cdspanel-csm.pdf

Noborio, H., & Schmidt, G. (1996). Mobile robot navigation under sensor and localization uncertainties. In *Autonomous Mobile Systems* (pp. 118–127). Berlin, Germany: Springer-Verlag. doi:10.1007/978-3-642-80324-6_11

Ogata, K. (1996). *Discrete time control systems* (2nd ed.). Upper Saddle River, NJ: Prentice Hall.

Pacheco, L., & Luo, N. (2006). Mobile robot experimental modelling and control strategies using sensor fusion. *Control Engineering and Applied Informatics, 8*(3), 47–55.

Pacheco, L., & Luo, N. (2007). Trajectory planning with a control horizon based on a narrow local grid perception: A practical approach for WMR with monocular visual data. In *Lecture Notes in Control and Information Sciences*. Berlin, Germany: Springer-Verlag.

Pacheco, L., Luo, N., & Cufí, X. (2010). *Control for navigation of a mobile robot using monocular data*. Saarbrucken, Germany: Lambert Academic Publishing.

Rao, C. V., Rawling, J. B., & Mayne, D. Q. (2003). Constrained state estimation for nonlinear discrete-time systems: Stability and moving horizon approximations. *IEEE Transactions on Automatic Control, 48*, 246–258. doi:10.1109/TAC.2002.808470

Rawling, J. B., & Muske, K. R. (1993). The stability of constrained receding horizon control. *IEEE Transactions on Automatic Control, 38*, 1512–1516. doi:10.1109/9.241565

Rimon, E., & Koditschek, D. (1992). Exact robot navigation using artificial potential functions. *IEEE Transactions on Robotics and Automation, 8*(5), 501–518. doi:10.1109/70.163777

Schäfer, H., Proetzsch, M., & Berns, K. (2007). Obstacle detection in mobile outdoor robots. In *Proceedings of the International Conference on Informatics in Control, Automation and Robotics*, (pp. 141-148). IEEE.

Schilling, R. J. (1990). *Fundamental of robotics*. Upper Saddle River, NJ: Prentice Hall.

Wilcox, B., Matthies, L., Gennery, D., Copper, B., Nguyen, T., & Litwin, T. … Stone, H. (1992). Robotic vehicles for planetary exploration. In *Proceedings of the IEEE International Conference on Robotics and Automations*, (pp. 175-180). IEEE Press.

KEY TERMS AND DEFINITIONS

Control Law: Implemented algorithms that constrain the output system values, within a predefined range, by using some related input and feedback signals.

Experimental Factor Tuning: Adjustments of the system parameter values based on statistical results, which are done by performing experimental test with high and low factor values.

Model-Based Control: Control system based on the previous knowledge of the system response as consequence of the previous states and sequence of produced inputs.

Odometer System: Sensorial on-robot system used for computing the incremental positioning of the robot as well as other kinematic parameters as velocities and accelerations.

Path-Following: Constrained robot navigation in which the robot should follow a determined trajectory or trajectories.

PID Controllers: Classic controller in which the control action is a function, proportional-integral-derivative, of the error between desired and obtained output values.

System Identification: Experimental methods performed with the aim of having the model of the system.

WMR: Wheeled mobile platform that has a certain degree for developing intelligent actions with autonomous capabilities.

Chapter 4
Mobile/Wireless Robot Navigation

Amina Waqar
National University of Computers and Emerging Sciences, Pakistan

ABSTRACT

Sensor-based localization has been found to be one of the most preliminary issues in the world of Mobile/ Wireless Robotics. One can easily track a mobile robot using a Kalman Filter, which uses a Phase Locked Loop for tracing via averaging the values. Tracking has now become very easy, but one wants to proceed to navigation. The reason behind this is that tracking does not help one determine where one is going. One would like to use a more precise "Navigation" like Monte Carlo Localization. It is a more efficient and precise way than a feedback loop because the feedback loops are more sensitive to noise, making one modify the external loop filter according to the variation in the processing. In this case, the robot updates its belief in the form of a probability density function (pdf). The supposition is considered to be one meter square. This probability density function expands over the entire supposition. A door in a wall can be identified as peak/rise in the probability function or the belief of the robot. The mobile updates a window of 1 meter square (area depends on the sensors) as its belief. One starts with a uniform probability density function, and then the sensors update it. The authors use Monte Carlo Localization for updating the belief, which is an efficient method and requires less space. It is an efficient method because it can be applied to continuous data input, unlike the feedback loop. It requires less space. The robot does not need to store the map and, hence, can delete the previous belief without any hesitation.

DOI: 10.4018/978-1-4666-2658-4.ch004

Copyright © 2013, IGI Global. Copying or distributing in print or electronic forms without written permission of IGI Global is prohibited.

INTRODUCTION TO NAVIGATION

Mobile Robotics is a vast topic and covers various strategies, and people have been working on it for a long time. It has some ancient as well as some modern techniques. Although a lot of work has been done in this field, robots cannot accept oral instructions by humans.

Navigation is the process of directing and monitoring a robot (as in our case) from one point to another. In addition, the extent or the range to which a mobile can navigate is called its resolution. An independent device should know its original position within its range (of sensors). For this purpose we have:

- Global Navigation;
- Local Navigation;
- Personal Navigation.

In Global Navigation, the robot itself determines its position relative to the environment and with the help of this location, it moves to the desired location. In Local Navigation, the robot determines its position relative to the objects in the environment and hence makes decisions to move after communication with them. In Personal Navigation, the robot is aware of the positioning. This case involves the monitoring of the individual robot and anything in contact with it.

Two technologies are used for Mobile Robot Navigation:

- Satellite-Based Global Positioning System

The first satellite-based navigation system was created in 1973 by the American Defense Department as a joint service between US Navy and Air Force. Limited access was given to non-military users. Presently, we have 24 space segmented satellites around the world. Each position on the earth requires 3 satellites to secure it. GPS provides accuracy 95% of the time. However, it may not work indoors because of some reasons like signal blockage and multipath interference.

- Image-Based Positioning System

This system works exclusively using optical sensors and the range is usually found by laser-based range finder, while the photometric cameras use CCD arrays. Extraction of the required information from these range finders and photometric cameras is extracted as follows:

- Environment representation.
- Sensing the environment.
- Applying Localization algorithm.

Techniques used for vision-based positioning are:

- Landmark-based positioning.
- Model-based approaches (3-D model or digital map).
- Feature-based positioning.

We can use vision-based techniques combined with other techniques like ultrasonic and laser-based sensors.

AIM OF THE CHAPTER

The main emphasis of this chapter will be to introduce mobile robot navigation rather than its tracking. Tracking is simple and can easily be implemented using a feedback loop (Phase locked loop). Kalman filter is one such example, which uses the phase locked loop and takes the average of the real and estimated positions of the robot to give the final position of the robot. Kalman Filter, however, is less accurate because it applies a weighted average methodology on the observed outcome of the next measurement. In addition, the weights are calculated with the help of covariance resulting in less accuracy. Kalman Filter basically tries to estimate a random state **'x'** of a discrete-time controlled process which can be represented by the following equation:

$$X_t = F_t x_{t-1} + B_t u_t + w_t \qquad (1)$$

where 'F' is the transition model, 'B' is the control input model and 'w' is the process noise.

In this chapter, we will try to solve this problem of less precise tracking occurring because of the use of a phase locked loop with the help of Markov Localization and Monte Carlo Localization.

The difference between navigation and tracking is that navigation monitors the user and directs the user from one point to another (as mentioned above), whereas tracking just updates the current location of the user on a given map. This concept motivates us to proceed to navigation.

LOCALIZATION

In Localization, the mobile robot extracts its position via global localization or via position tracking. In Global Localization, the robot needs to determine its original position with respect to the environment with the help of its sensor's readings. While in the case of position tracking, the initial position is already known to the robot. This position will be the initial belief of the robot.

WORKING

The robot has a belief, which is updated in the form of frames of space with fixed unit lengths, depending on the range of sensors attached to it. This belief is updated after every unit time interval as discussed above.

BELIEF UPDATE USING MARKOV LOCALIZATION

First of all, the robot will build a map as its belief and the map built in accordance with the range available is within its range (that is, in unit square area). The number of sensors affects the accuracy of mobile robot positioning. If there is more than one robot, then proper communication must take place between them. They should continuously be checking if they have entered into the range of any other robot. Now, as robots are performing multi tasking, the speed of the position computation and navigation should be maintained. In addition, the greater the number of sensors, the more computation time will be taken by the processor placed within the mobile robot. Hence, all the factors are dependent on the processor's efficiency (Berg, 2004).

Various methods can be used to update the belief of the robot. We use the Markov Localization, which is a pure probabilistic algorithm that will help us a lot in updating the robot's belief in the form of its probability density functions. The mobile robot stores its previous belief as well as calculates its recent belief. It then convolves the initial belief of the robot with the one, which has been updated after the unit time interval to get the final belief of the robot. In this time interval, the robot may or may not move, depending upon the user requirements and instructions given to the robot.

We will use Monte Carlo Localization to work in multi modal space. Multi modal space is useful for global localization as it uses more than one mode of space also useful for more than multiple instances. Monte Carlo Localization is applied on the received probability density function of each sensor attached on the robot, which then processes all of them and reproduces a final probability density function which is called the final belief of the robot.

The Robot finally makes a final decision about its next action based on this final belief (probability density function).

WORKING OF MONTE CARLO

Monte Carlo is an algorithm that is very helpful when the input is usually undeterministic. The input from the sensors of the robot is coming as a ray reflected from water, wall, or any such possible hindrance. Hence, Monte Carlo handles a high degree of freedom.

Since we are using integrals in our case as the range of the sensors, it can solve any kind of boundary condition on the integral. The Monte Carlo Algorithm has a general pattern that is stated below (Lee, 2009):

1. Declare all the possible inputs.
2. Produce the inputs with the help of a probability distribution (Markov Localization in our case) for the given range.
3. Perform an algorithm on the inputs that informally predict an output.
4. Unite all the results.

For Example Monte Carlo is used in games like bingo, connect four and battleship, where the best shot is estimated by first defining the range where the shots can be made. Then random shots are made. But possible shots are limited so selective shots can be made (Figure 1).

TRACKING AND KALMAN FILTER

Kalman Filter named after Rudolf E. Kalman works on the principle of recursion and weighted average. Kalman Filter has been used in many electronic communications equipment for the guidance of the earth vehicles as well as spacecraft (where sensors take the position and velocity as input for the location update on the map).

Kalman Filter uses discrete data and feeds the filter with a sampled input after every defined time interval. However, at the same time, it is useful as it can predict the future, also with the help of this recursion. Recursion is helpful in minimizing the mean error. One of its applications is FM radio that also uses Phase Locked Loop (PLL). Kalman Filter is basically used for tracking purpose and the main issue in our case is tracking and avoiding noise from the measurements as it gives output as the average of all previous values.

It follows the rule of "predict" and "update" in its phase locked loop. It tries to predict the noise.

Figure 2 shows how the Kalman filter works. The black boxes show the original position. Green stars show the estimated position and red crosses show the modified position by taking averages of both and trying to remove the error which is still present in the measurements. Therefore, there are more chances of error because of the average procedure.

Although Kalman filter is not a good choice because of error and noise present in the readings, Nathan Funk (2003) proved that Kalman filter can be used as a probabilistic prediction technique, making it more robust.

Figure 1. Bingo

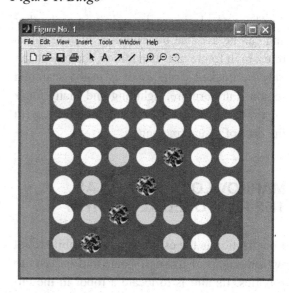

Figure 2. Tracking using Kalman filter

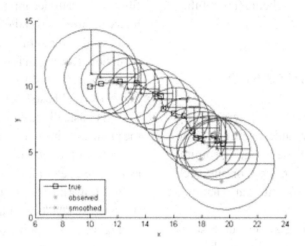

The results are improved and show the difference from simply guessing the values (by taking average.). However, the object might not stay in the region because of the following reasons:

- The object is moving too fast.
- The frame rate is too slow (Keeping the concept of belief frame in mind as explained above).
- The searched region of the sensors is too small.

These problems are dependent on each other and can be avoided by a number of constraints like a high frame rate. In addition, the object can only be tracked if the robot is in the searched region. Furthermore, lightning and many other factors can affect the appearance of the object (in the case of visual tracking).

MARKOV LOCALIZATION AS THE PROPOSED ALGORITHM

Localization is the estimation of a mobile robot with the help of its sensor data and is basic to robotics. Its aim is to locate a robot in the en-

vironment in which it is present, given the map of the environment, which is built by the robot itself as the first step of robot navigation. The environment may or may not be static. Hence, the method used for localization depends on the scenario. We have global techniques, which we use for global uncertainty, that are applicable in all scenarios and environments. For example, these techniques are applicable even for the kidnapped Mobile Robot problem, in which the robot starts from an arbitrary unknown position (Fox, Burgard, & Thrun, 1999).

We would use Markov Localization for the robot to estimate its position in any environment. Markov Localization is a purely probabilistic algorithm and maintains a probability density function over the entire space available and for all poses of the robot. We did this by using four sensors, one on each side face of the robot.

It mainly caters to the problem of state estimation with the help of the sensor values of the robot. It does not consider only a single investigation output. Rather, it processes all the sensor data and maintains a single probability density over the range available (As was explained in 'Monte Carlo'). Global localization is carried out because of the probabilistic advance we are using in this algorithm. In addition, because of the probabilistic algorithm, we can represent the hypothesis in a mathematical way; hence, the belief of the robot is a probability density function and can easily be understood by the humans. The main concept is:

$$Bel\,(l) = \int P(l|l',a_t)Bel(l')dl' \qquad (2)$$

Previous research was based on the static environment of the robot, which is not the case always. These researches failed in the dynamic environment because people around the robot are moving. Our method is highly accurate as it produces a probability density function according to every object placed in the surroundings of the robot and because the probability density functions are being updated after every unit time.

MATHEMATICAL FORMULA

Markov Localization is a recursive localization strategy that basically uses all the available data to estimate the posterior distribution over L_a (Russell & Norvig, 1995).

$$P(L_a = l|s) = P(L_a = s0, s1, s2, s3) \qquad (3)$$

with the help of Baye's rule for any sensor s_a,

$$\frac{P\left(s_a \mid s0, s1, s2, s3, \ L_a = 1\right) \ P\left(L_a = 1 \mid s0, s1, s2, s3\right)}{P\left(s_a \mid s0, s1, s2, s3\right)} \qquad (4)$$

This can now be reduced with the help of the Markov assumption,

$$\frac{P\left(s_a \mid L_a = 1\right) \ P\left(L_a = 1 \mid s0, s1, s2, s3\right)}{P\left(s_a \mid s0, s1, s2, s3\right)} \qquad (5)$$

where the denominator can be replaced by a constant 'k' because it does not depend on the position L_a. Hence, we can now write:

$$P(L_a = l|s) = k.P(s_a|l)\,P(L_a = l|s0, s1, s2, s3) \qquad (6)$$

or

$$Bel(L_a = l) = k.P(s_a|l)\,Bel(L_{a-1} = l) \qquad (7)$$

Algorithm (Basye, Dean, Kirman, & Lejter, 1992): Let us denote the action of the robot by 'a_t'. So if the robot moves, then its belief which is a uniform distribution initially would be modified.

Initialization: Let the initial position be l' and its belief is $Bel(L_0 = l')$, which is $P(L_0 = l')$. The modified position is l and its belief is $Bel(L_a = l)$, which is $P(L_a = l)$.

Do Forever: (Where k approaches 0) And it will forever perform the following given an action is performed:

$$Bel(L_a = l) = P(s_a|l)\,Bel(L_{a-1} = l) \qquad (8)$$

Both of the beliefs would be convolved as stated above.

For each Location l: (normalizing the belief)

$$Bel'(L_a = l) = k^{-1}.Bel(La = l) \qquad (9)$$

To the Robot:

$$Bel(L_a = l) = \int P(l|l',a_t)Bel(L_{a-1} = l')dl' \qquad (10)$$

EXAMPLE AND WORKING

- Assumptions:
  - We are considering one and two dimensional spaces.
  - The robot does not know its original position.
  - The initial position of uncertainty is determined by the Markov Localization. (Initially as a uniform probability density function.)

• **Sensor Functionality:** Sensors are now prompted and probability density functions are now updated as a peak if a door comes near the robot (within its range) and a relatively lowered probability density function in the case of the wall. The Robot has a multimodal belief and this shows that the information is insufficient for global localization.

• **Belief Modification:** This new probability density function is now multiplied with the previous uniform belief to obtain the final belief. After the multiplication, the new probability density function tells the robot about its confident position as most of the probability would be centered at a point (Cox, 1991).

• **Role of Noise:** Noise in the sensor reading may alter the belief of the robot (its probability density function). If only one of the sensors is showing door in the probability density function, it is not necessary to ignore the possibility of being next to the door. This noise leads to a loss of the information. In addition, this information loss may cause the new probability density function to miss some points on its graph, making it more smooth but indecisive at the same time.

MONTE CARLO LOCALIZATION

Monte Carlo Localization is used to determine the position of the robot given the map of the environment built up by the Markov Localization. It is a form of particle filter which estimates the sequence of hidden parameters x_k (where k can be any value), given y_k.

$$p(x_k|y_0,y_1,\ldots,y_k) \tag{11}$$

In this method a large number of imaginary current configurations are originally randomly sprinkled in the space of all possible positions that the robot systems can occupy. With each sensor update, the probability of each imaginary pattern is correct.

Figure 3. Graphs

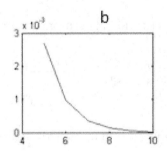

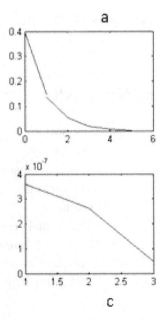

In Monte Carlo Localization, the space is digitized into some random samples (k). As discussed above, multi modal representation is used because of the global localization. Because of this reason, it requires less memory and is computationally more capable. Grid-based approaches were also used, but more memory was required because of 3-D figures (Oxford, 2012).

OUR MODEL

We have four sensors attached on the robot. They all emit a signal and wait for whether the radiation is reflected back or not. If the signal reflected back n gives a '1' if it is a wall and if there is no reflection, it is read as '0'. This data is for the range of unit length square around the robot (Chung, 1960).

In Figure 1, the belief of the robot is shown as it is moving away from the wall. In Figure 2, again, it is moving away from the wall towards a door but shifted scale because of the range of sensors.

Figure 3 indicates the convolved probability density function, which is now the final belief of the robot.

CONCLUSION

Monte Carlo Localization is trouble-free to put into practice and occupies a smaller amount of memory. A smaller amount of memory is occupied because the belief is updated rather than saved. Only a single last belief is required to calculate the new belief by convolution. Its results are more precise, accurate, and as error free than the feedback/phase locked loop of the Kalman Filter.

This statement can be proved by the experiment we did as shown in the Figure 3. The graphs are not so smooth because of accuracy and because maximum number of points were observed in the experiment. Laser range finder or ultrasound sensors can be used for error determination.

REFERENCES

Andrieu, C. (2003). An introduction to MCMC for machine learning. *Machine Learning, 50*, 5–43. doi:10.1023/A:1020281327116

Basye, K., Dean, T., Kirman, J., & Lejter, M. (1992). A decision-theoretic approach to planning, perception, and control. *IEEE Expert, 7*(4). doi:10.1109/64.153465

Berg, B. A. (2004). *Markov chain Monte Carlo simulations and their statistical analysis*. Singapore, Singapore: World Scientific. doi:10.1142/9789812700919_0001

Binder, K. (1995). *The Monte Carlo method in condensed matter physics*. New York, NY: Springer. doi:10.1007/3-540-60174-0

Chung, K. (1960). *Markov chains with stationary transition probabilities*. Berlin, Germany: Springer. doi:10.1007/978-3-642-49686-8

Cox, I. (1991). Blanche—An experiment in guidance and navigation of an autonomous robot vehicle. *IEEE Transactions on Robotics and Automation, 7*(2). doi:10.1109/70.75902

Fox, D., Burgard, W., & Thrun, S. (1999). Markov localization for robots in dynamic environments.

Funk, N. (2003). *A study of Kalman filter applied to visual tracking*. Retrieved from http://www.njfunk.com/research/courses/652-probability-report.pdf

Lee, E. A. (2009). *The problem with threads*. Retrieved from http://www.eecs.berkeley.edu/Pubs/TechRpts/2006/EECS-2006-1.pdf

Oxford. (2012). *An improved particle filter for non-linear problems*. Oxford, UK: Oxford University Press.

Russell, S. J., & Norvig, P. (1995). *Artificial intelligence: A modern approach*. Upper Saddle River, NJ: Prentice Hall.

ADDITIONAL READING

Kalman, R. (1960). A new approach to linear filtering and prediction problems. *Transactions of the ASME–Journal of Basic Engineering, 82*, 35–45.

Kanazawa, K., Koller, D., & Russell, S. (1995). Stochastic simulation algorithms for dynamic probabilistic networks. In *Proceedings of UAI 1995*. UAI.

Kitagawa, G. (1996). Monte Carlo filter and smoother for non-Gaussian nonlinear state space models. *Journal of Computational and Graphical Statistics, 5*(1).

Koller, D., & Fratkina, R. (1998). Using learning for approximation in stochastic processes. In *Proceedings of ICML 1998*. ICML.

Kortenkamp, D., Bonasso, R., & Murphy, R. (Eds.). (1997). *AIbased mobile robots: Case studies of successful robot systems*. Cambridge, MA: MIT Press.

Leonard, J., & Durrant-Whyte, H. (1992). *Directed sonar sensing for mobile robot navigation*. Dordrecht, The Netherlands: Kluwer Academic. doi:10.1007/978-1-4615-3652-9

Maybeck, P. (1979). *Stochastic models, estimation and control*. New York, NY: Academic Press.

KEY TERMS AND DEFINITIONS

Convolution: Mathematical Operation on two functions *f* and *g*, producing a third function that is typically viewed as a modified version of one of the original functions.

Markov Assumption: The Markov assumption, sometimes referred to as static world assumption, specifies that if one knows the robot's location, future measurements are independent of past ones (and vice versa).

Noise: Fluctuation in a signal is called noise.

Particle Filter: Usually estimates Bayesian models. Uses probabilistic theorem to determine the output of the process.

Phase Locked Loop: Is a control system that generates an output signal whose phase is related to the phase of an input signal.

Probability Density Function: Relative likelihood of a random variable to occur at a given point.

Tracking: A process starting with determining the current and past locations and other status of property in transit.

Chapter 5
Control Architecture Model in Mobile Robots for the Development of Navigation Routes in Structured Environments

Alejandro Hossian
Universidad Tecnológica Nacional, Argentina

Gustavo Monte
Universidad Tecnológica Nacional, Argentina

Verónica Olivera
Universidad Tecnológica Nacional, Argentina

ABSTRACT

Robotic navigation applies to multiple disciplines and industrial environments. Coupled with the application of Artificial Intelligence (AI) with intelligent technologies, it has become significant in the field of cognitive robotics. The capacity of reaction of a robot in unexpected situations is one of the main qualities needed to function effectively in the environment where it should operate, indicating its degree of autonomy. This leads to improved performance in structured environments with obstacles identified by evaluating the performance of the reactive paradigm under the application of the technology of neural networks with supervised learning. The methodology implemented a simulation environment to train different robot trajectories and analyze its behavior in navigation and performance in the operation phase, highlighting the characteristics of the trajectories of training used and its operating environment, the scope and limitations of paradigm applied, and future research.

DOI: 10.4018/978-1-4666-2658-4.ch005

Copyright © 2013, IGI Global. Copying or distributing in print or electronic forms without written permission of IGI Global is prohibited.

INTRODUCTION

Early research in the field of robotics were made based on development environments working with cells fixed in their positions for the robot to develop its work, as in the case of the welding robot in an assembly plant.

The significant technological advances in various industrial environments were one of the main reasons to provide adequate capacity to move the robot systems beyond its work cells. This discipline has been called, in due course, "Robotics Mobile" and is one of the greatest challenges addressed by the scientific community that works within the extensive and rich field of robotics.

In this context, it should be noted that successive investigations have worked to provide a level of autonomy to these mobile systems with the aim of robots that can navigate its operating environment and react to situations that have not been considered in scheduling its control. To make navigating the robot to meet the requirements that were specified, the robot must have a cognitive architecture on which to establish links between its sensory system and its system of performance in the operating environment, and therefore, the robot is able to achieve objectives.

Through the navigation, it is possible to guide the course of a mobile robot through an environment with the presence of obstacles. Different schemes are known to carry out this task, but all have the common goal of directing the vehicle to your destination in the most secure and efficient manner possible. The capacity of reaction that may possess the robot when faced with unexpected situations should be its most distinctive quality to function effectively in the environment where it should operate, which indicates the degree of autonomy that it has.

CONTROL ARCHITECTURE

In the context of the design of robot systems, a control architecture is some description of a system established on the basis of components of a structural nature and how it is assembled together to be able to shape a coherent and consistent set with respect to its "structure and function." In the context of mobile robotics, which is the branch of Robotics that concerns us in the present work, include the following sentence belonging to Javier De Lope Asiaín (De Lope, 2001):

Mobile robot control architecture is a software system that provides the actions or movements that should make the robot from the acquisition and processing of sensory information and of the objective or objectives that have been identified.

Given this reasoning, in which case it is intended to design and build a mobile robot, one of the key issues to solve this problem must be related the following to obtain a satisfactory design:

- Appropriate sensors to perceive the robot system operating environment.
- Engines, to facilitate the mobility of the robot to act on its environment.
- CPU (Central Processing Unit) to exercise control over this mechanism of relationship between sensors and motors.
- Command of tasks. The commands are ordered tasks the robot can complete.

The schematic shown in Figure 1 expresses the idea of integrating the four elements mentioned above to find a mechanism by which the robot can use environment information obtained from its sensory system, transforming it into actions and movements in this environment, according to the tasks and objectives to be achieved by the robot.

Figure 1. It illustrates an integrated outline of the aforementioned elements

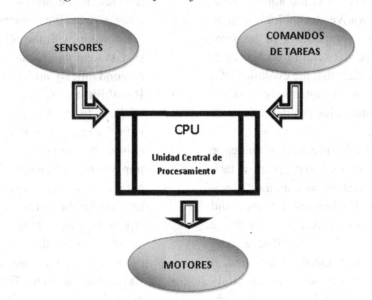

Components of the basic scheme, the CPU is the cornerstone which allows processing data collected by the sensor system to transform it into action on the environment of operation of the robot. Based on the foregoing, it is necessary to design a specific software for the purposes for the control of the robot. Even so, it should be noted that throughout history different control mechanisms have been supported on the basis of the hardware; indeed, initially the first mobile robot systems possessed no software in your internal configuration control. However, if it is true that today there are systems without software, it is also true that the solution focuses on the control software, the most extensive and flexible.

Therefore, dealing with the design of a mobile robot system process, one of the fundamentals to consider is the control software that allows the processing of the perceptions gathered by its sensory system and to translate them into actions carried out by its system of actuators or engines on the environment of transaction always depending on the objectives to be met by the robot. It is important to note concerning the main features of this control software, although it can be relatively simple, it is also possible to reach a

high level of complexity; as a result, it is not the same complexity that presents the functionality of a crawler robot with a robot humanoid of the last generation.

A satisfactory design of the architecture of a robot controller should conform to certain requirements deemed appropriate, such as (Foka & Panos, 2002):

- **Modularity:** It is the capacity that you must have the architecture to abstract over the spectrum of levels of behavior, facilitating the work of the designer and promoting the re-use of the software used. This feature is highlighted in those systems based on behavior.

- **Ease of Programming:** This requirement refers to the ability to perform multiple tasks and plan actions according to the current state of the robot and its operating environment.

- **Robustness:** It refers to the "fault tolerance" of the control system. This is critical in the design of architecture for the robot to continue working even when there has been a failure in any component.

- **Ability to Evolve:** It is the ability to incorporate innovative technologies, new equipment, and components to improve the performance weekend of architecture. Therefore, it is clear that the evolution of a control architecture is potentially more effective in an innovative profile projects or demonstration.

- **Efficiency:** This feature makes reference to the possibility of carrying out a task evaluating parameters such as the "precision" with which takes out the same and the "runtime," as well as, in relation to the resources used such as "hardware" and "software" required. On the other hand, it is noteworthy that efficiency is intimately linked with the robot behaviors that are consistent with the goals and plans to be developed.

- **Autonomy:** The vast majority of intelligent control architectures seek a high level of autonomy, in order to ensure consistency between the tasks the robot must perform and the environment where it must operate. It is important that the designer consider a transition from progressive and gradual, going from a typical control teleoperated to autonomous behavior characteristics.

- **Flexibility:** This feature represents the robot's ability to operate in environments with different characteristics and has only limited knowledge. This requirement reflects the importance of the robot's ability to respond to unexpected situations, modifying behavior to analyze the situations faced in designing a control architecture. Thus, the robot can select the most appropriate to consider reactions based on changes that may occur in its operating environment.

- **Reactivity:** By what has been said previously, the reactivity is determined by capacity possessing the robot to detect events in time and make decisions based on the environment of operation and the charac-

teristics of the task to be carried out. In other words, the "reactivity" is in close relationship with the efficiency of its sensory system and the ability of the robot to respond quickly and effectively.

- **Reliability:** It is very important that the reading from the sensors is consistent with the reading of the actuators. For this reason, it is appropriate in many cases to implement redundant functions, setting out different ways to accomplish the same task. As for the sensor system, the effect of redundancy helps to reduce the uncertainty associated with the process of perception of the robot in its actions.

- **Execution in Real Time:** This requirement describes the ability of the robot to generate responses at the moment it is necessary to do so, and not moments later.

It is important to note that the described characteristics could not always be satisfied simultaneously when designing a control architecture. It might be difficult to have a robust, flexible, and reactive features architecture, and, in turn, be highly efficient to carry out a particular task.

In summary, the current design of control architectures requires consideration of the commitment that must exist between efficiency in the design (e.g., plans led systems robots) and the flexibility and the ability (e.g., data-driven robots raised the world of action), and this has been the trigger for different types of control architectures for robots. At present, it may happen that the benefits to be provided by mobile robots, requiring a high degree of complexity in relation to its design, and consequently, it is essential to consider splitting the problem into a series of levels of abstraction that will address a problem so that this is more manageable. This shows the need to define a control architecture for the robot, usually understood as a hierarchical system of levels of abstraction to be implemented on the system control software robot.

In general you can say that the history of the field of robotics and the design of the architectures of control are subdivided into three main lines: deliberative, reactive, and hybrid architectures.

DIFFERENT APPROACHES TO CONTROL ARCHITECTURE

The mechanisms for autonomous robot navigation and artificial intelligence have been closely related, so autonomous robotics was stimulated, although somewhat limited by the premises and principles of Artificial Intelligence.

The framework of the research activity in the field of Robotics in recent years outlines some approaches on which underlie the different architectures of control for mobile robots.

Deliberative Approach

Architectures supported by this approach are characterized by reasoning techniques to decide actions to follow by robot based on a model the environment, and involve structures of reasoning in high level, which allow you to meet requirements of navigation of high complexity.

As stated, symbolic artificial intelligence was used in a first attempt to achieve autonomy in robots. The idea behind substantial symbolic artificial intelligence is that artificial intelligence is inherently a computational phenomenon, and therefore it can be studied and put into practice in different systems, even though the systems are out of its environment.

According to the characteristics of the present approach, its root is based on a breakdown of the processes that the robot should carry out, in independent tasks, which are then unified. These tasks can be identified as interpretation of the surveyed data from sensors, modeling of the environment in which the robot will operate, model on which a set of actions is planned to carry out a particular task, and finally, these actions are performed through the implementation module. Figure 2 illustrates the traditional approach with its corresponding tasks.

One of the main disadvantages of this approach is that for some of these processes, for example, the sensing and planning processes are carried

Figure 2. Traditional knowledge-based approach

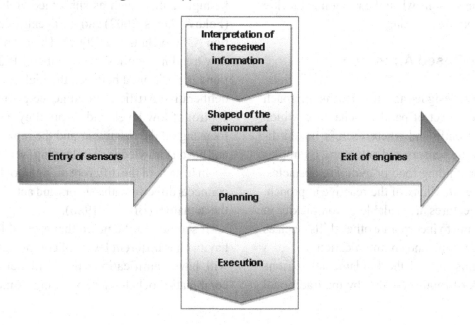

out through modules developed without considering the other existing modules in the architecture. In short, through this approach the designer defined by means of symbolic entities (e.g., predicate logic first order), a model of the environment that enables the development planning of the robot's movements to achieve a specific goal. However, what happens is that the environment turns out to be very difficult to formalize given that it varies in function of time; and even assuming that all cases could be defined, the model of world that would be too large to be processed. In other words, this kind of strategy turns out to be very little robust when the robot system must operate in dynamic environments. This difficulty in Artificial Intelligence is called "the frame problem" and focuses on the formal representation of the environment and its possible changes.

Reactive Approach

The architectures that fall within this type of approach used control strategies (e.g., a collection of pairs "conditions-actions") and do not make planning processes because it do not operate on an internal model of the environment. These architectures are faster in the processing of information, but it has limitations when the robot must tackle tasks that require planning.

Behavior-Based Approach

Architectures designed according to this approach are based on a set of parallel behaviors, which runs concurrently and where each is linked directly with the sensors and actuators. Examples of behavior: finding food or avoiding obstacles. Keeping the simplicity of the reactive approach, these architectures are scalable in complexity in robots seeking to incorporate other skills, such as memory, learning, and communication.

Given this, one of the fundamental reasons for the lack of robustness that by the traditional

approach is not considered in the design stage, the interaction between the different modules. For example, it is not possible to carry out a module of planning on a model of environment defined symbolically, then make the module corresponding to the sensory interpretation to the symbolic entities that are used in planning, and also in an isolated manner, configure the module for the implementation of the operators who took in planning on the actuators. Look at three key points to consider when it comes to pass to the design of an architecture of cognitive type (Harvey, 1996), namely:

Decomposition of a robot control system is not easy to perform in all cases. In many cases, the interactions between modules are determined by the environment and as a result, are more complex that simple links between these modules. The interactions between modules increases exponentially, as the complexity of the system grows. As result of these inconveniences caused by the design isolated individual modules, devoted greater efforts for the development of cognitive architectures with a greater degree of interaction between modules, i.e. the different modules are configured simultaneously and in a coordinated manner. Thus, in the mid-1980s various research in the field began to be highlighted, such as subsumed architectures (Foka & Panos, 2002) and the agents of competition (Ollero Baturone, 2007). Thus conforms "a new trend in robotics" (Braitenberg, 1987). This trend is positioned between the high level of the deliberative Artificial intelligence planning and control of low level, and inspired by nature and focusing on the behaviour and the rapid reaction and not on knowledge and planning (see Figure 3).

In it, each of the different modules of behavior connects directly with sensors and act directly on the actuators (Brooks, 1986).

It should also be noted that each of these behaviors has different levels of complexity, which will have ramifications that will define what combination of behavior governing the movement

Figure 3. Schematic of a behavior-based robotics architecture approach

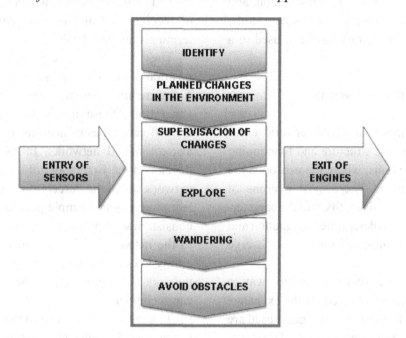

of the robot at each instant of time. This set of behaviors is called the "cognitive architecture" or "control architecture" of the robot.

Hybrid Approach

Considering the difficulties of the previous approach, design isolated individual modules in architectures that follow the approach based on knowledge and lack of planning with designs that are part of the approach based on behavior. The scientific community in this field began to devote greater efforts for the development of architectures cognitive with a higher degree of interaction between modules, that is to say that, the different modules are configured in a simultaneous and coordinated manner. Accordingly, in the mid-1980s are beginning to charge relevance different investigations in the field, such as subsumed architectures (Brooks, 1986) and agents of competition (Maes, 1991). This new trend is among high-level planning of artificial intelligence and deliberative control low-level and at the same time inspired by nature and highlights

the behavior and reactive power rapidly, leaving a little knowledge and planning (Santos & Duro, 2005).

Hybrid architectures show a relationship between the purely reactive and plans-oriented. This type of architecture aims to obtain a proper harmony between the so-called traditional symbolic methods of Artificial Intelligence and the response capacity and flexibility of pure reagents systems. In this way, by means of these architectures of character "hybrid" is a possible reactive control systems providing the ability to reason about the fundamental behaviors, optimizing the performance of navigation in complex operating environments.

A CASE STUDY OF MOBILE ROBOT NAVIGATION IN STRUCTURED ENVIRONMENT

This section presents a case study of navigation of a mobile robot with a type of architecture "sensor-motor," which fits within the reactive approach presented in the previous section, and in a structured environment. For this case study, neural networks technology has been used in a profitable manner.

Artificial Neural Networks

The theory and shape of artificial network neuronals is inspired by the structure and functioning of the nervous systems, where the neuron is the fundamental element. There exist neurons of varied forms, sizes, and lengths, which characteristics are important to determine the function and usefulness of the neuron. In Figure 4, we observe a model of it.

The concept of network becomes the way in which these are interconnected for the exchange of signals with information. The signals used are of two types: electrical and chemical. The signal generated by the neuron and transported along the axon is an electrical impulse, compared to the signal that is transmitted between the terminals of a neuron Axon and the dendrites of neurons following chemical origin; in other words, performed by molecules of transmitting substances (neurotransmitters) that flow through special

contacts, called synapses, which have the function of receiver and are located between the axon terminals and the dendrites of the next neuron.

The generation of electric signals and the composition of the cell membrane are closely linked. There are many complex processes related to the generation of these signals. As stated earlier, there are different types of neurons and consequently also of neural networks. In 1958, Rosenblatt developed the first model of an artificial neural network called the Perceptron with the ability to learn to recognize simple patterns. A Perceptron consists of several linear neurons receiving inputs to the network and an output neuron with the ability to decide whether an input that enters the network belongs to one of the two classes that can recognize.

To analyze the behavior of Perceptrons, using a technique through which is shown on a map, the decision regions created in the multidimensional space of inputs to the network, showing in these regions which patterns belong to a class and which another. The Perceptron regions separated by a hyperplane whose equation is dependent on the weights of the connections and the threshold value of the activation function of the neuron.

Figure 4. Biological model of a neural network

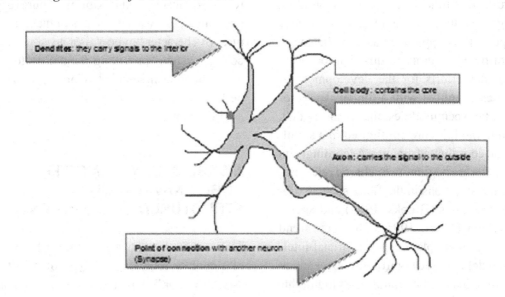

Such values of the weights may be fixed or adjusted by means of various training algorithms of the network.

However, as the Perceptron only has one input layer and one output, with a single neuron, is capable of producing a representation quite limited, discriminating very simple pattern, linearly separable. Look at Figure 5.

The Perceptron learning algorithm possesses a rate monitored, which means that it requires that the results are evaluated and, if necessary, make the corresponding modifications of the system.

A neural network model most advanced known as "Backpropagation," as shown in Figure 6. This model uses a rule that can be applied in network models with more than two layers of neurons, and learning that manages an internal representation of the knowledge that is able to organize in the intermediate layer of cells to achieve any correspondence between the input and output of the network. As we have seen, the search for the weights suitable for achieving a correspondence between the input and output is a very difficult task when there are no hidden layers, but the objective can be met with one of them.

In simple terms, the operation of a neural network of this kind is basically based on learning a set of predetermined pairs of output that are given as an example, using a propagation-adaptation cycle, consisting of two phases: first, stimulating the first layer of neurons in the network through a pattern of entry, which spreads through all layers to generate a result that is compared with the desired and estimates a value of error for each output neuron. Then, errors are transmitted to the rear, towards all the intermediate layer neurons that contribute directly to the output, receiving the error rate closer to the participation of the neuron intermediate in the original release. This process is repeated, layer-by-layer, until all neurons in the network have received an error indicating its contribution to the total error. In accordance with the received error, the weights of each neuron connection are reset, so the next time you submit the same pattern, the output is closer to the desired; in other words, the error is reduced.

This quality is what distinguishes this model. Its ability to adjust the weights of the neurons in the hidden layers, in order to learn the relation-

Figure 5. Perceptron neural network model

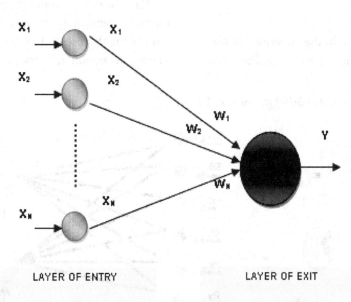

LAYER OF ENTRY   LAYER OF EXIT

Figure 6. Backpropagation neural network model

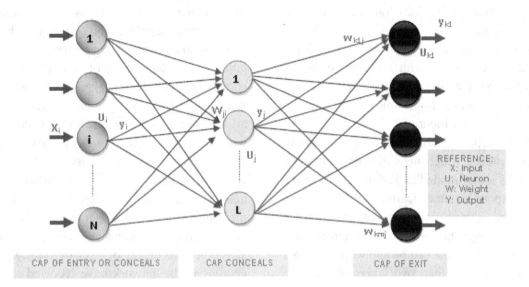

ship between an example of given patterns and corresponding outputs. Its objective is to find an internal representation that allows you to generate the desired exits when it receives entries from training, and that these can be applied to unknown entries, allowing us to classify them according to the characteristics and comparing them with the examples that were acquired during the training.

Resources

For the accomplishment of the experiments presented in this work, there was in use hardware

Pentium 5 and the package of software MATLAB (version 7.0) with support of the correspondent Toolbox of network neuronals.

Figure 7 shows the robot designed for experimentation. This robot has four proximity sensors for obstacle detection and two position sensors to locate the robot within the environment. It also has two wheels (motors) responsible for its movement when it establishes its system of sensors.

The browsing environment has implemented obstacles identified in it, and within a development environment presented as a menu that allows you to train the neural network proposal and evaluate

Figure 7. Design of robot. Model of the network used.

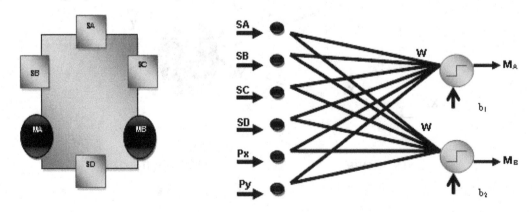

the robot in operation phase. Figure 8 shows a scheme for the simulation environment to navigate the robot. You can see the information about the location of the robot (provided by the coordinates X and Y, in this case corresponding to 0 as X and 6 according to Y), its orientation (the given for North, South, East, and West), the environment with the location of its obstacles and options to train the network, restart the process when it is necessary to simulate the steps of the path followed by the robot.

As you can see, the spaces occupied by black cross grids are places considered obstacles, which may not be occupied by the robot, and therefore should be avoided for the same reason not to collide, while the rest of the boxes, considered to be free, are suitable for its displacement. In addition, the limits of the grid are considered an obstacle to the robot. Each box can be individualized by a coordinate system defined by the numbers in the margins of the grid. All this representation is correlated with the X-axis and Y-axis coordinates of the conventional system.

Px and Py expressions are position sensors that contain both horizontal and vertical coordinates, respectively, corresponding to the position of the robot in the environment. Sensors SA, SB, SC, and SD are those that must detect the proximity of objects or obstacles, taking the value 1 in the presence of an object (whereas near an object when it is two lockers or two positions or less on the location of the sensor within the environment of operation); or the value -1 to the proximity of objects (whereas an object is not nearby when it is more than two boxes or positions on the sensor within the environment of operation). Similarly, the engines (MA, MB) acquire an ordered pair of values that reveal a certain effect, namely:

- **(-1, -1):** The robot must reverse a position.
- **(-1, 1):** The robot advances one position to the left.
- **(1, -1):** The robot advances one position to the right.
- **(1, 1):** The robot advances one position ahead.

Figure 8. The environment

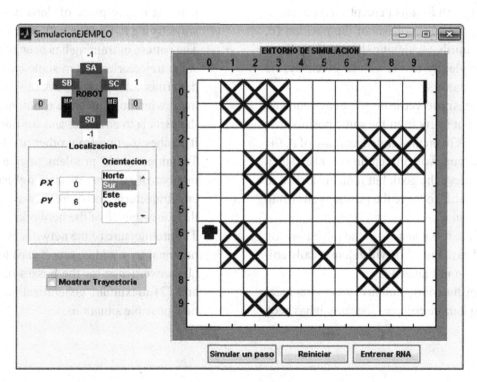

Consequently, the experimental method is the development of a training course determined from a mapping "sensor-motor" of the robot that responds to that path. Then you must train the network to that path and verify in the operation phase that the robot learned the path, and finally, test the robot with other paths and evaluate its performance.

The Experiments

Two kinds of experiments were carried out: the first used a network configuration corresponding to a model of simple Perceptron with a layer of input and an output layer, while the second class of experiments was conducted on the basis of a hidden type of more advanced architecture with the addition of a layer of neurons applying the model backpropagation. The application of this model allowed the mobile robot to develop more advanced behaviors for navigation in complex operating environments. We note that in both cases the process begins to train the network, as shown in Figure 9.

As you can see in Figure 10, the robot carries out a simple path for this Perceptron type neural network model, on the basis of the position (0, 6), facing north, achieving the goal without greater complexity to the values given by the sensors and motors.

The robot is observed in the same environment. Now the robot starts from the same position but facing south. Given the limited learning of it, the robot cannot cope with this environment and then does not achieve the goal, but remains locked in the same, unable to leave the positions shown in Figures 11 and 12.

Now, if the robot hand is in the same position and headed east, the robot will automatically collide, as shown in Figure 13.

Changing the configuration of the environment and making it more complex, because it has been shown that the previous model is very limited for certain trajectories, it can be seen that a neural network model a bit more advanced in its capacities resolves certain difficulties that the model presented with greater simplicity could not could achieve this.

In Figures 14 and 15, you can see that the world has changed its complexity, and with a network model more complex, it can solve the problem starting from different points to achieve the goal.

Analysis of Results

Experimental method is explained with the application of neural networks technology, the relationship with the environment, the type of robot, and how it interacts. It warns the different behaviors of the robot. When it starts from different points along the course of training or when it does start from the same point but with different orientation for the same course of training.

The following conclusions were obtained from research and experiments performed:

- Can see that the robot performs well depending on the place of departure and always trying to find the path of training.
- The pattern of training has been enough for some trajectories, but in some cases such a pattern is not representative of the environment which operates the robot, as most of the exits is to advance, and for this reason, it crashes starting from other positions.
- To mitigate this problem, increasing the number of patterns of training using several trajectories or more, always monitor the convergence of the network.
- The architecture of the network is modified neuronal (one adds a secret cap) to obtain: (1) convergence for the bosses of training and (2) maximum generalization, for the new possible situations.

Figure 9. Trained network

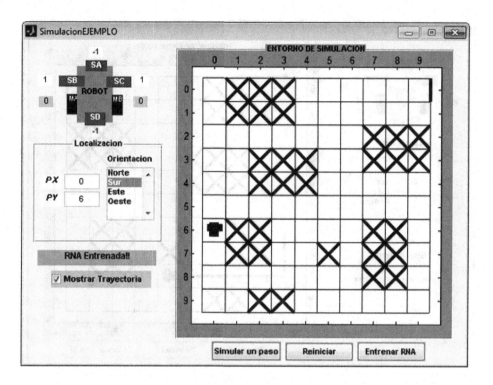

Figure 10. Simple path

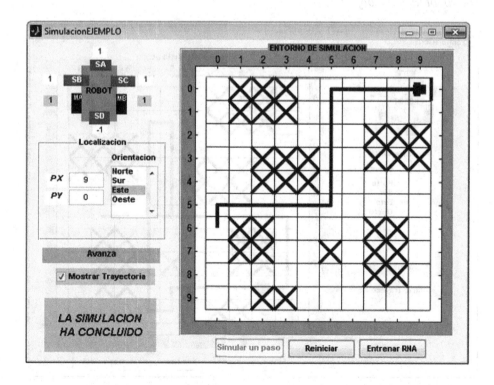

Figure 11. The robot fails the objective

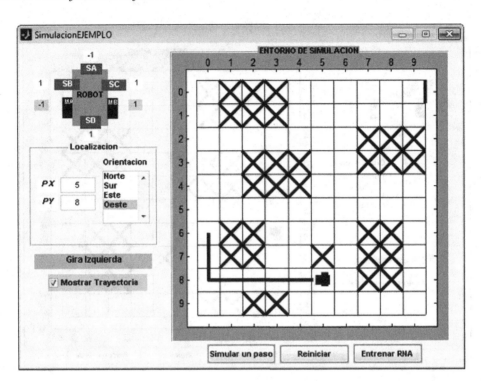

Figure 12. The robot fails the objective

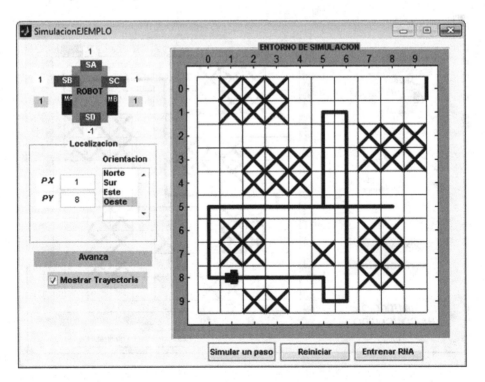

Figure 13. The robot collides

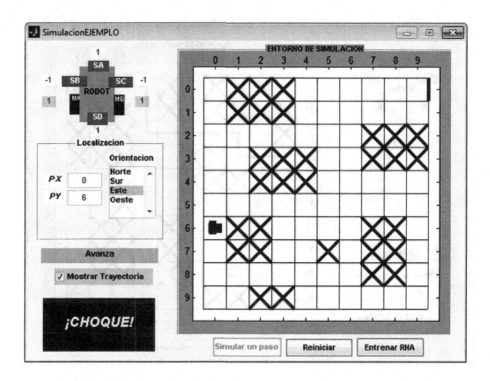

Figure 14. The robot achieves the objective on the basis of the position (0, 5)

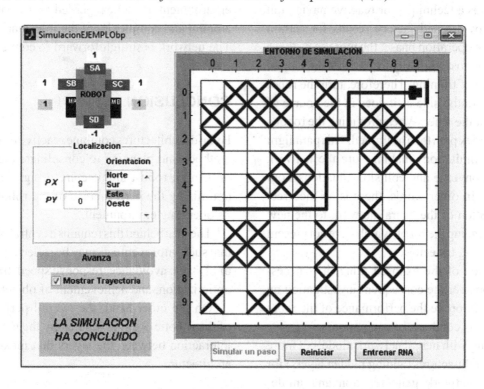

Figure 15. The robot achieves the objective on the basis of the position (0, 3)

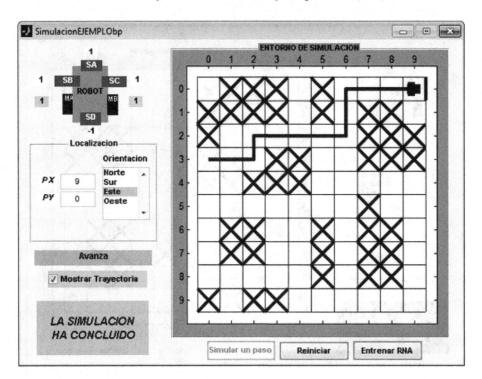

In summary, the use of Artificial Neural Networks as a technique for reactive navigation feature provides satisfactory results for certain paths in the operation phase, the more so as these paths are at most similarity to the robot developed at the stage of training. Therefore, turn the robot look more to the side as it does in the course of training for the other. When providing the robot a more complex path than he trained. This paradigm shows its limitations making sure the network does not converge and collision situations from occurring. In other words, there is an incorrect generalization of the neural network to the new situations facing the robot, which were not present in the training trajectories.

By virtue of the exposed thing, it is necessary to consider the consequent investigation on the aim to improve the performance of the robot regarding his conducts of navigation.

Working with neural network models of more complex architectures (adding hidden layers) for maximum network generalization and imple-ment trajectories representing robot navigation environment, must be regarded as increasing the amount and complexity of the information used. The network is straightforward to converge.

CONCLUSION

Hybrid architectures combine reactive processing methods and planning to consolidate the effects of a more robust architecture design, aimed at optimizing the performance of the robot within its operating environment.

The idea behind this remains a central objective that such improvements may be reflected in terms of obstacle avoidance, response speed, trajectory optimization, and achievement of objectives.

On the other hand, the overall performance of the robotic agent is defined by the continuous interaction between the layers that make up the architecture.

In synthesis, hybrid architectures, through a combination of both approaches (knowledge-based and behavioral-based), allow the reconfiguration of the reactive control systems through its ability to reason about the fundamental behaviors.

REFERENCES

Braitenberg, V. (1987). *Vehicles: Experiments in synthetic psychology*. Cambridge, MA: The MIT Press.

Brooks, R. (1986). *Achieving artificial intelligence through building robots*. Cambridge, MA: The MIT Press.

Foka, A. F., & Trahanias, P. E. (2002). Predictive autonomous robot navigation. In *Proceedings of the IEEE/RSJ Internacional Conference on Intelligent Robots and Systems (IROS)*, (pp. 490-495). IEEE Press.

Harvey, I. (1996). Artificial evolution and real robots. In *Proceedings of International Symposium on Artificial Life and Robotics (AROB)*, (pp. 138-141). Beppu, Japan: Masanori Sugisaka.

Maes, P. (1991). A bottom-up mechanism for behavior selection in an artificial creature. In *Proceedings of the First International Conference on Simulation of Adaptive Behavior (SAB90)*, (pp. 238-246). The MIT Press.

Mataric, M. J. (1992). Designing emergent behavior: From local interactions to collective intelligence, from animals to animat 2. In *Proceedings of the Second International Conference on Simulation of Adaptive Behavior (SAB92)*, (pp. 432-441). Cambridge, MA: The MIT Press.

Ollero Baturone, A. (2007). *Robótica, manipuladores y robots móviles*. Barcelona, Spain: Alfaomega-Marcombo.

Santos, J., & Duro, R. (2005). *Evolución artificial y robótica autónoma*. Mexico City, México: Alfaomega - Ra - Ma.

Scillato, A. E., Colón, D. L., & Balbuena, J. E. (2012). *Tesis de grado para obtener el grado de ingeniero electrónico*. Neuquén, Argentina: IEEE.

KEY TERMS AND DEFINITIONS

Artificial Intelligence (AI): AI aims to achieve autonomy in robots, focusing on the fact that intelligence is inherently a computational phenomenon, and as a result, it is possible to study and apply it in different systems, while these are outside the environment's performance.

Artificial Neural Network: Artificial neural networks are models that try to reproduce the behavior of the brain on the basis of the most important characteristics of biological neurons.

Control Architecture: System given structure and assembled with interacting components to establish the actions or movements to be performed by the robot with the acquisition and processing of information provided by its sensors and related to or stated objectives.

Deliberative Approach: Corresponds to the approach that supports the architectures whose reasoning techniques to decide the actions to be followed by the robot, are based on a model of the environment and need a high level of reasoning structures to deal with navigation of high complexity requirements.

Hybrid Approach: Corresponds to the approach that relates purely reactive architectures and oriented plans. Thus, through these architectures of character "hybrid," it can be supplied to reactive control systems of the ability to reason about the fundamental behaviors, optimizing the performance of navigation in environments of high complexity operation.

Mobile Robot: System robot that contains what is needed for its movement and displacement (power, control, and navigation system).

Reactive Approach: Corresponds to the approach whose architecture is that the road between sensors and motors, provides a response system real-time robot to possible changes that might occur in the environment where it operates. It is based on biological models to explain the behavior observed in individual organisms and implements strategies of control as a collection of pair conditions—does not perform planning processes because these do not operate on an internal model of the environment. The response to stimuli is reflexive, unregulated by deliberative processes of any kind and the movements of the robot are guided solely based on the information that is present at that time.

Structured Navigation Environment: It is the environment that can be defined with certainty in its configuration and location of obstacles, and that it does not undergo modifications over time. These changes may be predictable, and therefore be formalized from a computational point of view.

Chapter 6
A Swarm Robotics Approach to Decontamination

Daniel S. F. Alves
Instituto de Matemática, UFRJ, Brazil

Renan S. Freitas
Escola Politécnica, UFRJ, Brazil

E. Elael M. Soares
Escola Politécnica, UFRJ, Brazil

Thiago M. Santos
Escola Politécnica, UFRJ, Brazil

Guilherme C. Strachan
Escola Politécnica, UFRJ, Brazil

Vanessa C. F. Gonçalves
UFRJ, Brazil

Guilherme P. S. Carvalho
Escola Politécnica, UFRJ, Brazil

Luiza M. Mourelle
UERJ, Brazil

Marco F. S. Xaud
Escola Politécnica, UFRJ, Brazil

Nadia Nedjah
UERJ, Brazil

Marcos V. B. Couto
Escola Politécnica, UFRJ, Brazil

Nelson Maculan
UFRJ, Brazil

Rafael M. Mendonça
UERJ, Brazil

Priscila M. V. Lima
Instituto de Ciências Exactas, UFRRJ, Brazil

Felipe M. G. França
UFRJ, Brazil

ABSTRACT

Many interesting and difficult practical problems need to be tackled in the areas of firefighting, biological and/or chemical decontamination, tactical and/or rescue searches, and Web spamming, among others. These problems, however, can be mapped onto the graph decontamination problem, also called the graph search problem. Once the target space is mapped onto a graph G(N,E), where N is the set of G nodes and E the set of G edges, one initially considers all nodes in N to be contaminated. When a guard, i.e., a decontaminating agent, is placed in a node i ∈ N, i becomes (clean and) guarded. In case such a guard leaves node i, it can only be guaranteed that i will remain clean if all its neighboring nodes are either clean or clean and guarded. The graph decontamination/search problem consists of determining

DOI: 10.4018/978-1-4666-2658-4.ch006

Copyright © 2013, IGI Global. Copying or distributing in print or electronic forms without written permission of IGI Global is prohibited.

a sequence of guard movements, requiring the minimum number of guards needed for the decontamination of G. This chapter presents a novel swarm robotics approach to firefighting, a conflagration in a hypothetical apartment ground floor. The mechanism has been successfully simulated on the Webots platform, depicting a firefighting swarm of e-puck robots.

INTRODUCTION

Conflagrations are very delicate situations in which one or multiple heat sources cause an uncontrolled fire in a specific area. The spreading fire can seriously threaten human life and natural environments. They can also cause great destruction due to the risk of ignition when the fire is near explosive substances, like those found in chemical plants and hospitals. In spite of possible causes, a conflagration is defined as a contamination-like situation, in which an infected area can quickly infect neighboring areas.

In a burning apartment, it is extremely difficult and dangerous for the residents to try to extinguish the fire, especially if the area is already almost fully burning. A minimum number of humans have to be placed in specific rooms to use the decontamination agent, e.g., a water jet, and to guarantee that heated spots do not turn into fires again. The risk of this task is high for firefighters because of the unpredictability of the situation: the structure may fall apart, flammable products can cause a fatal explosion, and the smoke can disorientate and asphyxiate people. Therefore, it would be very appealing to have this phase of firefighting performed by non-living agents, such as robots.

Although the case study approached here deals with a small-scale problem, it is the possibility of a large-scale conflagration that motivates the conception of a distributed mechanism able to overcome the inability of tackling such situations with centralized strategies. This work presents a novel and distributed way of performing swarm robotics over problems that can be mapped into a graph decontamination or search problem. The *Scheduling by Edge Reversal* (SER) graph dynamics has been applied to many different resource-sharing problems and the graph decontamination has also been recently identified as such.

A pre-established number of robots will enter a hypothetical apartment and perform a cooperative task. They must extinguish the fire of the entire area, room by room. Each robot behaves like a cleaning agent. So, the presence of one of them in a room is assumed to be sufficient for its decontamination, as long as this agent does not leave the area. The act of cleaning is understood as the robot's ability to throw a water jet around the room and its readiness to extinguish any possible fire source that arises. The robot *has to* stay in the room to be decontaminated at least until all neighboring rooms are also *clean*. If this is not done, the cleaned area can be *infected*, i.e., catch fire, again.

DEFINITIONS

In order to present the problem and its respective solution in a coherent fashion, some concepts that will be use used in this chapter need to be presented. Given a connected graph G and two nodes v and u, which belong to G, if an edge e connects v and u, then v and u are *neighbors*. This implies that they have a relation of adjacency. If two nodes are *neighbors* in G and D results from an orientation of G, then those two nodes are also *neighbors* in D, regardless of the orientation of the edges. The reverse of an orientated edge e from v to u is an orientated edge f from u to v.

For the graph contamination problem, nodes of a graph C are in one of three states: contaminated, clean, and guarded or clean. Contaminated is the initial state of all nodes. A (clean and) guarded node is a node that has received a decontamination agent, which is a tool of the decontamination algorithm and can have different properties regarding creation of new agents and mobility. After leaving a node, the decontamination agent becomes clean. If a clean node has a contaminated neighbor, it becomes contaminated again. This event is called recontamination.

A sink decomposition is a distribution of the nodes of an acyclic connected digraph D in numbered layers. The layer of a node is the length of the longest simple path to any sink, or zero if the node is itself a sink. This also results in the property that a node in layer k, for $1 \leq k \leq l$, where 1 is the highest layer, has always at least one neighbor in layer $k - 1$.

DISTRIBUTED AND PARALLEL DECONTAMINATION

A decontamination graph approach can model solution strategies to conflagration problems. However, a centralized control is not reliable due to limitations of the scenarios, e.g., the size of the graph. Therefore, a distributed and parallel approach presents itself as a better and more natural alternative.

The decontamination problem, also known as node-search problem (Moscarini, Petreschi, & Szwarcfiter, 1998), is the process of changing all nodes of G from *contaminated* to *clean*, usually trying to use the smallest possible amount of decontamination agents. This is usually accomplished by planning the movements of the decontamination agents.

As the agents move, they might clean the transmission channel or not. If they clean the channel, their original node will not face recontamination risk from that direction. Therefore, it is safer to leave the node. If the channel is not cleaned by the agent, then the infection might run across it, after the node is left unguarded, and recontaminate the node.

Previous works either use controlled flood (Flocchini, Nayak, & Schulz, 2007) or explore the properties from certain graph structures (Luccio & Pagli, 2007) to attack the problem in a more efficient manner. SER-based decontamination constitutes a structure-independent and efficient approach (Gonçalves, Lima, Maculan, & França, 2010), which is the subject of this section.

Scheduling by Edge Reversal (SER)

The sharing of resources in a distributed system is a well-known problem. SER has been proposed as a way to maximize concurrency, thus enabling an optimized use of resources (Barbosa & Gafni, 1989). However, it can also be used to minimize concurrency in order to obtain the effect of delaying the repetition of system states.

In this method, a graph G models the processes and shared resources of the system, with nodes representing processes and edges linking processes that share a resource, so that each shared resource is represented by a clique, i.e., a fully connected subgraph. The edges are oriented from a process with lower priority to one with higher one, resulting in a digraph D. The orientation is given according to the needs of the problem, for instance, targeting higher or lower concurrency. The nodes can be organized as a sink decomposition scheme, with the sink nodes at level 0 and the others at the layer that corresponds to the length of the longest simple path from them to any sink.

A process can only operate if it has a higher priority than all its neighbors, i.e., if it is at the lowest layer. In other words, in case the node is a sink. After some time of operation, the process reverts its edges ceding priority to other processes. The sink decomposition is then modified. As the nodes in the first level (after level 0) become sinks, other non-sink nodes drop one layer and each sink goes to the highest layer that at least one of its neighbors had, producing a new orientation *D'* for *G*. The way the new sink decompositions are created shows that all the nodes will be sinks given enough reversals.

The successive reversals of edges always create another acyclic orientation, which is easily verifiable since the original orientation was acyclic and the reversal of edges of a sink will not create a cycle. Based on these properties, no deadlock or starvation will occur in the SER operation. There will be always a sink to operate, i.e., there is always an operating node. The time that a node

waits before it can operate is proportional to the length of its longest path to a sink. As this path is always finite, all nodes eventually operate. The sink decomposition can be related to the coloring of the acyclic graph, with one color per layer, with similar NP-hardness difficulty for optimizations of maximal or minimal concurrency. Figure 1 illustrates the process.

Barbosa et al. (1989) have proved that, under the SER, all nodes will operate the same number of times in a long enough interval of time, avoiding deadlocks and starvation. The interval of time between operations of a node and the number of

Figure 1. SER dynamics on a small graph with sink decomposition of the first acyclic orientation

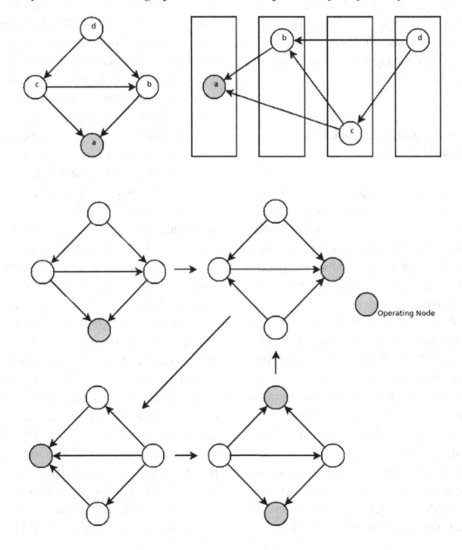

simultaneous nodes operating is highly dependent upon the initial acyclic orientation of the graph *G*. However, finding the initial acyclic orientation that maximizes the concurrency is an NP-hard problem with the reciprocal of the resulting concurrency limited by the chromatic and multichromatic numbers of the graph. This problem is similar to the vertex-coloring problem, which is also NP-hard. Therefore, attempts to maximize the number of concurrent operating processes have this difficulty.

The SER algorithm has been applied to distributed control of traffic intersections (Carvalho, Protti, Gregorio, & França, 2005), synchronous-to-asynchronous conversion of cryptographic circuits (Cassia, et al., 2009), Web graph decontamination (Gonçalves, et al., 2010), and job shop modeling (Lengerke, et al., 2008).

SER-Based Decontamination

In the SER-based decontamination (Gonçalves, et al., 2010), the use of the SER properties solve the graph contamination problem ensuring that all nodes will operate (be cleaned) and that there is no recontamination. Figure 2 illustrates a SER-based decontamination dynamics.

The decontamination starts with an acyclic orientation of the contaminated graph and guards (decontamination agents) positioned at the sink nodes. As the nodes are cleaned, the edges are reverted, new sinks appear, and the agents move to those in order to clean them, leaving a copy to protect the older sink from recontamination, if necessary. Decontamination agent stops guarding it when the node is not under recontamination risk. The edge reversions and agents' movements continue until the graph is clean.

In order to minimize the number of cleaning agents, the acyclic orientation of the graph must have a small number of nodes operating (being cleaned) simultaneously, i.e., it needs to lower the concurrency. As previously mentioned, this is related to a maximal coloring of the graph, an NP-hard problem (Gonçalves, et al., 2010). Therefore, the use of heuristics is more efficient.

Figure 2. SER-based graph decontamination

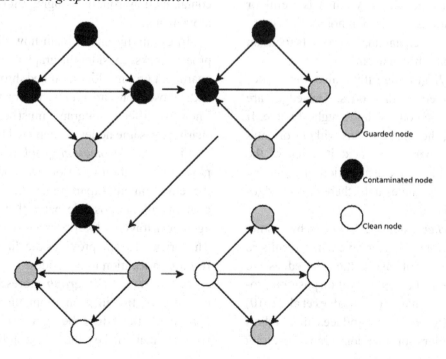

Some heuristics were analyzed for use with the SER and SER-based decontamination, such as *Alg-Neighbors*, *Alg-Colors*, and *Alg-Edges* (Arantes, França, & Martinhon, 2009). With *Alg-Stretcher* (Gonçalves, et al., 2010), an algorithm to enlarge the sink decomposition created by *Alg-Edges*, the amount of concurrency of the resulting SER dynamics decreases, hence the number of required agents.

The first attempts used a "coin" to separate the nodes. It started with a set of nodes N containing all the nodes of G. All nodes would receive a value 1 or 0 from the coin. A node i was a winner if it had 1 from the coin and all its neighbors in N had 0. The node i would then leave N. The process continued until N was empty. *Alg-Neighbors* expands this idea using f-sided dice, thus constituting a generalization of the previous idea, which had $f = 2$. This generalization accelerates the process of finding winner nodes.

Alg-Colors, instead of directly orienting the edges, gives the winner vertex the lowest color not used by a neighbor. Then it creates an orientation from the highest to the lowest color when the coloring process is finished. This algorithm finds the highest concurrency, which is useful for some applications of SER, but not for this use in the SER-based decontamination with its focus on the smallest number of agents.

The *Alg-Edges* algorithm also distributes random numbers to the nodes. The edges are orientated to the node with the highest value. If there is a tie, the edge remains without orientation and take part in the next iteration of the algorithm, with a new distribution of numbers. This process continues until there are no edges without orientation.

Alg-Stretcher extends *Alg-Edges* by testing moving nodes to a higher layer. If it results in a higher number of layers, then the edges are adjusted to reflect the new layer disposition producing a new orientation. Gonçalves et al. (2010) show that *Alg-Stretcher* produces the highest number of colors for a random graph, resulting in the lowest concurrency and lowest number of decontamination agents. Figure 3 illustrates *Alg-Edges* and *Alg-Stretcher*.

Mobility-Constrained SER-Based Decontamination

A mobility-constrained system, such as a firefighting situation, raises new considerations for SER-based decontamination. The impact of those constraints on the algorithm must be analyzed and the algorithm will need some modifications to cope with the changes.

Agents are now limited by quantity and localization, so they cannot appear or disappear on demand. The infection properties also suffer modifications and now their infection strength and speed must be studied, as well as their effects on the model, such as the agent's capacity of leaving a node earlier or for a short interval of time before it comes under risk of recontamination. The environment is also important: in a conflagration scenario, the robots have limited movement, while the fire has more freedom, ignoring barriers for the robots. This differentiates the subgraph for contamination from the subgraph for the agents' movements.

To exemplify and explain how this new approach works, consider the graph mapping of the apartment ground floor assumed throughout this text, whose rooms are represented by their nodes. Then, two different subgraphs must be constructed sharing the same nodes, as depicted by Figure 4.

Firstly, the *locomotion graph* represents the physical allocation of resources (robots) inside the environment (apartment), i.e., this graph contains all the possible paths that the mobile agents can traverse, like doors between the rooms. This oriented graph provides the direction of the resource allocation to the nodes.

Secondly, the *fire graph* represents the occurrence of fire in each room throughout the apartment, the edges being the channels for recontamination, i.e., the edges indicate which

Figure 3. Example of application of Alg-Edges, with the numbers used to order near the nodes, followed by application of Alg-Stretcher; node d is moved into a new layer

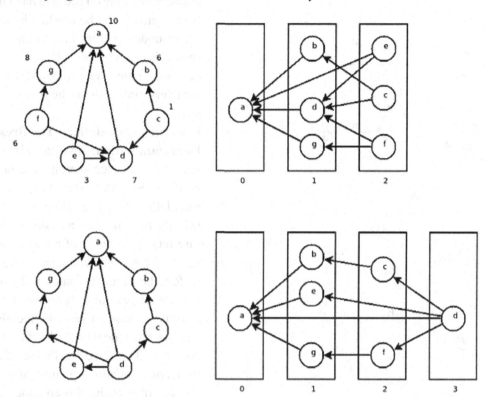

node (room) can infect another. The infection is possible if the rooms are adjacent, e.g., rooms separated only by one door or wall, or multiple nodes in the same room, representing a large area. Thus, if an agent stays at a node for the cleaning, it cannot leave the node until all the other neighbor nodes are cleaned too, otherwise, contamination may arise again.

As a premise for the proposed algorithm, let us consider that every mobile agent has access to the two graphs, i.e., that the apartment ground floor structure is known from the beginning. As soon as the robots enter the apartment, they have to follow a two-phase procedure:

Phase 1: Locomotion Graph Orientation:
- Number all the nodes in a sequence starting from 0 (zero);

- Choose the *entrance node* (room with the entrance door) and set it as a sink. All the edges must point toward that node;
- Search for the next sink nodes after an edge reversion, i.e., the nodes in which the above-orientated edges will point toward after one reversion. Orientate the remaining associated edges toward those nodes. In case of tie, choose the node with the lowest sequence number;
- Repeat the process for all the nodes.

Phase 2: Collaboration Algorithm Execution:
The algorithm returns, at every iteration, a two digit *allocation sequence* that displays the mobile agent dynamic allocation through the nodes of the graphs, e.g., the sequence "3/4" shows that some robot went from node 3 to node 4. If the fist digit is equal to "-1,"

Figure 4. Directed locomotion graph and undirected fire graph

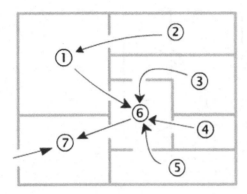

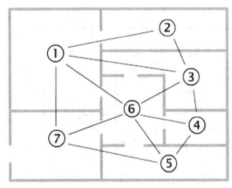

a new agent entered the apartment to get to a room because, at that moment, no robot inside the apartment could leave its position. It consists of the repetition of four steps until there are no nodes on fire anymore.

1. **Robots Release:** Verifies if any robot can leave a node. To do this, the algorithm evaluates if for every guarded node with a robot, all the respective adjacent nodes are also decontaminated. If so, the robot is released from that node and reallocated (ties of free robots are broken choosing the lower numbered node). Otherwise, a new robot enters the apartment.

2. **Decontamination Candidates Selection or Edges Reversion:** Verifies which nodes must be decontaminated. To carry out this function, the algorithm identifies the nodes,

which match the following conditions: have a sink, is catching on fire, and has a free path from which it can be reached by the robot origin node (or from the entrance node in case it is a new robot). If no node can be decontaminated, it performs edge reversion for every sink node in the graph and skips steps 3-5.

3. **Priorities Calculation, Tie Break, and Decontamination:** Analyzes all the nodes possible to be decontaminated. In case of multiple possible nodes, the tie is broken calculating their respective *priorities*. The priority *fire ratio* of a node is given by: Fire ratio = (number of adjacent nodes on fire / total number of adjacent nodes). The node to be decontaminated is the one with the lowest *fire ratio*, meaning the highest priority. In case of new tie, the algorithm chooses the node with the lowest number in sequence. This node is called the "D" (node in decontamination). It then allocates the chosen robot to the chosen node "D" to be decontaminated. For example, if one uses the robot in node 4 to decontaminated 1, the algorithm returns the allocation sequence 4/1. Likewise, if one uses a new robot to decontaminated 5, it returns the allocation sequence -1/5.

4. **Neighbor Sinks Decontamination:** For a previously decontaminated node "D," all of its adjacent sinks must also be decontaminated during this step, since it is expected that the latest allocated robot should be set free as soon as possible, while remaining "D" adjacent sinks still contaminated:

    a. It calculates the priority of all still contaminated "D" adjacent sink nodes. Chooses the node to be decontaminated.

    b. Runs Step 1 only to assign a free robot (or new robot) to be reallocated.

    c. Allocates robot and performs decontamination.

    d. Goes back to Step 4.

If the decontamination for "D" and for its adjacent nodes is finished, the algorithm goes back to Step 2 to check for more possible sinks to be decontaminated.

A SWARM ROBOTICS APPROACH TO FIREFIGHTING

The previous formulation of the SER-Based Decontamination algorithm to mobility-constrained scenarios is an interesting approach that can be demonstrated in practice. The problem of extinguishing a conflagration inside an apartment is an illustrative situation that is suitable for this demonstration.

The real problem can be tackled by using a mobile robot and/or fireproof machines focus, i.e., the decontamination agents. These robots belong to a cooperative network, that is, they are able to communicate with each other and share information. These features are assumed in the implementation of the *Mobility Constrained SER-Based Decontamination* algorithm. In a real case, mobile agents have to be versatile and fast enough to enter an apartment, search for a room and keep it clean of any fire focus until all neighboring rooms also get cleaned. While guarding that room, a robot must be aware of the environment around it and instantaneously extinguish all possible new fires which may arise. The logic of allocation and routing of each robot inside the apartment is given by the proposed algorithm.

Experimental Setup

A demonstration of the proposed *Mobility Constrained SER-Based Decontamination* in real scale requires a real apartment, fire, and possible expensive machines to perform the task. For this reason, the reduction of the problem into a small scale is a suitable option. The real apartment, which has its floor plan depicted by Figure 4, is represented by a respective Styrofoam mock up of its ground floor, with walls and doors constructed in proportion of the actual measures of this apartment.

As well as the environment, the mobile agents must have dimensions of the same proportions as the apartment they will decontaminate. In addition, these agents must have some essential features so that the algorithm can be properly executed: proximity sensors, camera vision, communication channel, expanded mobility and versatility, open hardware for programming and reduced size.

A solution that matches all these requirements is the e-puck education robot (Mondada, et al., 2009). It is a differential wheeled small round robot (7 cm of diameter), capable of reaching a maximum speed of 13 cm/s, with movement autonomy of 2 hours, which is especially designed for educational applications and demonstrations. It also includes: 3D accelerometers, 3 microphones, 1 loudspeaker, 10 LEDs, a 640x480 color camera, 8 infrared proximity sensors, 2 step motors, a 144 KB flash memory, 8 KB RAM and a dsPIC 30 CPU at 30 MHz (15 MIPS).

As previously mentioned, the *Mobility Constrained SER-based Decontamination* algorithm attempts to minimize the number of agents that must be used for the decontamination process. For this specific ground floor (see Figures 7 and 8), the graph requires four agents; so, in this project, four e-pucks perform the demonstration. In addition, the Webots 6.0 software runs a computer simulation of all robots performing the target task.

Trajectory Issues

While executing the algorithm, each robot in the network has to move inside the rooms. In order to make each robot reach some desired point inside the apartment, as well as avoiding obstacles and walls, a trajectory strategy had to be considered in the implementation. In this case, the use of pre-programmed potential fields is considered in each robot. Potential fields are functions that map all the points of a specific area into a specific value, thus generating a surface with a global minimum that represents the final target, and local maxima, that represent all obstacles.

The potential field is the sum of two fields: *repulsion* and *attraction* potential fields. The attraction field is given by:

$$U_{atr} = \frac{1}{2} * K_p * (x - x_d)^2$$

where K_p is a constant gain parameter, x is the position of the robot at some instant, and x_d is the target position. On the other hand, the repulsion field is given by:

$$U_{rep} = \frac{1}{2} * K_d * \left(\frac{1}{p}\right)^2$$

where K_b is a constant gain parameter and ρ is the distance from the robot to the obstacle. Considering that this distance can be zero at some time during the algorithm execution, a negligible constant is added to ρ in order to solve the problem of division by zero. According to the number of obstacles to be considered, the repulsion field can be given by the sum of multiple terms, each one with a respective ρ for each obstacle distance. Finally, the resulting force acting in the robot is given by:

$$F_{res} = -\nabla U.$$

This equation represents the force that arises in the robot according to its position over the surface generated by the use of potential fields. In other words, if a robot is near a peak, i.e., a local maximum representing an obstacle, a force arises acting on the opposite direction, repelling the agent far away from that point, like a ball rolling down a mountain. If the robot is near the valley, i.e., the global minimum representing the target, a force

arises trying to push the robot toward this point, like a ball falling in a hole. So, for every point over the surface, a respective force is assigned with established intensity and direction.

In order to avoid the situation in which a robot gets stuck in possible local minima, virtual targets are added in each door of the apartment ground floor, because the robots have to pass by them to reach the actual final target. This strategy does not eliminate local minima, but avoids the situation of having robots stuck at them. Figure 5 illustrates the resulting potential fields generated in order to do so.

Since the proximity sensors of e-pucks have low performance, the potential fields are implemented together with cellular decomposition. In this approach, the apartment ground floor is partitioned in small areas (cells). In this work, square cells were considered with sides equal to 0.5 cm. For each part, a value is assigned according to the designed potential field equations. The negative sign of the force equation is neglected in each cell value, because there is a routine in the implemented algorithm that uses the neighboring cell with the lowest value, indicating the actual path that the robot will follow.

In spite of the modeled strategy for the robot trajectory, the physical model of each robot has to be considered to develop the desired motion. Since the robot is a complex and extended body, its movements are not simple, but coordinated by dynamic mechanical equations, which are not taken into account in the original potential field approach. Together with parameters, such as friction, this leads to undesirable error accumulation, hence making the robots deviate from the planned trajectory.

To solve this problem, this work uses a mathematical model of a robot with differential driven movement, according to the *odometry* approach used by Santana, Souza, Britto, Alsina, and Medeiros (2008). The analysis of those equations

Figure 5. Potential fields used by a robot to go through a door inside the target floor plan

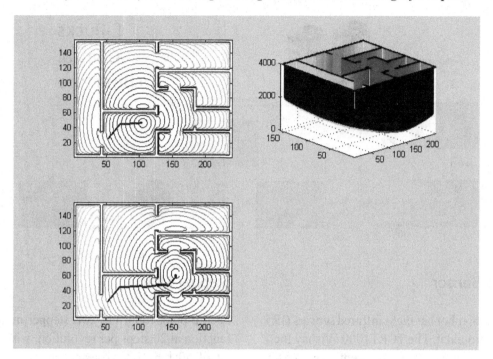

allows for the design of a controller for the projected robot angular and distance deviations from the desired trajectory through the potential fields approach. These variables errors are estimated by the computation of this model, after obtaining the encoder values of the wheels motors. Then, errors are fed into a PID controller, thus reestablishing an adequate path for each robot through the apartment ground floor. Figure 6 presents the integration between the trajectory of potential fields, the odometry model, and the controller.

The Webots platform allows for the simulation of all those equations together with other desired physical parameters, thus making it possible to better visualize the behavior of the real demonstration. In practice, this solution will not produce entirely accurate robot localization due to noise on the encoders of the stepper motors, but it is sufficient for the purposes of the experiment.

Collision Avoidance Issues

Microcontroller

There is a dsPIC microcontroller model 30F6014A, manufactured by Microchip (Microship Technology Inc., 2006), embedded into an e-puck's circuit board, which is responsible for managing all information between the robot devices. It has a total of 80 pins with 16 input registers and a unit of Digital Signal Processing (DSP) for analog signal processing. It also has a memory capacity of 8KB of RAM memory, 144KB of flash memory, and a maximum clock speed of 64MHz, provided by an external crystal oscillator. The microcontroller has five 16-bit timers that can compose two 32-bit counters.

Figure 6. Trajectory control scheme

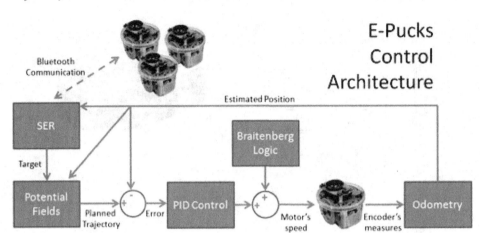

Infrared Sensor

Each e-puck robot has eight infrared sensors (IR) arranged around it. The TCRT1000 (Vishay Inc., 1999) device is an active exteroceptive IR sensor that indicates the presence of objects through a transmitter and a receiver arranged in the same direction of the object. The transmitter consists of a 950nm wavelength light infrared beam source at 80Hz frequency, which is reflected by some object and returns to the receptor. The receiver has a phototransistor, which the reflected waves reach changing the value of current flowing between the collector and emitter terminals, thus indicating the presence of the object.

Since the phototransistor is a semiconductor, it reacts with a non-linear response when faced with current and voltage changes. The TCRT1000 device guarantees the polarization of the phototransistor to make it operate in a region of high linearity, thus allowing a better match between both input and output values. The dsPIC microcontroller contains a group of registers that are in charge of the analog/digital (A/D) conversion and provide a twelve bit word, hence generating a dynamic range of 0 to 4095. All eight IR sensor measurements are sampled by the A/D converter at a frequency of 400Hz.

Stepper Motors

Each e-puck robot has two stepper motors that function at 20 steps per revolution, with an extra set of 50:1 reduction gear, hence generating a total of 1000 steps per revolution. This feature allows for the production of a more accurate and natural displacement. The motor speed is limited to one revolution per second. The movements of the motors are controlled by the association of the eight least significant bits obtained from the address 02D6 of the dsPIC. The four least significant bits are responsible for the movement of the left motor and the other four bits are responsible for the movement of the right motor.

Path Deviation with the Braitenberg Behavior

The deviation action is performed after the identification of an obstacle by the IR sensors. When approaching an obstacle, part of the emitted infrared signal is reflected on the obstacle surface and returns to the receptor. This signal varies the current between the collector and emitter terminals of the phototransistor, hence indicating the presence of the obstacle.

Figure 7. E-pucks inside the mock up

The process of A/D conversion is performed by the dsPIC microcontroller using the Digital Signal Processing unit (DSP). This unit contains a group of 12 registers at the output. They provide a digital signal for each IR sensor, which is represented by a value whose range varies from 0 to 4095. Due to the natural characteristic of the phototransistor and the process of A/D conversion, the signal is inversely proportional to the intensity of light received by the IR sensor. The variation in the digitalized signal of the IR sensors is further used to drive the two e-puck's stepper motors.

Simulation

The experiment consists of a specially configured simulation environment, representing the apartment ground floor in a Styrofoam mock-up, a finely tuned projector, and Web camera over the mock-up, 4 e-pucks with black paper discs attached over each one, and a computer linked to the camera and a Datashow projector. The camera and the projector provide a visual representation of the burning situation. A programmed application in the computer recognizes in real-time the mock-up image from the camera and constructs a colored image of the rooms, filling them with red to represent burning rooms and green for cleaned rooms. These images are real time updated and projected over the mock-up, as illustrated in Figure 7.

The application also recognizes the position of the robots through the black disc images available in real time provided by the camera. Using this information, the computer analyzes whether there is a robot inside a room, which would indicate that this room is decontaminated, hence coloring it in green. The program also runs the logic of contamination, i.e., if the application detects that a robot left a room with at least one still contaminated neighbor, that room will become contaminated again, hence colored in red, indicating the recurrent burning. The complete decontamination sequence obtained in the Webots environment is illustrated in Figure 8.

Communication Issues

For the operation of the distributed network of e-puck robots, a system of communication between them had to be implemented. The e-pucks must communicate to each other somehow and the

Figure 8. Complete decontamination sequence based on the Webots simulation environment

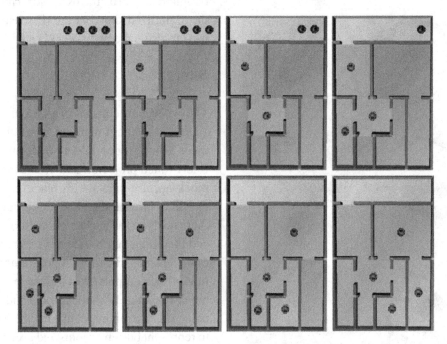

straightforward solution would be to adopt the Bluetooth library found on the official e-puck website. However, it was not possible to make it work outside the Webots simulation environment.

CONCLUSION AND FUTURE WORK

SER-based decontamination presents a suitable solution to conflagration scenarios. Even though the case study is small, its process scales to deal with larger situations. The robots can deal with environments that would endanger human lives in a more efficient manner, controlling both the fire and personal risks.

The modeling of the area as graphs gives the robots freedom to act in different environments without special adaptations. This versatility allows the application of the robots in different scenarios. Future work can develop the capability to map the conflagration environment and generate the needed graphs without previous knowledge of the area.

In addition, the dynamics of the contamination process can also affect the performance of the decontamination strategy. There are many parameters of contamination that can be more deeply studied to make modeling of more real situations possible. One of these parameters is the presence of a refractory period for recontamination. For this work, an extreme situation is considered, in which any room of the apartment is instantaneously contaminated as soon as any robot leaves it, and neighbor rooms are still catching on fire. Actually, the fire takes a delay to considerably spread again. This time lag to re-infect a node can be investigated to aid in developing a more complex but flexible decontamination algorithm, based on the presence of this refractory period of recontamination.

REFERENCES

Arantes, G. M. A. Jr, França, F. M. G., & Martinhon, C. A. J. (2009). Randomized generation of acyclic orientations upon anonymous distributed systems. *Journal of Parallel and Distributed Computing, 69*(3), 239–246. doi:10.1016/j.jpdc.2008.11.009

Barbosa, V., & Gafni, E. (1989). Concurrency in heavily loaded neighborhood-constrained systems. *ACM Transactions on Programming Languages and Systems, 11*(4), 562–584. doi:10.1145/69558.69560

Carvalho, D., Protti, F., Gregorio, M., & França, F. M. G. (2004). A novel distributed scheduling algorithm for resource sharing under near-heavy load. In *Proceedings of the International Conference on Principles of Distributed Systems (OPODIS),* (Vol. 8). OPODIS.

Cassia, R. F., Alves, V. C., Besnard, F. G.-D., & França, F. M. G. (2009). Synchronous-to-asynchronous conversion of cryptographic circuits. *Journal of Circuits, Systems, and Computers, 18*(2), 271–282. doi:10.1142/S0218126609005058

Flocchini, P., Nayak, A., & Schulz, A. (2005). Cleaning an arbitrary regular network with mobile agents. In G. Chakraborty (Ed.), *Distributed Computing and Internet Technology, Second International Conference,* (Vol. 3816, pp. 132-142). Berlin, Germany: Springer.

Gonçalves, V. C. F., Lima, P. M. V., Maculan, N., & França, F. M. G. (2010). A distributed dynamics for webgraph decontamination. In T. Margaria & B. Steffen (Eds.), *4th International Symposium on Leveraging Applications of Formal Methods, Verification and Validation,* (Vol. 6415, pp. 462-472). Berlin, Germany: Springer.

Lengerke, O., Carvalho, D., Lima, P. M. V., Dutra, M. S., Mora-Camino, F., & França, F. M. G. (2008). Controle distribuído de sistemas JOB SHOP usando escalonamento por reversão de arestas. In *Proceedings of the XIV Congreso Latino Ibero Americano de Investigación de Operaciones,* (pp. 1-3). IEEE.

Luccio, F., & Pagli, L. (2007). Web marshals fighting curly links farms. In P. Crescenzi, G. Prencipe, & G. Pucci (Eds.), *Fun with Algorithms, 4th International Conference,* (Vol. 4475, pp. 240-248). Berlin, Germany: Springer.

Microchip Technology Inc. (2006). *dsPIC30F family reference manual.* New York, NY: Microchip Technology, Inc.

Mondada, F., Bonani, M., Raemy, X., Pugh, J., Cianci, C., & Klaptocz, A. … Martinoli, A. (2009). The e-puck, a robot designed for education in engineering. In *Proceedings of the 9th Conference on Autonomous Robot Systems and Competitions,* (Vol. 1, pp. 59-65). IEEE.

Moscarini, M., Petreschi, R., & Szwarcfiter, J. L. (1998). On node searching and starlike graphs. IEEE. *Proceedings of Congressus Numerantion, 131,* 75–84.

Vishay Inc. (1999). *TCRT1000 datasheet.* New York, NY: Vishay, Inc.

ADDITIONAL READING

Choset, H. (2007). *Robotic motion planning: Potential functions.* Retrieved from http://www.cs.cmu.edu/~motionplanning/lecture/

Costa, T. A. de A., Ferreira, A. M., & Dutra, M. S. (2008). Parametric trajectory generation for mobile robots. In P. E. Miyagi, O. Horikawa, & J. M. Motta (Eds.), *ABCM Symposium Series in Mechatronics,* (Vol. 3, pp. 300-307). Brasília: Associação Brasileira de Engenharia e Ciências Mecânicas.

GCTronic. (2011). *E-puck – Gctronic wiki*. Retrieved October 7, 2011, from http://www.gctronic.com/doc/index.php/E-Puck#Software

Lengerke, O., Campos, A. M. V., Dutra, M. S., & Pinto, F. de N. C., & França, F. M. G. (2010). Trajectories and simulation model of AGVs with trailers. In V. J. De Negri, E. A. Perondi, M. A. B. Cunha, & O. Horikawa (Eds.), *ABCM Symposium Series in Maechatronics,* (Vol. 4, pp. 509-518). Brasília: Associação Brasileira de Engenharia e Ciências Mecânicas.

Oriolo, G. (2011). *Motion planning 3: Artificial potential fields*. Retrieved from http://www.dis.uniroma1.it/~oriolo/amr/

Pacheco, R. N., & Costa, A. H. R. (2002). Navegação de robôs móveis utilizando o método de campos potenciais. In M. T. S. Sakude & C. de A. Castro Cesar (Eds.), *Workshop de Computação WORKCOMP 2002,* (pp. 125-130). São José, California: SBC.

Santana, A. M., Souza, A. A. S., Britto, R. S., Alsina, P. J., & Medeiros, A. A. D. (2008). Localization of a mobile robot based on odometry and natural landmarks using extended Kalman filter. In *Proceedings of the VICINCO - International Conference on Informatics in Control, Automation and Robotics.* Madeira, Portugal: Funchal.

Zou, X., & Zhu, J. (2002). Virtual local target method for avoiding local minimum in potential field based robot navigation. *Journal of Zhejiang University Science A, 4*(3), 269–294.

Chapter 7
Path Planning in a Mobile Robot

Diego Alexander Tibaduiza Burgos
Universitat Politècnica de Catalunya, Spain

Maribel Anaya Vejar
Universitat Politècnica de Catalunya, Spain

ABSTRACT

This chapter presents the development and implementation of three approaches that contribute to solving the mobile robot path planning problems in dynamic and static environments. The algorithms include some items regarding the implementation of on-line and off-line situations in an environment with static and mobile obstacles. A first technique involves the use of genetic algorithms where a fitness function and the emulation of the natural evolution are used to find a free-collision path. The second and third techniques consider the use of potential fields for path planning using two different ways. Brief descriptions of the techniques and experimental setup used to test the algorithms are also included. Finally, the results applying the algorithms using different obstacle configurations are presented and discussed.

INTRODUCTION

Robotics is an area with vast prospects for industrial development that in a relatively short time could allow companies to improve production techniques, quality, and precision in many different processes.

For some years now, mobile robots have been introduced into various productive areas. This interaction has increased the need to provide autonomy to robots, which can interact beyond static environments. A basic example of this interaction is cleaning robots that currently are sold for domestic tasks and Automatic Guided Vehicles (AGV) which are often used in industrial applications.

Increasingly, robots must interact in environments where workspaces are not static, meaning that robots must have the tools to adequately perform their required tasks despite changes that can occur in cases where planned trajectories represent part of the solution.

DOI: 10.4018/978-1-4666-2658-4.ch007

Copyright © 2013, IGI Global. Copying or distributing in print or electronic forms without written permission of IGI Global is prohibited.

Path planning has been approached in different applications, including robotic manipulators, mobile robots, and underwater robots, among others. Similarly, different methodologies have been developed for both static and dynamic environments, which use different sensors for local and global planning of the workspace, as will be shown in the brief section review.

This chapter presents the development and implementation of three approaches for path planning. These are defined for working in a static workspace, but also can be applied to dynamic workspaces by updating the information of the position of all elements (obstacles and mobile robot). Specifically, in this case, an artificial vision system based on color algorithms is used for updating the position. The first technique involves the use of genetic algorithms which use a fitness function and the emulation of natural evolution to find the best trajectory; the results obtained are in function of the number of generations defined by the user. The second and third techniques include the use of potential fields as path planning method by means of two approaches; these algorithms define specific weights to the workspace for each element (obstacles and mobile robot) and compute a free path based on the potentials defined for the workspace. The chapter also includes a brief introduction to path planning and a brief state of the art that is focused on the methodologies presented. Additionally, each one of these techniques was proven in a physical system that consisted of a 1.50 m x 2.40 m wooden platform, an artificial vision system, two mobile robots, and some fixed obstacles. The results obtained are included.

BRIEF REVIEW ON PATH PLANNING

Research in path planning is an area of great interest in robotics because of the versatility it gives a robot to perform its work with reliability and autonomy. The domain of this ability gives the robot the option of avoiding collisions and obtaining different free collision paths for it to move in a workspace based on the user-defined criteria.

Currently, a large numbers of robots are equipped with sensors that allow them to obtain information from the environment. These sensors can provide global or local information to detect obstacles, which allows calculating free-collision paths using some planning strategies to move the robot step-by-step or to define a complete path in a workspace (Tibaduiza, 2008).

For more than 30 years, mobile robot autonomy has been one of the main motivations for developing path-planning strategies to provide the necessary tools for moving the robots through different environments and under difficult conditions. Below are some jobs mobile robots perform on a regular basis.

Lozano (1983) presents an approach for computing constraints on the position of an object due to the presence of obstacles. The algorithms presented allow persons to characterize the position and orientation of objects as a single point in a Configuration Space when the objects and obstacles are polygons or polyhedral.

Thorpe and Matthies (1984) propose a method called Path Relaxation which combines characteristics of grid search and potential fields. It works in two steps: a global grid search that finds a rough path, followed by a local relaxation step that adjusts to each node on the path to lower the overall path cost. The same year, Lumelsky and Stepanov (1984) describe one approach based on the continuous processing of incoming local information of the environment. A continuous computational model for the environment and vehicle operation is presented. Information about the environment (the scene) is assumed to be incomplete except that at any moment, the vehicle knows the coordinates of its target as well as its own coordinates. The vehicle is presented as a point; obstacles can be of any shape, with continuous borderline and finite size. Byung and Kang (1984) also present a method for minimum-time path planning in joint space subject to realistic constraints by testing the methodology in a Unimation PUMA 600 manipulator robot.

The next year, Crowley (1985) presented a navigation local method based on a dynamically maintained model of the local environment, called the composite local method, which integrates information from the rotating range sensor, the robot's touch sensor, and a pre-learned global model as the robot moves through its environment.

Habib and Yuta (1989) present a method for modeling a robot environment and structuring its free space in terms of a set of overlapped Prime Rectangular Areas (PRAs) which are suited for path planning and navigation tasks. Additionally, a connectivity table for the overlapped PRAs is constructed to provide the structure necessary for path-finding. In the same year, Elfes (1989) used occupancy grid, which is a probabilistic tessellated representation of spatial information. This approach estimates the occupancy state of the cells in a spatial lattice in order to define a map of the robot's world based on sensor information.

Rimon and Koditschek (1992) present a methodology consisting of an exact robot motion planning and a control unifying the kinematic path-planning problem with the lower level feedback controller design. Information about the free space and goal are encoded in an artificial potential function, and a navigation function gives the bounded-torque feedback controller information to the robot's actuators to avoid the obstacles in the path-planning process.

Goldman (1994) presented a proposal for path planning in the context of an airplane traveling to avoid circular danger zones. The proposal includes two aspects: (1) The use of the Collins decomposition algorithm to find an optimal path before the airplane even leaves and (2) an algorithm based on sensors to avoid danger zones.

In the same year, Hofner and Schmidt (1994) introduce a technique for planning and guiding a mobile floor-cleaning robot. Their methodology considers a kinematic and geometrical model of the robot and the cleaning units as well as a

2D-map of the indoor environment; the path is represented by a concatenation of two kinds of typical motion patterns. Each pattern is defined by a sequence of discrete Cartesian intermediate goal frames. These frames represent the position and orientation of the vehicle and must be translated into motion commands for the robot.

Changkyu, Sun, Jin, Il-Kwon, and Ju-Jang (1995) present a dynamical local path-planning algorithm of an autonomous mobile robot available for stationary obstacle avoidance using nonlinear friction. Together with the Virtual Force Field method (VFF), the path of the mobile robot is a solution of a path planning equation and the local minima problems in stationary environments are solved by introducing nonlinear friction into the chaotic neuron.

Sugihara (1997) presents an approach based on genetic algorithms for path planning in a mobile robot; this approach considers the use of danger zones. After Suguihara (1998) presented online path-planning based on genetic algorithms for an autonomous underwater vehicle, it was employed to integrate on-line path planning with off-line planning to make path planning adaptive.

Cao, Huang, and Zhou (2006) propose a methodology based on artificial potential fields for path planning in dynamic environments where the target and obstacles are moving. Here, a force function and the relative threat coefficient function are defined.

Han-Dong, Bao-hua, Yu-wan, Rui Hashimoto, and Saegusa (2007) use a genetic algorithm and propose a fixed-length decimal encoding mechanism. The obstacles are described as polygons based on the given grids. The path planning for the mobile robot is encoded into a string, which is a decimal encoding. This is equal to the sum of the numbers of all effective verteces of polygons in the path grids. In the chromosome, the decimal value of each nonzero bit is the code of vertex about corresponding obstacle for the path planning.

Yong, Lin, and Xiaohua (2008) present an algorithm for path planning in an unknown environment using potential field in order to generate the initial path population and genetic algorithms to optimize the path. The same year, Tibaduiza and Chio (2008) presented the description of two potential field methodologies applied to a mobile robot. Pinna, Masson, and Guivant (2008) used the Dijkstra algorithm to obtain the optimal path. The trajectory is determined by defining local regions and calculating some gates. The main idea is to move the robot through the gates in each region.

Bodhale, Afzulpurkar, and Thanh (2008) integrate potential fields and use Monte Carlo localization for navigation, obstacle avoidance, and mobile robot localization in a dynamic environment. The path planning algorithm is divided in two submodules, the first includes visibility graph with A* search method and the second is the local planning using potential fields.

Jianming, Liang, Qing, and Yongyu (2009) use D* algorithm for path planning in a partially known environment. A two dimensional model of the workspace with some obstacles expressed in grids is included after the robot movement is calculated by using the D* algorithm by searching in the neighbor grid cells on 16 directions.

Hassanzadeh, Madani, and Badamchizadeh (2010), propose a method based on Shuffled Frog Leaping (SFL) optimization that allows a mobile robot execute a trajectory in workspace with static obstacles. This approach transforms the problem of robot path planning to one of minimization and then defines the fitness of a frog based on the positions of the target and the obstacles in the environment. The position of the globally best frog in each iterative is selected and reached by the robot in sequence. In addition, the environment is partially unknown by the robot due to the limit detection range of its sensors.

Kang and Wang (2010) presented an algorithm based on PDE methods and some traditional solutions on path planning. The adaptive mesh is used to solve the PDE model for path planning.

PATH PLANNING

Path planning is one of the most important tasks regarding intelligent control of an autonomous mobile robot (Tibaduiza, 2006). The autonomy of a robot is evaluated in terms of the intelligence the robot displays by taking a decision such as going from one point to another without colliding with the elements presented in the workspace.

Robot movement is often divided into road planning and path planning, although they are independent of each other. Path planning consists of generating a free-collision path in an environment with obstacles and optimizing it, according to specified criteria, while the objective regarding road planning is to set the motion of a mobile robot along a planned path.

In general, path planning can be divided into two different approaches: online and offline planning. Online path planning is directly performed on the robot in real time, allowing it to respond to real-time changes in the workspace. Offline planning is normally used when obstacles are fixed and it is possible to generate a safe path that is not affected by any changes in the workspace.

In this section, two algorithms of path planning for a mobile robot will be explained. First, an introduction about genetic algorithms and its basic elements will be described. Later, an introduction to potential fields will be presented showing the basic elements to implement the approach. Specific elements regarding the implementation will be presented in the following section.

Genetic Algorithms (GAs)

Genetic Algorithms are an artificial intelligence technique inspired on the evolution theory. This technique is based on artificial natural selection mechanisms and genetics, and represents the solutions as an optimization problem. Biology shows how adaptations to the environment are transmitted to offspring by means of individual genes; these are a biological structure that contains information about the environment.

Evolution takes place in the chromosomes, where information is encoded (Hartl, 1998). The information is stored in the chromosomes and it changes from one generation to the next. In the forming of a new individual, the information in the parental chromosomes is combined, however this process is not perfectly understood. Importantly, there are many aspects that are not clear, although the scientific community widely accepts some general principles:

1. Evolution operates on chromosomes.
2. Natural selection is the process by which chromosomes with "good structures" are reproduced more often than others.
3. The reproduction process takes place by combining parental chromosomes.
4. Biological evolution has no memory in the sense that the formation of chromosomes only considers the previous phase information.

The implementation of a simple genetic algorithm implies defining some characteristics such as reproduction, crossover, and mutation (Goldberg, 1989). To perform these steps, the algorithm is encoded in strings called chromosomes, which consist of characteristic elements called genes. There are several ways to encode the data, according to what kind of symbol is used to represent the gene. These encoding methods can be classified as binary encoding, real-number encoding, integer or literal permutation encoding, and general data structure encoding (Mitsuo, 2000).

In general, it is necessary to consider the following to implement a genetic algorithm:

- A chromosomal representation, defined by means of an encoded route.
- An initial population.

- A measure of fitness. This objective function indicates the fitness of any potential solution (Fogel, 1994).
- A selection criterion to choose / eliminate chromosomes. This selection is probabilistic, not deterministic (Fogel, 1994).
- One or more operations of recombination.
- One or more mutation operations.

All these elements must be included by means of emulation in a computational program. In general terms, these elements can be represented in three steps: reproduction, crossover, and mutation.

Reproduction: Is the process by which individual strings are copied according to the value of its fitness function. Intuitively, it is possible to define this function as a measure of benefit, utility, or goodness that is necessary to maximize.

This operator is an artificial version of the Darwin's natural selection and defines who lives or dies. The reproduction operator can be implemented by various methods, but perhaps the most widely used is based on the roulette wheel, where each individual is selected according to its fitness.

Crossover: Is the fundamental mechanism of genetic rearrangement (Holland, 1992) for both real organisms and genetic algorithms. Some of the most frequently used crossover operators include:

- **One Point:** One breakpoint selected randomly in the parents is defined to exchange chromosomes.
- **Two Points:** Two breakpoints are chosen at random for exchange.
- **Uniform:** Each gene is chosen at random from one of the parents in order to contribute its gene to the child, while the second child receives the gene of the other parent.
- **PMX, SEX:** These are the more sophisticated operators and they can be combined with other operators.

Mutation: It is a change in the genome sequence. Normally, it is small and can be defined as random changes in the algorithm (Goldberg, 1989). Significantly, it has been found that the frequency of mutation is in the order of one mutation per thousand genes.

The mutation rate in natural populations is small so it is appropriate to consider it as a secondary adaptation mechanism of a genetic algorithm.

The mutation in a binary code consists of moving from one state to another. For example, the chromosome 00010011 might be mutated in its second gene to yield 01010011.

Potential Fields

The potential field method is known as a local method because it only calculates the movement for a given initial-final configuration. It starts at an initial position and moves the robot toward the goal in small steps. This method essentially considers the workspace as a force field by assigning a different value to each object on the workspace (Siegwart, 2004).

The goal point is considered a magnet of opposite polarity to the mobile robot, meaning that this point exerts an attraction force. Other important characteristic about this force is that it is different to the force exerted from the obstacles, which behave as magnets with repulsive forces. In general, the definition of a free-obstacle path is done by adding the potential obtained for the goal point and the potential for all the obstacles. In this way, the robot moves from one potential with a high value to another with a lesser value.

To understand how potential fields work, it is possible to review a simple example by assuming that a mobile robot needs to go from one point to another avoiding an obstacle between them. As will be understood, if no obstacles exist between the initial and the goal point, the best path would be defined by a straight line joining the two points. However, since there is an obstacle, this informa-

tion must be included in the algorithm in order to avoid this position.

Figure 1 shows a representation of the potential fields for the goal point and shows that it is possible to go to the goal point from any position in the workspace. To include the obstacle, the repulsive potential is generated from the obstacle position (Figure 2). Finally, the free-obstacle path (Figure 3) can be obtained by adding the results in Figures 1 and 2.

Figures 1, 2, and 3 show the attraction forces to the goal point, the repulsive forces for the obstacles, and the final trajectory when three obstacles at the same position are used.

IMPLEMENTATION OF ALGORITHMS

Genetic Algorithms

Algorithm implementation starts by defining data encoding. As previously discussed, there are different ways to code the path. For this work, we chose a binary encoding for chromosomes (Sugihara, 1997). This encoding defines each gene as a binary number (0 or 1) and the chromosomes as a combination of these numbers.

To define the encoding data, it is necessary to bear in mind the real dimensions of the workspace. In general, it is possible to consider the workspace as a grid with dimensions $X \times Y$, where X is the horizontal dimension and Y the vertical dimension. Objects in the workspace (mobile robot and obstacles) can be defined as a collection of cells in the grid, depending on their shape there are two types of obstacles: solid and hazardous (Suguihara, 1997). Solid elements correspond to the obstacles and the hazardous zones are risk zones. In real applications, it is necessary to include the hazardous areas in order to assure a free-collision path.

Some mobile robot elements such as their orientation in the workspace and their distance to the goal point should be included in the chromosome representation. The chromosomes chosen

Figure 1. Attractive forces to the goal point

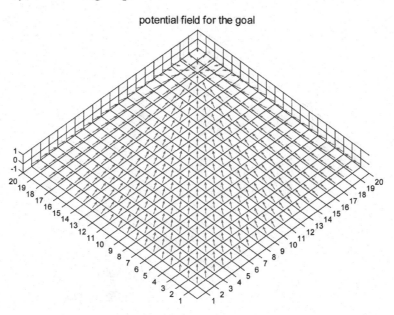

Figure 2. Repulsive forces for the obstacles

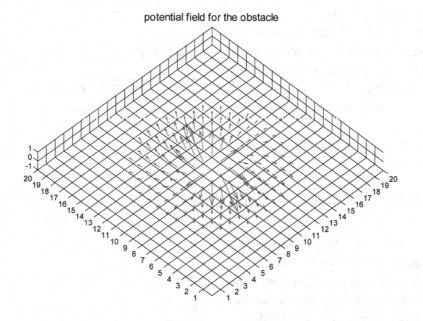

Figure 3. Final trajectory with 3 obstacles at the same place

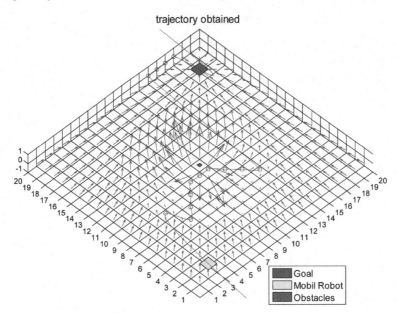

represent: monotony of the trajectory in x and y by mean of α chromosome; address: β chromosome, and distance: δ chromosome (Navas, 2005; Navas, 2006; Tibaduiza, 2006).

In a two-dimensional workspace, a path is said to be monotone with respect to its coordinate if no lines parallel to the x-axis cross the path at two distinct points (Suguihara, 1997).

The binary encoding defined is shows in Table 1.

The number of genes in the chromosome is given by the expression

$$1 + \log_2 N = GenesChromosome\delta \qquad (1)$$

where N is the number of cells representing the territory to be covered by the mobile robot.

Each direction-distance pair represents steps with variable distances (βδ couples). When the chromosome of the direction is encoded "00," the chromosome δ can be positive or negative. This means that the first gene of this chromosome is 0 for positive or 1 for negative. If it is positive and the individual is MX, it represents a vertical

positive up followed by a diagonal top right. If it is negative, this gene represents a vertical downward slash followed by a lower right (Figure 4).

Table 1. Encoding of chromosomes

CHROMOSOME	ENCODING	FEATURE THAT REPRESENTS
α	0	Monotony in X (MX)
	1	Monotony in Y (MY)
β	00	Vertical for MX and horizontal for MY
	01	Upper diagonal for MX Left diagonal for MY
	10	Horizontal for MX Vertical for MY
	11	Lower diagonal para MX Right diagonal para MY
δ	According to Equation 1	

Figure 4. Example of coded path

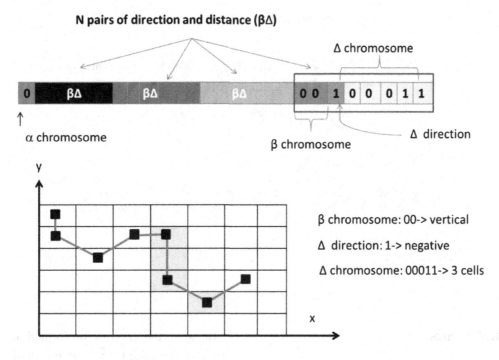

If this gene is positive and the individual is MY, this represents a horizontal positive to the right followed by a lower right diagonal. If this is negative, it represents a negative horizontal to the left followed by a lower left diagonal. Figure 4 shows an example of a section with the distribution of each chromosome using monotony in x.

Once the encoding is defined, the first parents are selected by random methods. We selected 6 parents in each generation which are organized by using the fitness function. This considers the maximum size of the workspace and the current distance from the robot mobile to the goal, for instance: in a workspace with 20×20 cells for going from the position (1, 1) to position (20, 20), in the worst case, the robot should move 399 cells. Knowing this information, the fitness function can be described as the difference between the total number of cells in the worst case (399) and the distance to the goal point. In order to find good paths, this function must be optimized with some criteria like, for instance, shorter trajectories and fewer obstacles.

The selection of chromosomes for crossover is performed using roulette wheel selection. This methodology gives each chromosome one space in the wheel in function of its fitness function. This means that chromosomes with high values have a greater probability of being selected. To implement the crossover, a single crossover point is selected (Figure 5). This operator randomly chooses a locus in the chromosomes and after an exchange between genes before and after of each parent is performed in order to obtain two new offspring.

In the offspring, a 10% rate of mutation is used. This means that only 10% of the offspring obtained from the crossing will be mutated. Similarly, the definition of which gene will mutate is randomly defined. The results obtained on this phase represent the new parents. Finally, all these steps are repeated until to determine a good trajectory.

Figure 5. Crossover method

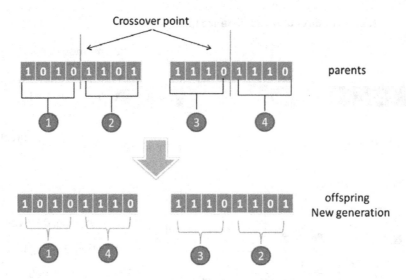

Potential Fields

The implementation of potential fields can be approached from different points of view. This chapter presents two approaches; the first makes use of quadratic functions and the second presents a linear approach to find the free-obstacle path.

First Approach

The algorithm implementation considers the following characteristics:

- The workspace around of the robot is divided in square zones.
- A matrix with the number of zones is defined.
- Each cell of the matrix contains a vector with the potential force associated to the environment.
- Each vector contains m magnitude and *d* direction.
- Each behavior generates a potential field *F*, whose forces are derived and determine robot movements.

Because the path is performed in a two-dimensional space (x, y), the potential gradient in *x* and *y* can be described as in Equation (2).

$$\nabla U = \begin{bmatrix} \dfrac{\partial U}{\partial x} \\ \dfrac{\partial U}{\partial y} \end{bmatrix} \qquad (2)$$

The potential field acting on the robot can be described as the sum of the forces of attraction and repulsion or rejection (Equation (3)).

$$U(q) = U_{attractive} + U_{rejection} \qquad (3)$$

Forces can be separated into attractive and repulsive forces as in Equations (4) and (5) (Dudek, 2000).

$$F(q) = F_{attraction}(q) - F_{rejection}(q) \qquad (4)$$

$$F(q) = -\nabla U_{attraction}(q) - \nabla U_{rejection}(q) \qquad (5)$$

Figure 6. Attractive potential field in the goal point

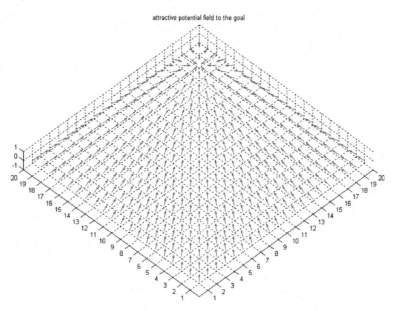

attractive potential field to the goal

Behavior in the Goal Point: The potential in the goal point corresponds to attractive forces since all trajectories must converge to this point. An example is seen in Figure 6.

The attractor force can be expressed as a parabolic function as shown in Equation (6).

$$U_{attraction}(q) = \frac{1}{2} k_{attraction} * \rho_{goal}^2(q) \qquad (6)$$

where $k_{attraction}$ is a positive scaling factor and ρ_{goal} (q) is the Euclidean distance $\left\| q - q_{goal} \right\|$.

This attractor potential is differentiable and leads to the attractor force of Equation (7). This equation can be re-expressed as in Equations (8) and (9).

$$F_{attraction} = -\nabla U_{attractive}(q) \qquad (7)$$

$$F_{attraction} = -k_{attraction} * \rho_{goal}(q) * \nabla \rho_{goal}(q) \qquad (8)$$

$$F_{attraction} = -k_{attraction}(q - q_{goal}) \qquad (9)$$

Behavior in the Obstacles: Obstacles should be considered to avoid the complications in the path-planning task. The simplest way to do this is to define repulsive forces (Figure 7). Equation (10) describes one way to formulate the reject potential.

$$U_{reject} = \left\{ \frac{1}{2} k_{rejection} \left(\frac{1}{\rho(q)} - \frac{1}{\rho_0} \right)^2 \quad si \quad \rho(q) \le \rho_0 \right. \qquad (10)$$

$$U_{reject} = \left\{ 0 \quad si \quad \rho(q) \ge \rho_0 \right. \qquad (11)$$

where $K_{rejection}$ is a scaling factor, $\rho(q)$ is the minimum distance between the point q and the object and ρ_0 is the distance of influence of the object. The reject potential function U_{reject} has positive or zero values and tends to infinite when the object contains the path.

Figure 7. Reject potential fields using 4 obstacles

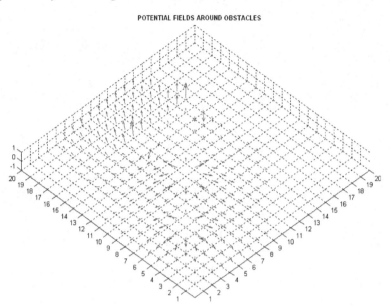

POTENTIAL FIELDS AROUND OBSTACLES

The repulsive force can be defined as in the Equation (12).

$$F_{rejection}(q) = -\nabla U(q) \qquad (12)$$

$$if \quad \rho(q) \leq \rho_o \qquad (13)$$

$$F_{rejection} = \left\{ k_{rejection}(\frac{1}{\rho(q)} - \frac{1}{\rho_0})\frac{1}{\rho^2(q)} * \frac{q - q_{obstacle}}{\rho(q)} \right\} \qquad (14)$$

and its value is "0" if $\rho(q) \geq \rho_o$.

Finally, the trajectory can be obtained by adding the repulsive and attractive forces. Figure 8 shows the result with four obstacles.

Results (Figure 8) shows that the trajectory obtained can move the mobile robot from an initial position to the goal point avoiding the four obstacles. In addition, it shows that the trajectory obtained is not optimal since it is possible to reach the target in fewer steps using a different trajectory. Because the objective of this work is to present the basics of path planning using this strategy, optimization strategies are not included.

Second Approach

In this approach, the potential fields are emulated using the same elements previously defined, but without the use of gradient functions. In general terms, the algorithm contains 5 steps (Tibaduiza, 2005):

- Initialize variables.
- Read and locate the most important points on the matrix.
- Define the potential around the goal.
- Define potential to each obstacle.
- Move the robot from the starting point to the goal point between the cells with values equal or less than the currently position.

To implement this method it is necessary to bear in mind the following considerations:

- There is a discretization of the workspace.

Figure 8. Final trajectory with 4 obstacles

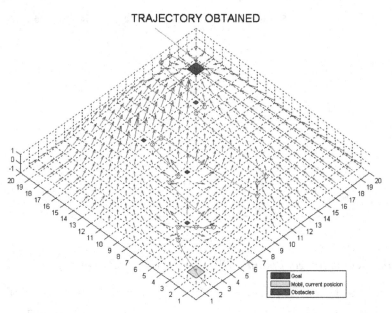

- Each point on the workspace is affected by a numeric value indicating the relationship of this to the point of arrival.
- According to the numerical distribution assigned, the path follows a downward potential.
- The algorithm can be online if the values from the sensors are updated regularly and the algorithm is calculated more than one time, allowing the robot to move in the workspace without collisions. The robot can also run off-line if it has the position of the static obstacles only at the beginning of the program and the algorithm is executed only one time.
- The algorithm is adjustable to the work-space. In general, this can be defined depending of the distribution of the sensors.

Initialization of Variables: This is the first stage of the algorithm and is used to obtain data such as the starting point, obstacle positions, and goal position. All these elements depend of the kind of sensors used in the implementation. In this work, a camera is used to obtain workspace

images, which are later processed and used to calculate the position of each element. This step can be performed by two ways, if the algorithm is on-line, all the data should be actualized in real time to determine the dynamic distribution of the workspace (Tibaduiza, 2006). However, if the algorithm is offline the information of the workspace should be updated one time at the beginning of the algorithm.

Once the workspace size is known, a matrix with zeros using this information should be defined. In this matrix, each element corresponds to a given area in the workspace, for instance if the real workspace has an area of 200 cm^2, this can be represented by a matrix of 20 rows and 20 columns, where each element corresponds to an area of 10 cm^2 (Figure 9). The area of each cell can be adjusted regardless of the dimensions of the workspace.

The next step to make all roads converge to the goal point is to generate the potential field for this objective. This is performed by assigning a numerical value of "0," from this point all the cells around it are increased until the borders of the workspace, as can be seen in Figure 10. The

Figure 9. Representation of a workspace of 20 x 20 cells

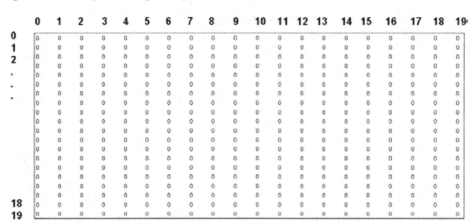

Figure 10. Potential field in the goal point x 20

3	3	3	3	3	3	3
3	2	2	2	2	2	3
3	2	1	1	1	2	3
3	2	1	0	1	2	3
3	2	1	1	1	2	3
3	2	2	2	2	2	3
3	3	3	3	3	3	3

Figure 11. Potential field in the obstacle

49	49	49	49	49
49	50	50	50	49
49	50	200	50	49
49	50	50	50	49
49	49	49	49	49

only point that remains is the starting point which is represented by another value greater than the potential defined for the goal point in the workspace.

In this configuration, the attraction force is warranted, because it ensures the robot can move to any place in the workspace by means of a descending sequence of numbers to the goal point.

Numerical Potential in the Obstacles: To include obstacles, it is necessary to bear in mind the following:

- The obstacle by itself has an effective area, depending of its dimensions, through which you cannot make a path because of a restriction on the real world. This obstacle is represented in the algorithm as a solid area with the number "200," although the value can be selected as desired. The important point is that this value should be greater than any other already defined in the workspace.

- When the obstacle has an irregular shape, a risk area should be considered to prevent a path defined in its surroundings, to define this area equally big numbers can be used as in Figure 11, where 200 define the solid obstacle and the other cells define the danger zones by means a downward numerical potential.

One kind of risk zone that very often occurs is when two obstacles are very close to each other. There is the possibility of the algorithm generating a path between the obstacles the robot cannot follow because of its physical dimensions. Especially in these cases, the danger zones are

Figure 12. General view including the potential field in the borders

50	49	50	50	50	50	50	50	50	50	50	50	50	50	50	50	50	50	49	50
50	49	49	49	49	49	49	49	49	49	49	49	49	49	49	49	49	49	49	49
50	49	13	13	13	13	13	13	13	13	13	13	13	13	13	13	13	13	49	50
50	49	13	100	12	12	12	12	12	12	12	12	12	12	12	12	12	12	49	50
50	49	13	12	11	11	11	11	11	11	11	11	11	11	11	11	11	11	49	50
50	49	13	12	11	47	47	47	47	47	47	47	47	47	10	10	10	10	49	50
50	49	13	12	11	47	48	48	48	48	48	48	48	47	9	9	9	9	49	50
50	49	13	12	11	47	48	49	49	49	49	49	48	47	8	8	8	8	49	50
50	49	13	12	11	47	48	49	50	50	50	49	48	47	7	7	7	7	49	50
50	49	13	12	11	47	48	49	50	200	50	49	48	47	6	6	6	6	49	50
50	49	13	12	11	47	48	49	50	50	50	49	48	47	5	5	5	5	49	50
50	49	13	12	11	47	48	49	49	49	49	49	48	47	4	4	4	4	49	50
50	49	13	12	11	47	48	48	48	48	48	48	48	47	3	3	3	3	49	50
50	49	13	12	11	47	47	47	47	47	47	47	47	47	2	2	2	2	49	50
50	49	13	12	11	10	9	8	7	6	5	4	3	2	1	1	1	2	49	50
50	49	13	12	11	10	9	8	7	6	5	4	3	2	1	0	1	2	49	50
50	49	13	12	11	10	9	8	7	6	5	4	3	2	1	1	1	2	49	50
50	49	13	12	11	10	9	8	7	6	5	4	3	2	2	2	2	2	49	50
50	49	49	49	49	49	49	49	49	49	49	49	49	49	49	49	49	49	49	49
50	49	50	50	50	50	50	50	50	50	50	50	50	50	50	50	50	50	49	50

Figure 13. Trajectory obtained

50	49	50	50	50	50	50	50	50	50	50	50	50	50	50	50	50	50	49	50
50	49	49	49	49	49	49	49	49	49	49	49	49	49	49	49	49	49	49	49
50	49	12	12	12	12	12	12	12	12	12	12	12	12	12	12	12	12	49	50
50	49	12	100	11	11	11	11	11	11	11	11	11	11	11	11	11	11	49	50
50	49	12	100	100	47	47	47	47	47	47	47	47	10	10	10	10	10	49	50
50	49	12	100	47	48	48	48	48	48	48	48	47	9	9	9	9	9	49	50
50	49	12	100	47	48	49	49	49	49	49	48	47	8	8	8	8	8	49	50
50	49	12	100	47	48	49	50	50	50	49	48	47	7	7	7	7	7	49	50
50	49	12	100	47	48	49	50	200	50	49	48	47	6	6	6	6	6	49	50
50	49	12	100	47	48	49	50	50	50	49	48	47	5	5	5	5	5	49	50
50	49	12	100	47	48	49	49	49	49	49	48	47	4	4	4	4	4	49	50
50	49	12	100	47	48	48	48	48	48	48	48	47	3	3	3	3	3	49	50
50	49	12	100	47	47	47	47	47	47	47	47	47	2	2	2	2	3	49	50
50	49	12	11	100	100	100	100	100	100	100	100	100	100	1	1	2	3	49	50
50	49	12	11	10	9	8	7	6	5	4	3	2	1	100	1	2	3	49	50
50	49	12	11	10	9	8	7	6	5	4	3	2	1	1	1	2	3	49	50
50	49	12	11	10	9	8	7	6	5	4	3	2	2	2	2	2	3	49	50
50	49	12	11	10	9	8	7	6	5	4	3	3	3	3	3	3	3	49	50
50	49	49	49	49	49	49	49	49	49	49	49	49	49	49	49	49	49	49	49
50	49	50	50	50	50	50	50	50	50	50	50	50	50	50	50	50	50	49	50

very important and can help to avoid a significant number of collisions.

The number of cells around the obstacle with a high numerical potential depends on the risk area to consider.

Potential Field on the Workspace Borders: This is one way to ensure that the robot never leaves the workspace and corresponds to a security method more than a necessity.

To consider the potential, the edges of the workspace are represented as a danger zone with a big potential as in Figure 12.

The defined structure for the movement is seen in Figure 12.

Once all the previous elements are considered, the movement of the robot is calculated; here, the goal point can be found moving the robot by the cells with same or minor value as is show in the Figure 13.

Figure 14. General view of the workspace

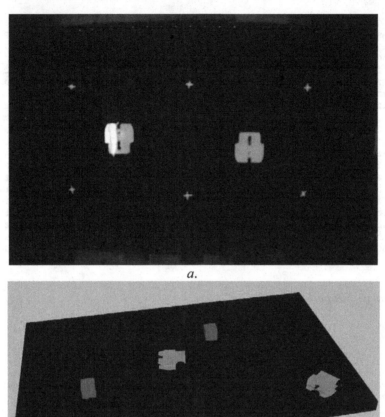

a.

b.

EXPERIMENTAL SETUP

The experimental setup includes two mobile robots with differential guided locomotion, a wooden platform, a camera, and two fixed obstacles (Figures 14(a) and 14(b)).

One of the robots is used to make the planned trajectory and the other robot is used as a mobile obstacle in order to change the workspace. The wooden platform is an opaque black color to avoid glare or shadows that can give false data to the vision algorithm.

Artificial Vision System

The hardware of the artificial vision system consists of an analog camera, a digitizer card, and a computer to process all the images captured from workspace. From the images, the information about the actual position of the mobile robot and the obstacles are obtained. The trajectory defined by the path planning method must be recalculated depending, for instance, on whether the mobile obstacle stops its execution. In the workspace, the stationary objects were identified with a blue color and the mobile obstacles with yellow.

Figure 15. Camera position

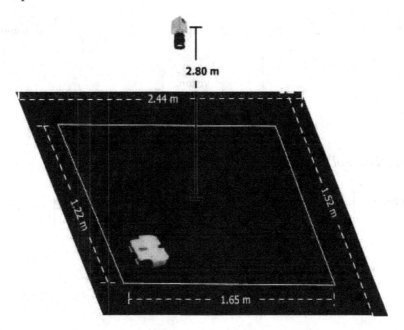

To obtain and digitize the images, an analog camera DFK50H13 / N with a 4 mm lens (Figure 15) and a 32-bit PCI DFG video compression card is used. In pre-processing, a median filter was also used (Amaya, 2005) to eliminate the noise present in the image without noticeably altering its visual quality.

Two types of approaches were implemented for detecting and segmenting movement in this workspace, (Amaya, 2005; Tibaduiza, 2006). The first approach is based on color, which can distinguish the color of objects of interest present in the scene segmentation. This methodology is performed by means of two methods (xyY and HLS). The xyY color model defines three imaginary colors: x, y, and Y to solve the problem inherent with the RGB model (Red-Green-Blue) where many colors perceived by the eye cannot be represented. The color information is only found in two of its components (x & y) since the Y component represents the clarity of the colors. The use of HLS method provides an advantage over the RGB model in which color information is contained in its three components. HLS color model uses three basic components of color: hue, saturation, and brightness (luminance).

The second approach for detecting and segmenting images includes gradient-based methods. This method provides a solution to the problem of motion estimation from the observation of changes in the brightness of an image sequence. Using this approach makes it possible to detect the edges of a moving element in any dynamic scene.

In motion estimation that uses gradient-based methods, one of the main problems is the presence of mobile elements that may be part of the scene and interfere with correct detection of the edges of the moving elements. Therefore, we also implemented an algorithm that is based on the Jain algorithm (Jain, 1995) which solves this problem greatly. This algorithm uses two gradients, the spatial and temporal, which are combined by a logical AND operation.

More details about of the vision algorithm implementation and results can be found in Amaya (2005) and Tibaduiza (2006, 2008).

Figure 16. *High level control strategy*

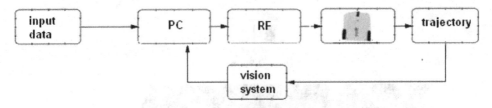

Figure 17. *Low level control strategy*

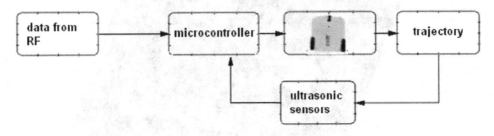

Control Algorithm

The control strategies include two different approaches divided as high-level control and low-level control (Tibaduiza, 2006). In the first, the free-obstacle path is calculated based on information from the artificial vision system and this is sent to the robot using RF communication (Figure 16).

In the low-level control (Torres, 2006; Tibaduiza, 2006), all the strategies associated with the specific movement of the robot are included. All the specifications about how the robot moves are defined by the microcontroller, where the information from the RF module concerning to the trajectory to execute required the assigned path and the signal from the ultrasonic sensors are processed (Figure 17).

Communication

The path is defined in the control host using 9 bits, which include the step-by-step definition for each chromosome. All the information is sent through

an RF module from the principal computer to the mobile in order to execute the trajectory.

The RSW433 and TSW433 radiofrequency modules are used; these modules work at a frequency of 433 MHz and use AM modulation for data transmission in serial protocol at a speed of 1200 baud.

RESULTS

Figure 18 show an example of the results in the artificial vision system using two static and one mobile obstacle. This figure also shows how the elements in the workspace are correctly detected and are an example of the inputs in the path planning algorithms.

More details and results of the artificial vision system can be found in Tibaduiza (2007).

Several tests were performed in order to test the genetic algorithm. Two of these tests are show in Figures 19(a) and 19(b). These examples use different scenarios and number of generations.

Because of the randomness in defining the first parents, the results can significantly differ.

Figure 18. Artificial vision system results

 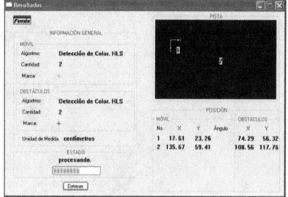

Figure 19. Genetic algorithm results: (a) 20 generations, (b) 110 generation

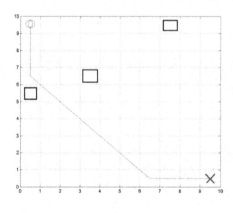 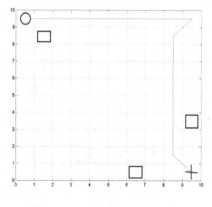

a.   b.

For example, Figure 19(a) provides a better path (least distance) when it is compared with Figure 19(b), despite having a smaller number of generations. To improve this situation, techniques like elitism (Goldberg, 1989) can be included in the crossover step; this element allows the new generation to conserve the most important genes of the ideal solution.

Equally, different tests were performed to evaluate potential field approaches and some examples are included. Figure 20(a) and 20(b) show the results obtained when the first approach of potential field is used.

Figure 20 shows the free-path obtained. Since these paths are not optimized, there are sections of paths that could be improved by using optimization methods.

Figures 21(a), 21(b), and Figure 22 shows the results of the second approach using potential fields.

Figure 20(a) shows the positions of the obstacles in the workspace, the goal position, and the current position of the mobile robot and in Figure 20(b) the trajectory obtained is presented. As can be seen, this is the shortest path for going between the start and goal points without unnecessary movements.

Figure 20. Potential field algorithm results with two different workspace configuration

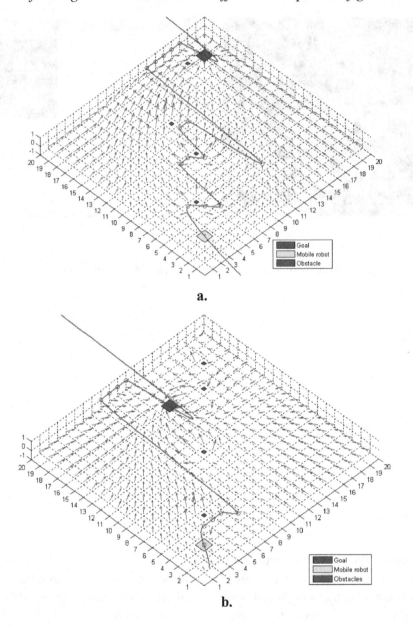

a.

b.

Figure 22 shows another example when there are 6 obstacles in the workspace using the second approach. Again, the algorithm again found an obstacle-free trajectory.

Comparing both potential field approaches makes it possible to see that the trajectory obtained by using the second approach generates more direct paths. This is a very important point because the robot can avoid performing extra movements to reach the goal point by means of this algorithm.

CONCLUSION

The approaches in path planning presented in the current chapter allowed us to calculate free-

Figure 21. Potential field algorithm results

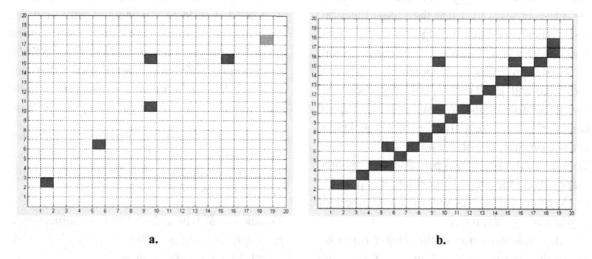

a. **b.**

Figure 22. Potential field algorithm with 6 obstacles

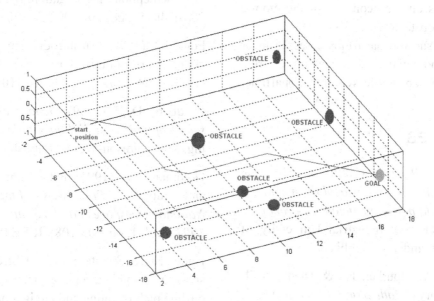

collision paths. These approaches were tested in different scenarios and some examples were included to provide some results. These methodologies allowed robot movement in both a static and dynamic workspace. In addition, we explained the differences between the implementation in static and dynamic environments.

The methodologies allowed implementing a genetic algorithm and two potential fields, which are adaptable to the dimensions of any workspace.

It was shown that genetic algorithms provide a satisfactory solution to the path-planning problem. The trajectory obtained depends on the combination of the variables, like the fitness function and the number of generations. This chapter only presented a binary codification, but there are different ways to define the fitness function. Real tests showed that the algorithm gives good results with a good balance in time consumption.

Two approaches using potential field were presented with good results. Both approaches allowed the mobile robot to move in the workspace considering the potential forces of the goal point and the obstacles to obtain a free-collision path. Comparing the algorithms permits us to conclude that by using the second algorithm, it is possible to calculate trajectories with a few numbers of movements for the robot.

The implementation of the danger zones allowed the algorithms to obtain trajectories with good results avoiding the collisions with the obstacles in the real tests. The definition of these zones depends on the geometry of the mobile robot and the obstacles.

The results showed that in most cases the linear approach of potential field creates more direct trajectories that consume less computational time.

REFERENCES

Amaya, Y., & Ruiz, J. (2005). *Localización dinámica de móviles y obstáculos en una escena controlada para aplicaciones en robótica.* (Bachelor`s Thesis). Industrial University of Santander. Santander, Colombia.

Bodhale, D., Afzulpurkar, N., & Thanh, N. T. (2008, February). *Path planning for a mobile robot in a dynamic environment.* Paper presented at IEEE International Conference on Robotics and Biomimetics, 2008. Bangkok, Thailand.

Byung, K., & Kang, S. (1984). An efficient minimum-time robot path planning under realistic conditions. In *Proceedings of the American Control Conference, 1984*, (pp. 296-303). IEEE.

Choi, C., Hong, S.-G., Shin, J.-H., Jeong, I.-K., & Lee, J.-J. (1995). Dynamical path-planning algorithm of a mobile robot using chaotic neuron model. In *Proceedings of the 1995 IEEE/RSJ International Conference on Intelligent Robots and Systems: Human Robot Interaction and Cooperative Robots,* (Vol. 2, pp. 456 – 461). IEEE Press.

Crowley, J. (1985). Navigation for an intelligent mobile robot. *IEEE Journal on Robotics and Automation, 1*(1), 31–41. doi:10.1109/JRA.1985.1087002

Dudek, G., & Jenkin, M. (2000). *Computational principles of mobile robotics.* Cambridge, UK: Cambridge University Press.

Elfes, A. (1989). Using occupancy grids for mobile robot perception and navigation. *IEEE Computer, 22*(6), 46–57. doi:10.1109/2.30720

Fogel, D. (1994). An introduction to simulated evolutionary optimization. *IEEE Transactions on Neural Networks, 5*(1). doi:10.1109/72.265956

Goldberg, D. E. (1989). *Genetic algorithms in search optimization, and machine learning.* Reading, MA: Addison-Wesley Publishing Company.

Goldman, J. A. (1994). Path planning problems and solutions. In *Proceedings of the IEEE 1994 National Aerospace and Electronics Conference, 1994,* (Vol. 1, pp. 105-108). IEEE Press.

Habib, M. K., & Yuta, S. (1989). Structuring free space as prime rectangular areas (PRAs) with on-line path planning and navigation for mobile robots. In *Proceedings of the IEEE International Conference on Systems, Man and Cybernetics,* (Vol. 2, pp. 557 – 565). IEEE Press.

Hartl, D., & Jones, E. (1998). *Genetic principles and analysis* (4th ed.). New York, NY: Jones and Bartlett Publishers.

Hassanzadeh, I., Madani, K., & Badamchizadeh, M. A. (2010). *Mobile robot path planning based on shuffled frog leaping optimization algorithm.* Paper presented at the IEEE Conference on Automation Science and Engineering (CASE). Toronto, Canada.

Hofner, C., & Schmidt, G. (1994). Path planning and guidance techniques for an autonomous mobile cleaning robot. In *Proceedings of the IEEE/RSJ/GI International Conference on Intelligent Robots and Systems 1994: Advanced Robotic Systems and the Real World, IROS 1994,* (Vol. 1, pp. 610 – 617). IEEE Press.

Holland, J. (1992). Genetic algorithms. *Scientific American, 267*(1), 66–72. doi:10.1038/scientificamerican0792-66

Liang, K., & Jian-Jun, W. (2010). *The solution for mobile robot path planning based on partial differential equations method.* Paper presented at the International Conference on Electrical and Control Engineering (ICECE) 2010. Wuhan, China.

Lozano-Perez, T. (1983). Spatial planning: A configuration space approach. *IEEE Transactions on Computers, C-32,* 108–120. doi:10.1109/TC.1983.1676196

Lumelsky, V., & Stepanov, A. (1984). Effect of uncertainty on continuous path planning for an autonomous vehicle. In *Proceedings of the 23rd IEEE Conference on Decision and Control,* (Vol. 23, pp. 1616-1621). IEEE Press.

Mitchell, M. (1998). *An introduction to genetic algorithms.* Cambridge, MA: MIT Press.

Mitsuo, G., & Runwei, C. (2000). *Genetic algorithms & engineering optimization.* New York, NY: Wiley.

Navas, O., & Ortiz, N. (2006). *Algoritmos genéticos aplicados al planeamiento de trayectorias de un robot móvil.* (Bachelor`s Thesis). Industrial University of Santander. Santander, Colombia.

Navas, O., Ortiz, N., Tibaduiza, D. A., & Martinez, R. (2005). *Algoritmos genéticos aplicados al planeamiento de trayectorias de un robot móvil.* Paper presented at I Simposio Regional de Electrónica y Aplicaciones Industriales. Ibagué, Colombia.

Pinna, J., Masson, F., & Guivant, J. (2008). *Planeamiento de trayectoria en mapas extendidos mediante programación dinámica de dos niveles.* Paper presented at V Jornadas Argentinas de Robotica-JAR 2008. Bahía Blanca, Argentina.

Qixin, C., Yanwen, H., & Jingliang, Z. (2006). *An evolutionary artificial potential field algorithm for dynamic path planning of mobile robot.* Paper presented at the International Conference on Intelligent Robots and Systems. Beijing, China.

Ramesh, J., Kastur, R., & Shunck, B. (1995). *Machine vision.* New York, NY: McGraw Hill Book Co. Guo, J., Liu, L., Liu, Q., & Qu, Y. (2009). An improvement of D* algorithm for mobile robot path planning in partial unknown environment. In *Proceedings of the Second International Conference on Intelligent Computation Technology and Automation, 2009,* (Vol. 3, pp. 394-397). ICICTA.

Rimon, E., & Koditschek, D. (1992). Exact robot navigation using artificial potential functions. *IEEE Transactions on Robotics and Automation, 8*(5), 501–518. doi:10.1109/70.163777

Siegwart, R., & Nourbakhsh, I. (2004). *Introduction to autonomous mobile robots.* Cambridge, MA: MIT Press.

Sugihara, K., & Smith, J. (1999). Genetic algorithms for adaptive planning of path and trajectory of a mobile robot in 2D terrains. *IEICE Transactions on Information and Systems, 82*(1), 309–317.

Suguihara, K. (1998). GA-based on-line path planning for SAUVIM. *Lecture Notes in Computer Science, 1416,* 329–338. doi:10.1007/3-540-64574-8_419

Thorpe, C., & Matthies, L. (1984). Path relaxation: Path planning for a mobile robot. [OCEANS.]. *Proceedings of OCEANS, 1984*, 576–581.

Tibaduiza, D. A. (2006). *Planeamiento de trayectorias de un robot móvil*. (Master Thesis). Industrial University of Santander. Santander, Colombia.

Tibaduiza, D. A., Amaya, Y., Ruiz, J., & Martínez, R. (2007). Localización dinámica de móviles y obstáculos para aplicaciones en robótica. *Colombian Journal of Computation, 8*, 93–120.

Tibaduiza, D. A., & Anaya, M. (2007). Campos de potencial aplicados al planeamiento de trayectorias de robots móviles. *Revista de Ingeniería, 1*, 23–34.

Tibaduiza, D. A., & Anaya, V. (2008). *Algoritmos de planificación de trayectorias para un robot móvil*. Paper presented at the 6th International Conference on Electrical and Electronics Engineering Research. Aguascalientes, Mexico.

Tibaduiza, D. A., & Chio, N. (2008). Metodologías de campos de potencial para el planeamiento de trayectorias de robots móviles. *Colombian Journal of Computation, 9*, 104–119.

Tibaduiza, D. A., Martínez, R., & Barrero, J. (2005). *Planificación de trayectorias para robots móviles*. Paper presented at the Simposio Regional de Electrónica y Aplicaciones Industriales. Ibagué, Colombia.

Tibaduiza, D. A., Martinez, R., & Barrero, J. (2006). *Algoritmos de planificación de trayectorias para un robot móvil*. Paper presented at the 2nd IEEE Colombian Workshop on Robotics and Automation. Bogotá, Colombia.

Torres, C. H., & Mendoza, E. Y. (2006). *Control de dos móviles en un entorno dinámico*. (Bachelor's Thesis). Industrial University of Santander. Santander, Colombia.

Zhang, H. D., Dong, B.-H., Cen, Y.-W., Zheng, R., Hashimoto, S., & Saegusa, R. (2007). *Path planning algorithm for mobile robot based on path grids encoding novel mechanism*. Paper presented at the Third International Conference on Natural Computation, 2007. Haikou, China.

Zhang, Y., Zhang, L., & Zhang, X. (2008). Mobile robot path planning base on the hybrid genetic algorithm in unknown environment. In *Proceedings of Eight International Conference on Intelligent Systems Design and Applications, 2008*, (Vol. 2. pp. 661-665). IEEE.

KEY TERMS AND DEFINITIONS

Chromosome: Is a set of genes, which contains information about the specific problem to solve by means of a encoding. The most common implementation is by using of a binary encoding.

Gene: Is the minimal unity of a chromosome in the Genetic Algorithms, this contains information about specific characteristics that define to an individual.

Genetic Algorithms: Genetic Algorithms are an artificial intelligence technique inspired on the evolution theory. This technique is based on artificial natural selection mechanisms and genetics, and represents the solutions as an optimization problem.

Mobile Robots: Is a type of robot whose movements are not limited to the sum of the length of each of its elements, this means that these robots don't have a typical workspace defined for the movement. The most common mobile robots are those that include wheels; depending on how these are configured (position, number) there are different ways and degrees of freedom for the robot motion.

Path Planning: The path planning is an important task in the definition of the movement of a robot, this consists of generate a free-collision path in an environment with obstacles and optimize it, according to specified criteria.

Potential Fields: Is a methodology for path planning which considers the workspace as a force field by assigning different values to each object on the workspace.

Road Planning: The objective regarding road planning is to set the motion of a mobile robot along a planned path.

Chapter 8
An Alternative for Trajectory Tracking in Mobile Robots Applying Differential Flatness

Elkin Yesid Veslin Díaz
Universidad de Boyacá, Colombia

Jules G. Slama
Universidade Federal do Rio de Janeiro, Brazil

Max Suell Dutra
Universidade Federal do Rio de Janeiro, Brazil

Omar Lengerke Pérez
Universidad Autónoma de Bucaramanga, Colombia

Hernán Gonzalez Acuña
Universidad Autónoma de Bucaramanga, Colombia

ABSTRACT

One solution for trajectory tracking in a non-holonomic vehicle, like a mobile robot, is proposed in this chapter. Using the boundary values, a desired route is converted into a polynomial using a point-to-point algorithm. With the properties of Differential Flatness, the system is driven along this route, finding the necessary input values so that the system can perform the desired movement.

DOI: 10.4018/978-1-4666-2658-4.ch008

Copyright © 2013, IGI Global. Copying or distributing in print or electronic forms without written permission of IGI Global is prohibited.

INTRODUCTION

Throughout time, mobile robots have acquired great importance because of a wide variety of applications arising from the potential for autonomy that they present; some examples include autonomous robots and transportation systems (AGV—Automated Guided Vehicles). The function of an autonomous robot is to carry out different tasks without any human intervention in unknown environments, in which transportation systems move objects from place to place without needing a driver. In these operations, the main task is to control the displacement of a robot through a given route. However, the main problem that exists in the control systems is precisely the performance of the system from one space to another, and the mobile robot is not an exception.

Nowadays, researchers in this field have developed several inquiries such as the application of chaotic routes for the exploration of uncertain spaces and the control of AGV systems in order to be applied in industry; the former highlights the application of flexible systems of manufacturing theories or FSF for the generation of routes in vehicles with trailers (Tavera, 2009; Lengerke, 2008).

Due to the existence of friction during displacement, this kind of system presents non-holonomic restrictions in its kinematic structural and, therefore, the mobility is reduced (Siciliano, 2009). It has also been shown that these systems are differentially flat; meaning that the system has a set of outputs called flat outputs, which according to the properties of flat systems, the outputs and their derivatives allow one to describe the whole system (Fliess, 1994). This paper highlights this property in the case of trajectory tracking, with a point-to-point steering algorithm (van Nieuwstadt, 1997; De Doná, 2009). The desired route is reformulated through a function in time and space. The parameterization, which is combined with the differential flatness systems concepts, determines a set of inputs that allows for the control of robot movement by means of such routes.

The methodology designed is conceived from a brief description of differential flatness systems concepts; afterwards the concept of parameterization of routes is introduced, with the tracking method from one point to another in order to be implemented in the system. This analysis provides different graphical simulations that show the outcomes. Finally, a discussion is opened about the advantages of the implementation and future possibilities for studies.

BACKGROUND

Flatness Systems

The differential flatness systems' concept was introduced by Michel Fliess, and his teamwork through the concepts of differential algebra (Fliess, 1994). They conceive a system as a differential field, which is generated by a set of variables (states and inputs). Later, Martin (1997) redefined this concept in a more geometric context, in which flatness systems could be described in terms of absolute equivalence.

According to Fliess (1994), a system is defined as differentially flat if there is a set of outputs (with the same number of inputs) called flat outputs. These outputs can determine the state of the system and its inputs without the need of integrations; this means that a system is flat when a set of outputs $Y \in$ can be found in the way:

$$y = h\left(x, u, \dot{u}, \ldots, u^{(r)}\right) \qquad (1)$$

Such that the states (x) and the inputs (u) could be defined by a set of outputs and its derivatives as:

$$x = f\left(y, \dot{y}, \ldots, u^{(q)}\right) \qquad (2)$$

$$u = g\left(y, \dot{y}, \ldots, u^{(q)}\right) \qquad (3)$$

As can be observed, in a flatness system the states of the outputs are in function of the flat outputs and a finite number of their derivatives. These kinds of associations are useful in problems that require an explicit generation of routes (Murray, 1995); it means that the system, in order to obtain a desired behavior, achieves it through the design of a route in the outputs. Later, based on Siciliano (2009), adequate inputs are determined in order to make the route. Different works present the use of differential flatness systems that include applications on mobile robots (Murray, 1995), mobile robot with trailers (Deligiannis, 2006; Lamiraux, 2000; Rouchon, 1993; Ryu, 2008), simplified cranes (Fliess, 1991), and robotic manipulators (Murray, 1995; Veslin, 2011).

Mobile Robot

The formulation is developed for a mobile robot with four wheels as shown in Figure 1. In this model, the front wheels perform as signal wheels and the back wheels steer.

Where P is the position of the back wheels in space, $(x(t), y(t))$ is defined by the inclination of the vehicle $\theta(t)$ respecting coordinated axes. The robot moves with a speed $v(t)$ and the orientation of the front wheels is defined by the variable $\varphi(t)$. As is supposed, the system does not present slipping during its movement, therefore the mobile robot kinematic model is described by Equation (4):

$$\dot{x}(t) = v(t)\cos\theta(t)$$
$$\dot{y}(t) = v(t)\operatorname{sen}\theta(t) \qquad (4)$$
$$\dot{\theta}(t) = \frac{v(t)}{l}\tan\varphi(t)$$

Equation (4) shows that a mobile robot is represent in a non-linear equation system, where l is the distance between the front and back wheels. The inputs of the system, $u_1(t)$ and $u_2(t)$ represents the vehicle system $v(t)$ and orientation of the front wheels $\varphi(t)$. It is a flat system and the flat outputs are the position of back wheels in two-dimensional space $(x(t), y(t))$. In this way,

Figure 1. Mobile robot proposed model

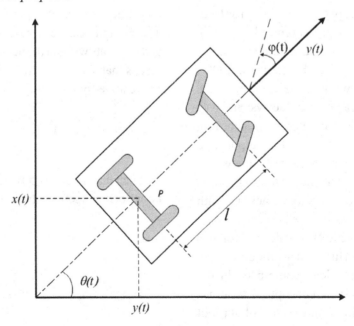

from geometrical considerations (Rotella,2008), the first input in function of flat outputs is defined (3) as follows:

$$u_1(t) = v(t) = \sqrt{\dot{x}(t)^2 + \dot{y}(t)^2} \qquad (5)$$

In which vehicle speeds in the space x and y are defined by the first derivative of the flat outputs. The variation of the angle versus time is determined by relating the components of the vehicle's speed by the Equation (6).

$$\theta(t) = \arctan\left(\frac{\dot{y}(t)}{\dot{x}(t)}\right) \qquad (6)$$

This answer achieves definition 2. In order to find the relationship between the second input u_2 (t), and the flat outputs $x(t)$ and $y(t)$, from Equation (4), the following is obtained:

$$\varphi(t) = \arctan\left(\frac{\dot{\theta}(t)l}{v(t)}\right) \qquad (7)$$

Deriving Equation (6) over time shows that:

$$\dot{\theta}(t) = \frac{\ddot{y}(t)\dot{x}(t) - \dot{y}(t)\ddot{x}(t)}{\dot{x}(t)^2 + \dot{y}(t)^2} \qquad (8)$$

Finally, replacing Equation (8) and Equation (5) in Equation (7) we obtain:

$$u_2(t) = \varphi(t) = \arctan\left(l\frac{\ddot{y}(t)\dot{x}(t) - \dot{y}(t)\ddot{x}(t)}{\left(\dot{x}(t)^2 + \dot{y}(t)^2\right)^{3/2}}\right) \qquad (9)$$

The performed operations demonstrate that this is a flat system according to (2) and (3). System transformation to the flat domain is completely geometric with respect to the system's properties and non-holonomic restrictions. The geometric property is one of the main advantages of flat systems, whereby it is not a linear approximation but a characteristic itself (Martin, 1997). A system is flat when it indicates that the non-linear structure is well characterized and can be used to design control algorithms to generate routes, to plan movements, and to introduce stabilization.

Routes Parameterization

Considering a flat system, it is possible to write desired routes $x_d(t)$ in terms of flat outputs and their derivatives. Supposing that the vehicle moves from $x_d(t_0) = x_0$ to $x_d(t_f) = x_f$. These values are known as the derivatives (the desired route is a $f(t)$ function which is derivative in time). Flat outputs (z) are parameterized as follows:

$$z_i(t) = \sum_j A_{ij}\theta_j(t) \qquad (10)$$

In which θ_j (t) is a set of basic functions are generally polynomial. This concept defines the search problem in a finite set of parameters defined by A_{ij}. Assuming that the initial position x_0 at time t_0 and the final position x_f at time t_f (they are separated by a time τ) are known, the functions that define the flat output can be calculated as:

$$z_i(t_0) = \sum_j a_{ij}\theta_j(t_0) \qquad z_i(t_f) = \sum_j a_{ij}\theta_j(t_f)$$
$$\vdots \qquad\qquad \vdots$$
$$z_i^q(t_0) = \sum_j a_{ij}\theta_j(t_0) \qquad z_i^q(t_f) = \sum_j a_{ij}\theta_j(t_f)$$
$$(11)$$

The procedure described in Equation (11) is repeated depending on the number of flat outputs. This model is a set of n equations whose magnitude depends on the number of derivatives q in the necessary flat outputs to determine the system. The higher the number of base functions of this parameterization, the closer it will follow the route. Bearing in mind that A represents unknown factors, the solution of Equation (11) is given by:

$$A = \Omega(t)^{-1} Z(t) \qquad (12)$$

In which Ω is the matrix of all basis function in Equation (11) and Z represents the outline conditions for the routes and their derivatives. The solution is a polynomial Equation (13) that represents the plane output that is parameterized according to the established boundary conditions. This procedure is repeated as shown in Figure 2 until the whole route is completed. Hence, this type of monitoring is known as point-to-point. One of the most important considerations is precisely the number of points that will exist in this route; if the number of points is small, the parameterization is more accurate and the system will perform better.

$$z_i(t) = a_{i1}x^n + a_{i2}x^{n-1} + \ldots + a_{in-1}x + a_{in} \qquad (13)$$

The next step of the route-tracking algorithm is the application of these polynomials. They will represent the flat outputs $(Z = x_d(t))$ that characterize the system and its derivatives.

The number of points and the polynomial degree that parameterizes the function are an optimization point that may vary according to the system objectives. A convergence point is a τ time extension among each high point (the order of 0.1 seconds), with a polynomial high order parameterization (for example, six or seven grade). Time extensions are obtained by determining the whole time and dividing it by the number of points, according to the results and the type of polynomial to parameterize, whose results can be modified. This calibration is important in order to carry out desired routes during long periods of time. Therefore, it is important to reduce the processing machine time.

FLATNESS TRAJECTORY TRACKING VEHICLE BEHAVIOUR

Point-to-point tracking technique is implemented in the proposed mobile robot system. Assuming that a desired route described as $x_d(t)$ and $y_d(t)$, to the coordinate axis x and y, is obtained from a movement planning system. For instance, it is desirable that the mobile robot move in a sinusoidal route, for example: $y_d(t) = 5sen(t)$, while in the space $x_d(t) = 4t-5$, which describes a linear route.

The equations' movements are parameterized according to the set of equations described in Equations (11) and (12). That gets a set of $z(t)$ flat output functions as sixth-order polynomials

Figure 2. Route parameterization; each point interprets a period of time in the last route

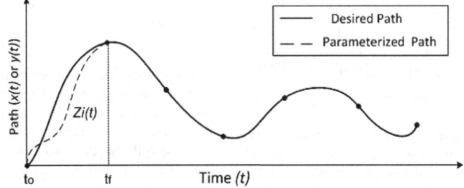

that describe the complete behavior defined by the path during a period of time τ. These equations are implemented in Equation (5) and Equation (9) in order to get the set of necessary outputs $(u_1(t), u_2(t))$ to follow the path. The results can be observed in Figures 3, 4 and 5, in which the simulation is performed by the numeric integration over time and through the implementation of input variables obtained due to the route parameterization. The test time is 30 seconds. Figure 3 represents the journey performed by the vehicle along the route.

Figures 4 and 5 describe the variation of u_1 and u_2. Values of variation are because of the trend of the route in which the speed increases in zero amplitude regions, while the variation is null. On the other hand, the speed slows down (without zero reduction) while the vehicle receives high orientation levels that can be positive or negative according to the direction of movement.

In another experiment, the desired route is a circular course along the space x and y, with a center in $(0, 0)$; the robot's initial position is outside the desired route. Figures 6, 7, and 8 illustrate the robot's movement along the route.

Because periods of time for each portion of the parameterized route are the same in Figure 7, the system has a high initial speed at the beginning of the route, which is then reduced until it stays constant over the circular route. The angle variation is similar (Figure 8) and it is observed that along the journey the control maintains wheel orientation; thus, the vehicle keeps the correct position with respect to the circumference.

Solutions and Recommendations

In both experiences, the initial conditions of the orientation in the $\theta(t)$ vehicle were given by Equation (6). The trajectory tracking through flatness systems theory gives enough conditions to obtain the guidance of the system according to the initial position of the vehicle, such as in the simulation shown in Figures 4 and 6.

Vehicle journeys as in Figure 3 and Figure 6 present a deviation with respect to the route; whereas tracking parameters are approximate to the desired route in both functions $x_d(t)$ and $y_d(t)$. There are persistent deviation conditions that tend

Figure 3. Sinusoidal route follow-up along the space

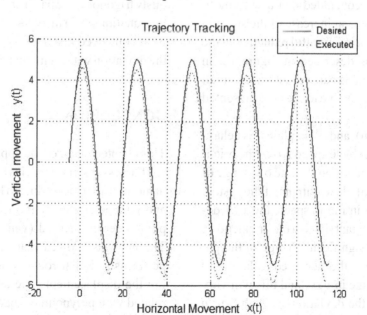

Figure 4. Vehicle input u1(t), the speed variation according to the route in Figure 3

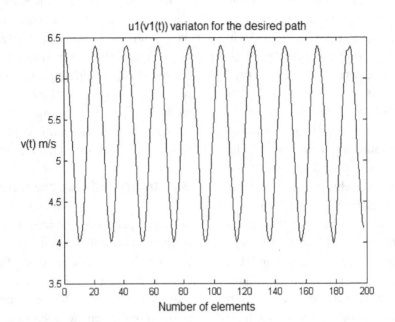

to increase the error in a linear manner, as shown in Figures 9 and 10, which increase error over time. Whether the route to follow is an exponential or quadratic function (Figure 11), the slope of the error becomes greater and it causes a significant deviation for the same period of time.

Based on the observations made in the quadratic route, it was concluded that according to the nature of the path, an increase of the slope in the equation allows a meaningful reduction in the route's deviation as described in Figure 12, in which the same route is carried out with a small degree of deviation relative to the case presented in Figure 11.

Equations (5), (6), and (9) explain this behavior. Therefore, the vehicle speed $v(t)$ as the orientation angle $\theta(t)$ depends on the rate of change of the first derivative of plane outputs; if the system accelerates in time in the y_d space, the $x_d(t)$ derivative will be a constant (if it is a linear function), it will not cause a significant change in the parameters. Therefore, the achieved values will depend on the change in $y_d(t)$, and this will vary over time causing the deviation, as is shown in

Figure 11. Nevertheless, if the $x_d(t)$ lineal equation has a slope which allows a rapid increase at the same time as its quadratic counterpart; this will contribute to obtaining the values of the outputs so that the resulting deviation will be reduced.

While these polynomials were an example, in routes generated by the other methods (the previously trigonometric path), the influences between the equations exist in the same way. In other cases, it was observed that a temporal route can reduce the deviation with an increase in wave amplitude.

CONCLUSION

This chapter analyzed the application of differential flatness systems theories to trajectory tracking in mobile robots according to Flies' studies (1994) and parameterizing flat outputs with a point to point steering method (van Nieuwstadt, 1997). The analysis showed that although the vehicle performs a similar route, a deviation exists during the displacement time and this deviation is related with polynomials' nature that defines the

Figure 5. Variation of the orientation of u2(t) front wheels according to described route in Figure 3

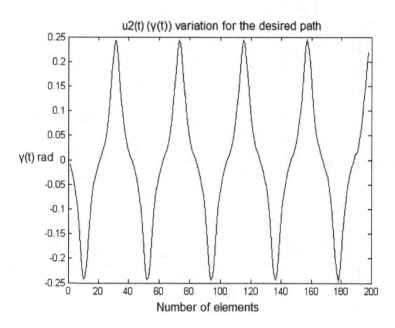

Figure 6. Circular route tracking, in which the system´s initial position is outside of the route

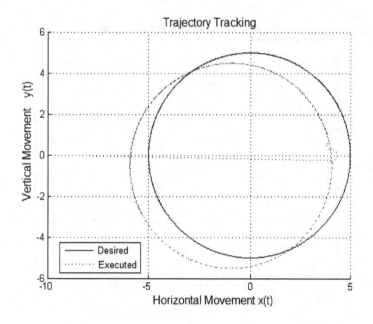

Figure 7. Input u₁(t), the speed variation of the mobile robot, according to the described route in Figure 6

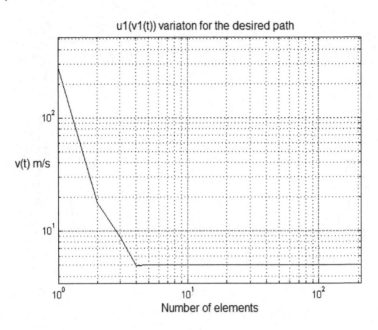

Figure 8. Variation of the orientation of the front wheels u₂(t) according to the route described in Figure 6

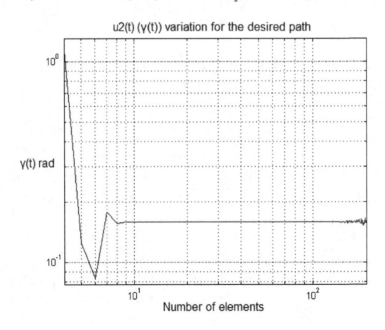

Figure 9. Route deviation respecting to the horizontal displacement in x(t) along the time for Figure 3's route

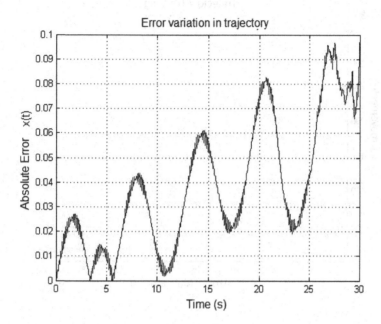

Figure 10. Route deviation respecting the vertical displacement in y(t) along the time for the Figure 3's route

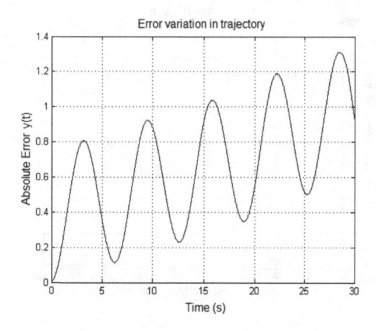

Figure 11. Quadratic route described in space

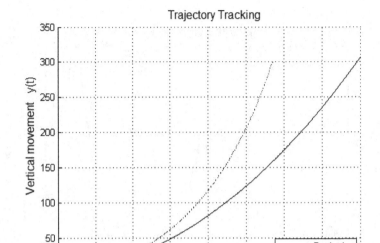

Figure 12. Quadratic route described in space with variation in lineal function slope that describes the route in x(t)

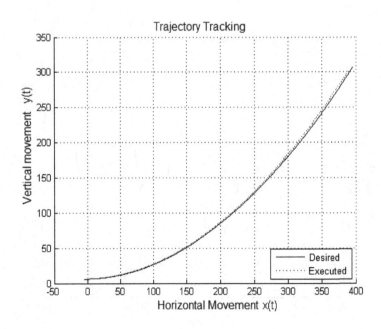

routes. Fortunately, there are methods to reduce this deviation and there is a need for implementing a control that reduces the distances of the route.

These results suggests a closed-loop control to ensure this system´s output. Through this comparison, the output values are modified by means of flat equations derived from the flat system. This control is important to obtain different behaviors that approach the desired objective in order to reduce the existing error.

FUTURE RESEARCH DIRECTIONS

In Veslin (2011), a PD controller is applied in an upper arm dynamic model that follows a trajectory parameterized through polynomials. The application of the controller allows the model to follow a path with minimal deviation in its movement. The methodology applied in that work will be implemented in the mobile robot, in order to reduce the error in the trajectory tracking that presents itself in the current system.

REFERENCES

De Doná, J. A., Suryawan, F., Seron, M. M., & Lévine, J. (2009). A flatness-based iterative method for reference trajectory generation in constrained NMPC. In Magni, L., Raimondo, D. M., & Allgöwer, F. (Eds.), *Nonlinear Model Predictive Control: Towards New Challenging Applications* (pp. 325–333). Berlin, Germany: Springer-Verlag.

Deligiannis, V., Davrazos, G., Manesis, S., & Arampatzis, T. (2006). Flatness conservation in the n-trailer system equipped with a sliding kingpin mechanism. *Journal of Intelligent & Robotic Systems*, 46(2), 151–162. doi:10.1007/s10846-006-9056-2

Fliess, M., Levine, J., Martin, P., & Rouchon, P. (1994). *Flatness and defect of non liner systems: Introductory theory and examples*. CAS.

Fliess, M., Lévine, J., & Rouchon, P. (1991). A simple approach of crane control via a generalized state-space model. In *Proceedings of IEEE Control and Decision Conference*, (pp. 736-741). IEEE Press.

Lamiraux, F., & Laumond, J. (2000). Flatness and small-time controllability of multibody mobile robots: Application to motion planning. *IEEE Transactions on Automatic Control*, 45(10), 1878–1881. doi:10.1109/TAC.2000.880989

Lengerke, O., Dutra, M., França, F., & Tavera, M. (2008). Automated guided vehicles (AGV): Searching a path in the flexible manufacturing system. *Journal of Konbin*, 8(1), 113–124. doi:10.2478/v10040-008-0106-7

Martin, P., Murray, R., & Rouchon, P. (1997). *Flat systems*. Paper presented at Mini-Course European Control Conference. Brussels, Belgium.

Murray, R. (1995). Non linear control of mechanical systems: A Lagrangian perspective. In *Proceedings of the IFAC Symposium on Nonlinear Control Systems Design (NOLCOS)*. IFAC.

Murray, R., Rathinam, M., & Sluis, W. (1995). Differential flatness of mechanical control systems: A catalog of prototype systems. In *Proceedings of the ASME International Mechanical Engineering Congress and Expo*. San Francisco, CA: ASME.

Rotella, F., & Zambettakis, I. (2008). Comande des systèmes par platitude. *Techniques de l'ingénieur, 450*.

Rouchon, P., Fliess, M., Lévine, J., & Martin, P. (1993). Flatness and motion planning: The car with n trailers. In *Proceedings European Control Conference*, (pp. 1-6). ECC.

Ryu, J. C., Franch, J., & Agrawal, S. K. (2008). Motion planning and control of a tractor with a steerable trailer using differential flatness. *Journal of Computation and Nonlinear Dynamics*, 3(3), 1–8.

Siciliano, B., Sciavicco, L., Villani, L., & Oriolo, G. (2009). *Robotics, modelling, planning and control*. London, UK: Springer-Verlag.

Tavera, M., Dutra, M., Veslin, E., & Lengerke, O. (2009). *Implementation of chaotic behavior on a fire fighting robot*. Paper presented at the 20th International Congress of Mechanical Engineering. Gramados, Brasil.

van Nieuwstadt, M. (1997). *Trajectory generation for nonlinear control systems. Technical Report CS 96-011*. Pasadena, CA: California Institute of Technology.

Veslin, E., Slama, J., Dutra, M., Lengerke, O., & Morales, M. (2011). Trajectory tracking for robot manipulators using differential flatness. *Ingeniería e Investigación, 31*(2), 84–90.

Section 2
Wireless Robotic Applications

Chapter 9
A Hierarchically Structured Collective of Coordinating Mobile Robots Supervised by a Single Human

Choon Yue Wong
Nanyang Technological University, Singapore

Gerald Seet
Nanyang Technological University, Singapore

Siang Kok Sim
Nanyang Technological University, Singapore

Wee Ching Pang
Nanyang Technological University, Singapore

ABSTRACT

Using a Single-Human Multiple-Robot System (SHMRS) to deploy rescue robots in Urban Search and Rescue (USAR) can induce high levels of cognitive workload and poor situation awareness. Yet, the provision of autonomous coordination between robots to alleviate cognitive workload and promote situation awareness must be made with careful management of limited robot computational and communication resources. Therefore, a technique for autonomous coordination using a hierarchically structured collective of robots has been devised to address these concerns. The technique calls for an Apex robot to perform most of the computation required for coordination, allowing Subordinate robots to be simpler computationally and to communicate with only the Apex robot instead of with many robots. This method has been integrated into a physical implementation of the SHMRS. As such, this chapter also presents practical components of the SHMRS including the robots used, the control station, and the graphical user interface.

DOI: 10.4018/978-1-4666-2658-4.ch009

Copyright © 2013, IGI Global. Copying or distributing in print or electronic forms without written permission of IGI Global is prohibited.

INTRODUCTION

Mobile robots have been found to be useful for a number of applications such as combat (Yamauchi, 2004; Barnes, Everett, & Rudakevych, 2005), oceanography (Bellingham & Rajan, 2007), space exploration (Bellingham & Rajan, 2007; Halberstam, et al., 2006; Schreckenghost, Fong, & Milam, 2008), as well as search and rescue.

In particular, attention has been focused towards developing technologies for using mobile robots for Urban Search and Rescue (USAR) and research interest into USAR robotics is currently at a high level. This is not surprising given the potential benefits of applying robots to the domain of USAR. For instance, robots can be used to gather information pertaining to victims, potential hazards as well as the condition of the mission environment (Tadokoro, 2009; Casper, Micire, & Murphy, Issues in intelligent robots for search and rescue, 2000). In addition, robots can even be used to deliver water and medical supplies to victims (Casper, Micire, & Murphy, Issues in intelligent robots for search and rescue, 2000). If sufficiently compact in size, a robot can also be employed to explore spaces in building rubble that are too constricted or treacherous for human or dog rescuers to enter (Burke & Murphy, 2004; Yanco, Drury, & Scholtz, 2004). In practice, mobile robots have been deployed at Ground Zero in New York City after the September 2001 attacks to search for victims. More recently, similar robots have also been applied to survey the damage caused by the March 2011 earthquake in Japan. Therefore, given the potential advantages that may be derived from deploying robots for USAR, it has been selected as the envisioned application of the system of multiple mobile robots presented in this chapter.

However, the deployment of robots for USAR is not a straightforward affair. Despite the evolution of sensing and computing technology in the past half century, mobile robots are still unable to perform entirely devoid of guidance from humans for long durations. This is because the state-of-the-art robotic technology is yet to be able to robustly cope with the challenges of highly dynamic real-world scenarios (Wong, Seet, & Sim, 2011). This fact is especially true for demanding applications such as USAR. Therefore, to ensure that the deployed robots perform as desired, humans must supervise mobile robots and from time to time intervene in the robots' activities so that qualities currently unique to humans such as judgment, experience, flexibility, and adaptability may be used to augment robot capabilities (Haight & Kecojevic, 2005).

The availability of wireless communication means that humans can provide the required supervision for the deployed robots and yet be able to establish increased separation (compared to when communicating using tethers) between themselves and the possible hazards inherent in the robots' mission environments (Shiroma, Chiu, Sato, & Matsuno, 2005). Wireless communication has also facilitated improved robot mobility since deployed robots do not have to be encumbered with a communication tether.

Yet, despite the larger separation between humans and deployed robots made possible through the use of wireless communication, standoff distances are typically not infinite because the strength and integrity of wireless signals can deteriorate with increased separation. Therefore, humans must always be within communication range of the robots and some degree of human vulnerability to mission environment dangers will still exist. To minimize overall human exposure to such hazards, the number of humans allowed in or near the mission environment must be low (Murphy, Blitch, & Casper, 2002) but this should preferably not result in only a small number of robots being allowed to work on the mission. Rather, technology should be advanced to allow multiple robots to be supervised or controlled by as few humans as possible. Given that supervision from humans is indispensible for robots deployed in complex applications such as USAR, it would be ideal then if multiple robots could be supervised by only a single human. In this manner, the advantages of fielding multiple robots[1] such as added redundancy and reliability as well as parallelism can still be harnessed without significantly increasing the number of humans required to field them. The preference for such a system is reflected in the rules of the RoboCup Rescue Physical Agent League Competition, which issues penalties to teams with more than one human individual in the simulated disaster site. Hence, this study has been geared towards the design and development of a Single-Human Multiple-Robot System (SHMRS). In a SHMRS, the robots together form a robot collective while the human acts as the sole supervisor to the deployed robot collective.

Unfortunately, the supervision of multiple robots via a single human is not easy as the role of the human supervisor is expected to demand high amounts of attention. Subsequently, this results in elevated levels of cognitive workload and consequently, degraded situation awareness (Kadous, Sheh, & Sammut, 2006; Wong, Seet, & Sim, 2011). To prevent this, deployed robots should perform certain aspects of the mission by themselves. In some works, robots provide assistance by performing navigation at some level of autonomy (Doroodgar, Ficocelli, Mobedi, & Negat, 2010), thereby reducing reliance on constant human attention required for teleoperation. However, in addition to efforts from individual robots to perform tasks autonomously, multiple robots can also assist in reducing human supervisor workload by autonomously coordinating their actions to work cohesively as a unit in reducing the supervisor's mental exertion. In an article by Scerri et al. (2010), coordination has, in fact, been identified as a suitable aspect of multiple robot deployment to automate due to the immense amount of attention resources demanded by coordinating robot efforts.

However, coordination should not only be designed to assist the human in deploying the robots but also to make efficient use of robot on-board computational power and communication bandwidth. Communication bandwidth influences the volume and rate of communication (Barrett, Rabideau, Estlin, & Chien, 2007) while the amount of computational power and computational storage on a robot can affect the maximum achievable level of robot autonomy or the sophistication of a robot's behaviours (Barrett, Rabideau, Estlin, & Chien, 2007; Parker, 2008). When robots are required to autonomously coordinate their actions, consideration for the limited amount of on-board communication and computational resources is necessary (Barrett, Rabideau, Estlin, & Chien, 2007; Anderson & Papanikolopoulos, 2008) and interactions between agents "must be well-structured to minimize resource consumption" (Fox, 1981).

If autonomous coordination of robot actions consumes exorbitant amounts of computational power and communication bandwidth, then it is probable that the deployed robots would be bogged down by such demands on their limited resources, possibly resulting even in a failure of the whole system. The likelihood of such a scenario is expected to increase as the number of autonomously coordinating robots rises. This is because each robot adds to the coordination complexity, consequently prolonging the time needed to compute the coordination (Burgard, Moors, Stachniss, & Schneider, 2005)(Burgard, Moors, Stachniss, & Schneider, Coordinated Multi-Robot Exploration, 2005)(Burgard, Moors, Stachniss, & Schneider, 2005). In addition, multiple agent systems populated with large numbers of coordinating agents may have to communicate with more entities (Fox, Organization structuring: Designing large complex software, 1979), increasing the possibility of agents incurring crippling communication overheads (Anderson & Papanikolopoulos, 2008; Sugihara & Suzuki, 1990; Sweeney, Li, Grupen, & Ramamritham, 2003).

The problem of limited robot computational resources is considered even when the latest computing solutions are available for multiple-robot computation. For instance, Vincent et al.'s (2008) study of coordinated multiple-robot exploration and map building factored the finite computational resources on board robots in the design of group-level behaviours despite the robots used being imbued with "state-of-the-art" laptops for computation. Given that robots used for USAR (such as the Microtracs robots described by Casper and Murphy[2003] (2003) (2003)) are not always large enough to accommodate laptops, the concern amongst roboticists for limited computational resources in USAR robots is not unfounded. Therefore, while autonomous coordination between robots can be tailored to mitigate supervisor cognitive workload levels and to promote situation awareness, it is important that methods for autonomous coordination be designed to optimally utilize the limited supply of computational power and communication bandwidth on each robot.

With these concerns in mind, a method for autonomous robot coordination has been designed in this research for application in a SHMRS. This solution has been inspired by the use of the hierarchical organizational structure in human organizations as a means to simultaneously yield results and minimize manpower costs. The focus on using organizational structure to address the concerns identified in the preceding paragraphs is due to the fact that structure can significantly impact a system's behavior and performance (Horling & Lesser, 2005). Furthermore, regardless of whether or not it is the explicit intention of the designer, any agent or robot collective will adopt some kind of structure (Wong, Seet, & Sim, 2011). As such, structure is an aspect of robot collective design that should not be regarded as an afterthought, but rather exploited to achieve the desired system performance.

Here, the key characteristic of the proposed method for autonomous coordination is the organizational structuring of the deployed robots into a hierarchy, resulting in a robot collective in which a single (Apex) robot atop the hierarchy is responsible for performing the majority of as well as the most complex aspects of the computation

necessary for autonomous coordination. This helps to reduce the computational sophistication collective members. In this study, such collective members which are not the Apex robot are referred to as Subordinate robots. By deferring the bulk of the computational responsibilities to the Apex, the Subordinate robots may be computationally simpler, possibly running on basic microprocessors. Furthermore, this technique also has the potential to produce savings in terms of communication bandwidth (Fox, Organization structuring: Designing large complex software, 1979). Yet, the use of a dedicated robot to perform coordination in a hierarchically structured collective implies that the loss of the Apex robot to the hazards of the mission automatically results in the loss in capability for autonomous coordination for the entire collective of robots (Horling & Lesser, 2005; Maturana, Shen, & Norrie, 1999; Parker, Multiple mobile robot systems, 2008). Therefore, extra attention and care may have to be provided in supervising the Apex.

In the research presented here, two techniques for autonomous coordination have been implemented on a physical SHMRS: (a) coordination with a hierarchically structured collective and (b) coordination with a horizontally structured collective. Unlike coordination achieved via the use of a hierarchical structure, the coordination performed within a horizontally structured collective requires all collective members to equally share the responsibility of the coordination. It is important to note that discussions pertaining to organizational structures in this chapter are limited only to the robot collective and do not refer to the overall SHMRS. Given superior capabilities for judgment, perception, and learning in humans compared to those in robots, an important assumption in this research is that the human supervisor is always ranked above the robots. Furthermore, it is necessary to note that current study does not envision the robots in the collective to adopt more than one collective structure during the duration of a deployment.

The aim of implementing two techniques for autonomous coordination is to compare one technique against the other. As such, an experiment involving the use of the developed SHMRS to perform a search task in a mock USAR scenario has been devised to demonstrate the feasibility of autonomous coordination using the hierarchical structure and to determine if diversion of supervisor attention to ensure Apex robot survival during the mission negatively impacts supervisor workload and situation awareness. The mission environment within this experiment, like real world USAR environments, has various hazards, which are detrimental to both humans and robots. Therefore, at times, during the mock USAR mission, robots may perish or become incapacitated. The experiment calls for robots to be used to comb a defined mission space in search of (simulated) victims prior to the entry of human rescuers. The robots are to help locate victims and point out hazards so that human rescuers can quickly and directly proceed to victim locations rather than spend longer periods exposed to the dangers of the mission environment while searching for victims. To assist their human supervisor, deployed robots coordinate their actions by dividing the entire mission area into smaller zones for each robot to comb. Furthermore, to facilitate progress tracking and coordination, individual robots keep track of waypoints, which they have visited or instructed by the human supervisor to skip.

The objectives of this chapter are threefold. Firstly, this chapter seeks to acquaint the reader (via the introduction section), with the challenges of multiple robot deployment using a single human supervisor. The second objective is to present the proposed solution for multiple robot deployment and coordination in the face of the identified challenges. The final objective for this chapter is to describe the practical aspects of implementing a system consisting of autonomously coordinating robots that are supervised by a single human. Hence, this chapter is decomposed into six sections. The first section which

follows this introduction is one which describes the methods for autonomous coordination used in this implementation of the SHMRS. The section after that presents the practical or physical aspects of implementing the SHMRS. To facilitate appreciation of how components of the SHMRS (including the human supervisor, the robots, the control station, and the graphical user interface) work together to contribute to USAR, the mock USAR mission used to evaluate the effectiveness of the SHMRS is also described. Following that, some research directions are presented before concluding remarks are offered.

AUTONOMOUS COORDINATION IN THE IMPLEMENTED SHMRS

In this section, the techniques for multiple robot autonomous coordination in performing an area coverage task are presented. As described in the previous section, each technique utilizes either the hierarchical or horizontal organizational structure for coordination. Both techniques are highly similar as each one involves the allocation of waypoints based on the K-means clustering algorithm. The K-means clustering algorithm (MacQueen, 1967) presented below is a method for partitioning data points in a Euclidean space into k number of clusters. The algorithm allocates each data point to the cluster that has the smallest distance between the point and the cluster's centroid (see Algorithm 1).

Still, despite the use of the K-means clustering algorithm by both techniques of autonomous coordination, each technique differs from the other in terms of whether or not robots rely on any particular member of the collective to perform the bulk to the computation required for coordination. The first technique presented is that of coordination performed with the hierarchically structured collective. Following that, coordination performed with the horizontally structured collective is described.

Algorithm 1. Waypoint allocation using K-means clustering

Input:
$X = \{x_1, x_2, ..., x_N\}$ = a set of N waypoints
K = Number of robots
$R = \{r_1, r_2, ..., r_K\}$ = a set of robot current position
$C = \{c_1, c_2, ..., c_K\}$ = a set of cluster centroids of K clustering
Output:
$P = \{p(x) | x = 1, 2, ..., N\}$ = a set of allocated cluster labels of X
1. $X \leftarrow$ Extract_Available_Waypoints_From_Current_Map ()
2. $C \leftarrow R$
3.   **repeat**
4.   **for each** $x_i \in X$ **do**
5. $P(xi) \leftarrow$ arg min Distance $(xi\ cj)$ where $j \in \{1...K\}$
6.   *for* **all** $j \in \{1...K\}$ **do**
7. $Cnew_{(j)} \leftarrow$ Average of X, where $p(x) = j$
8.   If $C \neq Cnew$ then
9. $C \leftarrow Cnew$
10.   *until* $C = Cnew$
11.   *re***turn** **P**

The Hierarchically Structured Collective

The deployment of multiple mobile robots for USAR using this organizational structure for the robot collective is the proposed solution in this research. As described earlier, the design for autonomous coordination using the hierarchical structure for the robot collective has been derived from the desire to make efficient use of robot onboard computational power and communication bandwidth while attempting to alleviate human supervisor cognitive workload levels and loss in situation awareness. The schematic of the SHMRS equipped with a hierarchically structured robot collective is illustrated in Figure 1.

An example of a hierarchically structured robot collective is the system developed by Seib et al. (2011). In that robot collective consisting of two robots, the Apex robot performed the role of instructing a Subordinate robot to navigate an unknown environment, with the objective of cleaning the environment. The Subordinate robot was imbued with minimal intelligence and the robot collective relied on the Apex robot's abilities to develop a SLAM (simultaneous localization and mapping) (Durrant-White & Bailey, 2006)

map and to localize the Subordinate robot in order to direct it to clean targeted areas of the environment. Burmitt and Stentz (1998) also proposed a hierarchically structured robot collective with the intention of using the robots to visit parts of partially known environments. The researchers devised a "central-mission-planner" to coordinate the actions of the robots. This planner resided within one of the robotic vehicles, effectively making it the Apex robot.

Figure 1. The SHMRS with a hierarchically structured robot collective

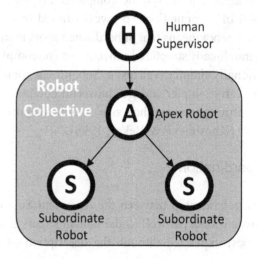

In the following paragraphs, the characteristics of the hierarchically structured collective are first described. Following that, some background information regarding autonomous coordination in this study is provided. To illustrate this technique of autonomous coordination, a detailed presentation of a coordination episode in the hierarchically structured collective is presented as well.

Characteristics of the Hierarchically Structured Collective

The hierarchically structured collective consists of two robot types: the Apex robot and the Subordinate robot. Within the collective, there is one Apex robot and possibly multiple Subordinate robots. As an example, the illustrated robot collective in Figure 1 consists of three robots. The number of Subordinate robots may be increased[2] or decreased according to mission needs.

The Apex robot shoulders most of the overall computational load performed within the collective. Therefore, the Apex robot should be imbued with adequate computational power to deal with the burden of handling most of the collective's computational needs, consequently increasing its level of sophistication. However, the enhanced capabilities brought on by increased sophistication are often accompanied by an increased possibility for breakdowns (Landauer, 1995; Wickens & Hollands, 2000). Yet, the comparatively higher level of computational prowess offered by the Apex robot means that Subordinate robots in the hierarchically structured collective can be computationally simpler. As a result, Subordinate robots can carry simpler and less powerful computers, possibly decreasing their size[3] and development costs (Khoshnevis & Bekey, 1998).

Coordination

The relationship between the Apex and Subordinate robots (as well as the human supervisor) is best explained through the description of a coordination episode. The coordination episode refers to the event involving the robots autonomously harmonizing their actions such that the likelihood of each robot redundantly performing actions which another robot is to perform (or has performed) is reduced or minimized. In this implementation, this refers specifically to the division of the mission area into smaller zones such that each robot is allocated its own zone to search.

Yet, prior to actually performing coordination, the Apex robot must be imbued with knowledge about the mission area's layout. Here, this is achieved by providing the robot with a 160×160 pixel image file of the map, corresponding to a 16m×16m real world mission area. The layout of the mission area is represented with black pixels denoting occupied space such as walls while white pixels represent unoccupied space. From this image, the robot identifies cells of 10×10 pixels, consisting only of white pixels. Each cell is then designated as a waypoint. As such, when all the waypoints are compiled into a list, they provide the robot with a representation of all the points on the mission area that the deployed robots must visit. This list of waypoints is termed as the global waypoints list. An example of this process is shown in Figure 2.

Subsequently, coordination in a robot collective of *k* number of robots is possible when the waypoints within the global waypoints list are divided into *k* separate and smaller waypoints lists using the K-means clustering algorithm. Each newly generated waypoints list resulting from a coordination episode denotes the allocated zone for one member of the robot collective and is termed as the local waypoints list. At this juncture, it is good to note that in addition to having the knowledge about the mission area's layout, the Apex robot must also be aware of the number of active (or still surviving) robots within the collective.

The process of deriving a local waypoints list from a representation of the mission area based on black and white pixels appears fairly

Figure 2. The process of generating a global waypoints list

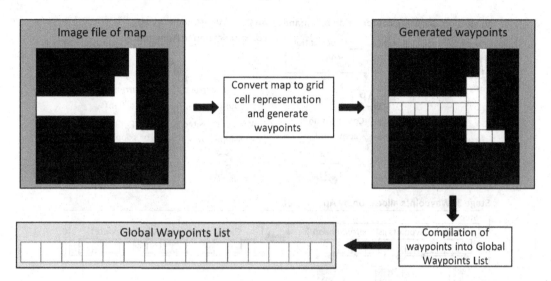

straightforward and is, indeed, a brief description of the recipe for coordination at the beginning of a mission when none of the waypoints have been visited or skipped by any robot, as well as when none of the robots have perished. However, the coordination at the start of the mission is a special case. The following sub-section offers a description of the general case of autonomous inter-robot coordination in this work. This general case handles the added complexity resulting from wanting to prevent waypoints that have already been visited or skipped from being reallocated to robots, as well as from keeping track of which robots are capable of continuing the mission.

Coordination Episode in a Hierarchically Structured Collective

It has to be said that the process for autonomous coordination in different hierarchically structured collectives are likely to vary, depending on the mission that the robots are meant to address. For the hierarchically structured collective presented in this chapter, which is designed for combing a partially known area, the process of autonomous coordination has been simplified to five key stages. Each stage is described below. In addition, each step is illustrated in Figure 3 using a 3-robot

collective as an example. In Figure 3, the Apex robot is denoted by the initials "AP," while the two Subordinate robots are designated "S_1" and S_2."

Stage 1: *Propagating Coordination Command.* In Stage 1, the command for the robot collective to coordinate is provided by the human supervisor (via the control station) to the Apex robot. Along with this command, the Apex robot also receives from the human supervisor, information pertaining to the health of each Subordinate robot. Specifically, this refers to whether or not the human supervisor has deemed any of the Subordinate robots to be inactive or to have perished during the course of the mission. After this, the Apex robot propagates the command for coordination to the Subordinate robots.

Stage 2: *Subordinate Robots Communication to Apex Robot.* Next, in Stage 2, active or surviving Subordinate robots transmit their current position (in two-dimensional Cartesian coordinates) to the Apex Robot. In addition to relaying its current position to the Apex robot, a Subordinate robot will also transmit a list consisting of waypoints, which it has visited or skipped since the previous coordination episode.

Figure 3. The coordination episode in a SHMRS with a hierarchically structured collective

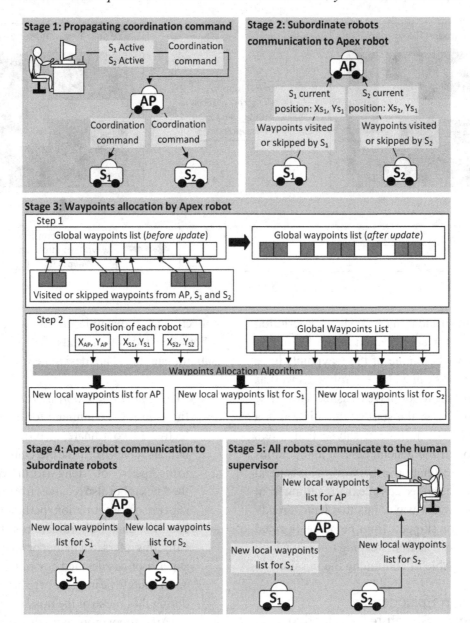

Stage 3: *Waypoints Allocation by Apex Robot.* Following the transmission of the relevant information from Subordinate robots to the Apex robot, the Apex robot executes Stage 3. Stage 3 can be further decomposed into two steps.

In the first step, the Apex robot updates its global waypoints list by using the lists consisting of visited or skipped waypoints received from each Subordinate robot as well as the Apex's own list of visited or skipped waypoints. In Figure 3, such waypoints are represented by shaded squares. Once this is accomplished, the Apex is equipped with a global waypoints list, which reflects the mission's progress in terms of area coverage. More importantly, the list reveals which waypoints are yet to be explored by the robots. Such waypoints are termed here as normal waypoints and are shown in Figure 3 as white squares. The second step in Stage 3 involves applying a waypoints allocation algorithm based on the K-means clustering technique on the normal waypoints. If there are k number of robots participating in the coordination episode, the algorithm attempts to divide the normal waypoints into k clusters, thereby allocating one cluster to each robot. A cluster is paired with a robot based on the proximity between the robot and the center of the cluster. Waypoints in each cluster are compiled into new local waypoints lists for each robot.

Stage 4: *Apex Robot Communication to Subordinate Robots.* In the Stage 4, each Subordinate robot receives a new local waypoints list from the Apex robot. However, if a cluster has been assigned to the Apex robot, then wireless communication is not necessary.

Stage 5: *All Robots Communicate to the Human Supervisor.* In the final stage, each robot transmits its newly allocated waypoints to the control station to indicate to the human supervisor that it has successfully coordinated.

The Horizontally Structured Collective

An alternative to achieving autonomous coordination is to distribute the computation and communication required for coordination amongst the deployed robots more evenly. Figure 4 provides an overview of this alternative. By attempting to equally distribute the responsibility for coordination, the robots become peers of one another, resulting in a horizontally structured collective where no particular robot is ranked higher or lower than the others.

Figure 4. The SHMRS with a horizontally structured robot collective

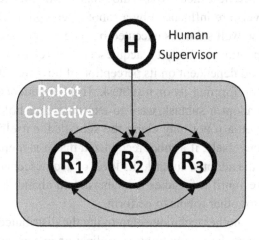

Such a method is in contrast to the dedication of a single robot to perform the bulk of the computation and communication within a hierarchically structured collective. In a horizontally structured collective, robots are all capable of executing the waypoints allocation algorithm by themselves. This reduces the likelihood of a total loss in the capability for autonomous coordination when a particular robot fails in the field. Yet, this also implies that each robot requires information such as the map of the mission environment as well as the position and visited or skipped waypoints from each member of the collective. Furthermore, as Figure 4 indicates, this means that each robot must establish communication links with all other robots within the collective. Due to these reasons, all robots in a horizontally structured collective must have sufficient computational power and communication bandwidth to provide, procure, and process the necessary information for coordination.

An example of a horizontally structured robot collective is the ALLIANCE architecture presented by Parker (1998). In that framework, the robots autonomously coordinated their actions by being allocated motivation level for each of the many predefined subtasks necessary to achieve a global goal. Each robot's motivation to perform a subtask would be influenced by its impatience or acquiescence towards the subtask. Impatience levels are influenced by a robot's perception of how well subtasks executed by other robots are performed. A robot's acquiescence level is instead dependent on its perception of how well it is performing its own subtask. If its impatience to adopt a subtask were to exceed a threshold level, a robot may claim the subtask for itself. Conversely, if a robot recognized that it is inept at the adopted subtask; its level of acquiescence rises until it forsakes the subtask and abandons it for other robots to perform.

Another technique employing the distributed approach to autonomous coordination in a group of robots is the Broadcast of Local Eligibility (BLE) (Werger & Matarić, 2000). With BLE, each robot in the group is imbued with an identical set of behaviors and each behavior corresponds to the execution of a particular subtask. A robot assigns a value to each behavior in its behavior set. This value corresponds to the robot's self-assessed ability (or local eligibility) to perform the given subtasks. Each robot may communicate its values for each of the imbued behaviors to other peer robots. The robot with the highest value for a certain behavior claims the right to execute the behavior and then inhibits other robots from executing the same behavior. The failure of a particular robot means also that the inhibition of other robots to perform the subtask originally performed by the failed individual is ceased. Other robots are therefore free to adopt the failed robot's task.

Coordination Episode in a Horizontally Structured Collective (General Case)

For the horizontally structured collective presented in this chapter, autonomous coordination occurs in four key stages. The process is briefly described in the paragraphs below.

Stage 1: *Propagating Coordination Command.* In Stage 1, the coordination episode begins with the human supervisor providing the command. This command does not have to be communicated to a specific collective member but instead can be provided to any robot. Together with the command is information about which robots have been rendered inactive and which ones are still active. The robot which receives the coordination command from the human supervisor propagates the command (and the relevant information about which robots are still active) to the other active members of the robot collective.

Stage 2: *Inter Robot Coordination.* In Stage 2, each robot sends its current position and the waypoints that it has visited or skipped (since the previous coordination episode) to all other active robots in the collective.

Stage 3: *Waypoints Allocation by all Participating Robots.* Stage 3 involves all active robots executing the waypoints allocation algorithm much like what the Apex robot does in the hierarchically structured collective. As each robot in the horizontally structured collective is capable of generating new local waypoints lists for all robots participating in the coordination episode, a robot would only have to search amongst the self-generated local waypoints lists for the one corresponding to its own identification tag[4]. From this list, a robot can infer for itself its allocated waypoints. The use of the same waypoints allocation algorithm as well as identical sets of input parameters to the algorithm ensures that robots do not produce conflicting results, meaning that robots do not claim for themselves waypoints meant for another robot.

Stage 4: *All Robots Communicate to the Human Supervisor.* The ability for every robot within the collective to derive for itself the waypoints which it should visit therefore eliminates the need for a robot to have to depend on another to provide its list of allocated waypoints. As a result, after Stage 3, each robot can transmit its newly allocated waypoints to the control station to indicate to the human supervisor that it has successfully coordinated.

The Pros and Cons of the Hierarchically Structured Robot Collective

Based on the reviewed literature and the coordination techniques presented in this chapter, the current chapter attempts to identify the benefits and the drawbacks of both types of robot collective. However, as the benefit of one collective structure typically translates to the drawback of the other, only the pros and cons of the hierarchically structured robot collective are discussed.

By comparing the proposed coordination techniques for both types of robot collective, it is possible to see that in order for a Subordinate robot in the hierarchically structured collective to take part in a coordination episode, it needs only to be aware of its own position and the waypoints, which it has visited or skipped. It does not need to maintain any representation of the larger mission area nor even be aware of the existence of other collective members besides the Apex robot. This simplicity translates into a number of advantages. Firstly, it reduces the likelihood of computer glitches or bugs on the Subordinate robot. Secondly, it means that the necessary computation performed within the Subordinate robot can quite likely be achieved with small and inexpensive microprocessors (rather than larger and more costly CPUs). This, in turn, implies that numerous Subordinate robots can be easily deployed and inexpensively replaced.

Also, as Subordinate robots in the hierarchically structured collective communicate only with their Apex robot. Inter-robot interactions are limited to only between Subordinates and the Apex. As such, the use of the hierarchical organizational structure allows for close coordination between robots while at the same time possibly facilitating reduced communication between robots (Fox, 1981). At the time of writing this chapter, the experiment to verify this claim is being conducted.

However, a disadvantage to structuring the collective hierarchically is that the Apex robot has to be more capable than its Subordinate robots in terms of computational prowess and available bandwidth. This is especially true if the number of Subordinates reporting to the Apex is large. Therefore, the added simplicity in coordination offered by structuring the collective hierarchically has a tradeoff. When designing the collective, it is therefore important not only to limit the number of robots populating the collective from the standpoint of preventing the human supervisor from being overloaded cognitively but also to impose a threshold based on consideration for the Apex robot's computational power and communication bandwidth.

Yet, by far the biggest drawback of the hierarchical structure is its reliance on a single individual to perform autonomous coordination, resulting in the robot collective's propensity for single-point-failure (Horling & Lesser, 2005). Therefore, the human supervisor may have to pay particular attention to the Apex robot, possibly limiting the risks that the robot is exposed to and ensuring its survival. Consequently, this may induce a higher level of cognitive workload for the human supervisor. One of the main objectives of this research is to determine if this is indeed true.

PHYSICAL IMPLEMENTATION OF A SHMRS WITH HIERARCHICALLY STRUCTURED COLLECTIVE

The aspects of physically implementing the SHMRS are presented in this section. Each aspect corresponds to one of the major components of the physical SHMRS. Not including the human supervisor, the SHMRS may be considered to consist of three other major components: (1) the robots, (2) the control station, and (3) the Graphical User Interface (GUI). Hence, in the paragraphs that follow, the robots used are first described. This is followed by a presentation of the control station and before the design of the GUI is described.

The Robots

The implemented prototype SHMRS currently supports up to three robots, regardless of the chosen robot collective organizational structure. The robots have been designed and developed to meet a certain criteria. For instance, each robot must be mobile and be able to move about untethered. As such, each robot needs to be capable of wireless communication. Robots must also be able to sense obstacles and visually assess their surroundings. Additionally, each robot is required to be capable of self-localization as well as processing commands from humans and its own sensors.

To satisfy this set of requirements, this research looked to the ATRV series of robots. Each robot has been imbued with a CPU running the Linux operating system. The software running on the robot has been developed in-house and designed to run based on lower level functionalities provided by the Carnegie Mellon Robot Navigation Toolkit (CARMEN). A laser scanner and a video camera have been installed on each robot to enable obstacle avoidance and visual assessment of the robot's environment. To facilitate untethered communication, every robot makes use of a wireless video transmitter to relay video imagery to the control station. In addition, communication of sensor feedback data and commands between robots and control station is achieved using a wireless Ethernet station adaptor on board each robot. Therefore, each robot connects wirelessly to the Ethernet access point at the control station. Ethernet connection in this implementation is achieved via TCP/IP in an ad hoc wireless network. Self localization on the robot is based on the Monte-Carlo Localization method (Thrun, Fox, Burgard, & Dellaert, 2001), making use of data from its laser scanner and odometer. By considering data from both of these sensors, a probabilistic approach to position is possible (Montemerlo, Roy, & Thrun, 2003). Figure 5 shows one of the deployed robots as well as lists the robot specifications in greater detail.

The physical differences between each of the three ATRV robots used are small and each robot has the same basic equipment suite of sensors as well as communication devices shown above in Figure 5. Therefore, by adjusting a few parameters on each robot, the same three robots can be jointly deployed either as a hierarchically structured or horizontally structured collective. In real-world deployments however, Subordinate robots in a hierarchically structured collective may require only small microprocessors to perform the limited computing required of them. Therefore, it is likely that Subordinate robots can be much more compact than the ATRV robots used here in this implementation. The rationale for enabling the same robots for use to deploy as a part of a collective organized either hierarchically or horizontally is so that meaningful comparisons can be made when considering the effects caused by deploying either collective type. These comparisons are outside the scope and, therefore, are not presented in this chapter.

The Control Station

The control station shown in Figure 6 is the platform, which houses the hardware components such as the communication and input devices

Figure 5. General specifications of a SHMRS robot individual

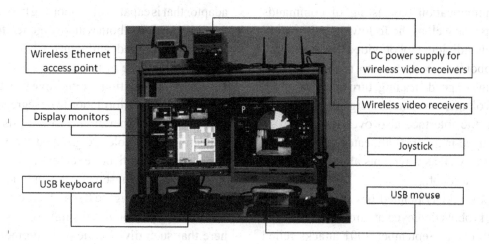

Physical Specifications
- Robot dimensions: 0.77m x 0.64m x 0.55m
- Weight: Approximately 50kg

Power System
- Supply: 2 x 12V rechargeable batteries in series

Mobility System
- Motor: Two brushed DC servo motors
- Drive: Differential drive
- Maximum Speed: 1.7m/s

Computing System
- Processor: Pentium III 800MHz
- Operating System: Linux Red Hat
- Motor Controller: iRobot rFlex system

Sensors
- Localization: Odometer (incremental optical motor encoders)
- Vision: SONY EVI-D30 pan, tilt and zoom camera
- Laser Scanner: SICK LMS 200

Wireless Communication Devices
- Ethernet Communication: BreezeNet SA-10 Station Adaptor PRO.11 (IEEE 802.11b wireless Ethernet)
- Video transmission: 1.2GHz wireless video transmitter

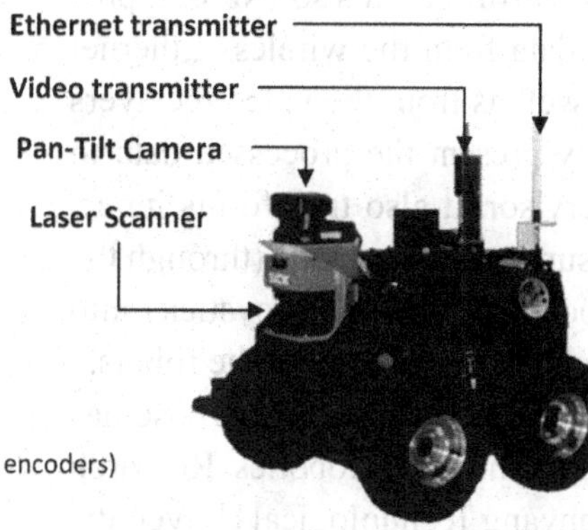

Ethernet transmitter

Video transmitter

Pan-Tilt Camera

Laser Scanner

Figure 6. The control station of the SHMRS

Wireless Ethernet access point

DC power supply for wireless video receivers

Display monitors

Wireless video receivers

Joystick

USB keyboard

USB mouse

necessary for the human to supervise the robots. In addition to that, the control station also houses the computer, which runs the graphical user interface necessary for interaction between the human supervisor and deployed robots.

The control station hardware consists of a PC with dual display monitors, a USB joystick, a wireless Ethernet access point, a USB keyboard as well as a USB mouse. For each robot in the collective, a wireless video receiver and a video capture device are also required at the control station.

For the mock USAR missions planned for the experiments in this research, the control station is situated some distance away from the robots to emulate the required separation between human rescuers and the hazardous environment in which the robots perform their mission. Because of this separation, there is no line of sight between human supervisor and robots during the experiment.

The Graphical User Interface (GUI)

The GUI is the control station's software. It processes received data from the wireless Ethernet access point as well as from the video receivers and then visually present the processed data to the human supervisor. It also transforms inputs that the human supervisor has made (through the use of the keyboard, mouse, and joystick) into commands which are transmitted to the robots.

The software employed for this purpose has been developed within the Robotics Research Centre of the Nanyang Technological University. It was developed using the C++ programming language and OpenGL (Open Graphics Library) and runs on the Microsoft Windows-XP operating system. Information displayed by the GUI include the map of the mission area, elapsed mission time, text communication (consisting of commands and alerts), as well as the following information from each individual robot: video imagery, robot status, robot location in the map, robot laser scanner readings, speed, heading, turn rate, remaining battery voltage, and wireless signal strength. In addition, the interface also overlays on top of the displayed map, areas allocated to each robot to search as well as waypoints already visited or skipped by a robot.

Murphy (2004) pointed out that human operators of robots deployed at Ground Zero in the aftermath of the September 2001 attacks relied entirely on using their visual sensory channel to extract information about the robots. As such, to enhance interaction and to prevent the human supervisor from being visually overloaded due to the volume of information presented visually, some information is also provided via the supervisor's auditory sensory channel. In this implementation, the GUI has been designed to present alerts in audio. Such alerts include coordination failures, the detection of victims, and the inability of a robot to autonomously plan a path to its goal waypoint. However, to make sure that the supervisor may still perceive an alert after it has been displayed in audio, the same alert is shown visually on screen as well. If the alert is not acknowledged or if the situation causing the alert is not rectified, the same audio alert is played to the supervisor once more thirty seconds after the previous audio alert.

During the early stages of the interface's design, it was realized that video imagery and laser scanner readings from each robot as well as the map of the mission area required large portions of the screen if the user is expected to view and comprehend the displayed information efficiently and easily. Therefore, rather than attempting to cram all the required information for display on a single computer monitor, the GUI displays the information on two windows which should be maximized on the monitor for effective use. Hence, the control station computer uses a display adaptor that is capable of supporting two computer monitors such that both windows generated by the GUI can be viewed simultaneously.

Having the use of two screens to display the necessary information for multiple robot deployment enables the information to be segregated such that information pertaining to the mission and collective as a whole are grouped for display on one screen (the collective screen shown in Figure 7) while the other screen (the individual robot screen shown in Figure 8) can be used to display information pertaining to single robots. It is felt here that such division helps to promote clarity and user friendliness. To switch between viewing information from another robot, a user can select the desired robot by clicking on its icon on the map (which is found on the collective screen) or on the display of the robot's camera view on the collective screen.

Each component described in this section has been integrated to form the SHMRS. To best illustrate how these components work together as a single system, a description of a typical mission to which the implemented SHMRS may be applied is offered next.

Figure 7. The collective screen

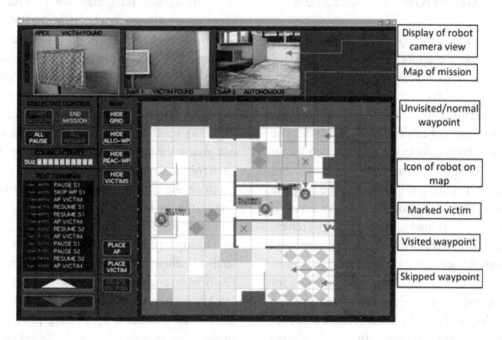

Figure 8. The individual robot screen

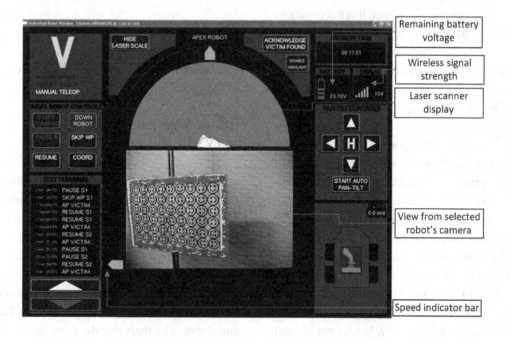

THE DESCRIPTION OF A MISSION

In this section, the mock USAR mission designed to demonstrate the applicability of the SHMRS is presented. This discussion serves to justify the implementation effort and describes how the human supervisor, the control station, and the robots function cohesively as a SHMRS. The mission has been developed as a test bed for an experiment, which is currently being conducted at the time of writing this chapter. The purpose of the experiment is to compare human supervisor cognitive workload and situation awareness when deploying either a hierarchically structured or horizontally structured collective. The aim of the experiment is also to make comparisons for demonstrating the effectiveness of coordination using the hierarchical structure in reducing robot computational and communication loads. Therefore, experiment runs conducted with the mock USAR mission sometimes feature hierarchically structured robot collectives and at other times, feature horizontally structured robot collectives. Before the mission is discussed, it is necessary to describe the background of the mission.

Mission Background

The goal for the human supervisor in the SHMRS is to deploy the robots to search for and locate all the victims in the mission area as quickly as possible. The robots may move about autonomously but such autonomous navigation can be temporarily suspended on the human supervisor's command. If necessary, the human supervisor can remotely pilot the robot via a joystick at the control station.

To simulate victims amid the disorganization and clutter of a typical USAR environment, printed boards of about 40×30cm in size and each consisting of an array of up to 54 smiley faces are used. Around the array of faces is a border about 2cm thick and orange in color. This brightly colored border helps the robot to use its on-board camera to autonomously recognize and alert the human supervisor to the presence of the simulated victim.

The human supervisor in this mock USAR scenario is to locate frowning faces in each array of smiley faces. A board may contain zero or up to two frowning faces. This makes the difficulty in confirming the presence of a victim more realistic as spotting a board from a distance does not guarantee the discovery of a victim. Rather, each robot must approach the board up close and examine the board in detail. Figure 8 shows the view from the individual robot screen of the GUI while the robot is examining the board.

During a typical USAR deployment, hazards such as debris and large holes in the ground are likely to be encountered. Therefore, in this mock USAR scenario, the presence of such hazards have been simulated by placing piles of wood in the mission area to replicate debris and black plastic boards about $1m^2$ in area to denote holes. If any robot crashes into the simulated debris or runs over the simulated holes, it is deemed to have perished[5].

The Mission

The following paragraphs describe the key aspects of the mock USAR mission.

Before Mission Start

The robots are to begin the mission at the entrance to the mission area. Presumably, the human supervisor is located outside of and away from the possible dangers of the mission area. Prior to entering the mission area, the Apex robot would have been provided with a map of the mission area[6]. While the map provides a layout of the walls, partitions, and rooms in general, it does not indicate the positions of non permanent objects such as furniture and doors. The map also does not indicate where hazards such as holes and debris are.

Once inside the mission area, each robot is localized with the help of the human supervisor, thereby providing the robot with its 2D position coordinates. After localization, the system is ready to start the mission.

Mission Start

The mission starts when the human supervisor clicks on the "Start mission" button at the GUI. This sends a command to the Apex robot and initiates the first coordination episode. Subsequently, each robot is allocated a series of waypoints to visit. This series or list of waypoints represents the zone within the mission area, which the robot is to comb.

After the coordination episode, the human supervisor can instruct each robot individually to begin combing the zone, which it has been allocated. This prevents the robots from beginning to move all at once while they are still in close proximity to one another.

Visiting and Skipping Waypoints

To visit waypoints and comb the area via autonomous behavior, robots plan paths from their current position to their goal waypoint. A new goal waypoint is allocated to the robot once its current goal waypoint has been reached or once the robot has been instructed by the human to skip that waypoint. A robot ascertains its goal waypoint by considering its allocated waypoints yet to be visited or skipped. The waypoint closest to the robot is designated as its goal waypoint and a path is planned towards it using the CARMEN path-planning feature.

As a robot performs area combing in its allocated zone, an increasing number of waypoints allocated to the robot are visited. The robot reports its arrival at each waypoint to the control station. On the GUI, this is indicated by shading the cor-responding waypoint on the map display in grey. Such visited waypoints may be seen on the screen capture of the collective screen shown in Figure 7.

If the human supervisor determines that a particular allocated waypoint is not critical and may be bypassed, then it is possible for a robot to be instructed to skip that waypoint. This means that while a robot may still traverse to the position of the skipped waypoint (on its way to another position), the robot will not deliberately make plans to move to the skipped waypoint. As a result, this can help to reduce the number of waypoints that a robot has to visit and shorten mission duration. Furthermore, this feature can allow robots to bypass areas, which are occupied by obstacles such as debris and furniture. Skipped waypoints can also be seen in Figure 7.

Each robot keeps track of which of its allocated waypoints has been visited or skipped. As earlier explained, this information is necessary for coordination episodes performed mid way through the mission. Such mid-mission coordination may be required when a re-allocation of waypoints can allow the robots to comb the mission area more efficiently. For instance, coordination may be called for when a particular robot still has much of its allocated zone unexplored while other robots have already finished combing their zones.

Victim Discovery

In this implementation, the robots may assist the human supervisor to discover victims by performing color detection. This involves analysis of each pixel of the video from each robot's on-board camera to determine if the pixel has the properties necessary to trigger a detection. If the number of pixels fitting the detection criteria exceeds a threshold, then the robot alerts the human supervisor (via the GUI) to the detection of a victim. The human supervisor can assess if the robot is correct by viewing the video from the relevant robot. If indeed the robot has accurately detected a victim, the human can use the GUI to mark the victim's position on the map display. The position of a marked victim is shown in Figure 7.

It has to be noted that autonomous victim detection using color detection is not the most practical method of detecting the presence of a victim in an USAR scenario. However, the application of this method here in this implementation is meant to mimic the use of heat detection technology to help locate victims through their body heat. It should also be noted that attempting to detect victims using such technologies (both color and heat detection) does not always produce robust results. With color detection, the robot may fail to pinpoint the location of a victim if ambient lighting is insufficient. It may also falsely detect a victim if other environment features have the same color. Similarly, if heat detection technology is applied, the heat from other sources may interfere with accurate detection. The inability for body heat to be detected by the robot (due perhaps to thick clothing worn by a victim) may also lead to failed detection. As such, total reliance on robotic technology for victim detection is not prudent and such technology should serve only to augment the system's capabilities and not to replace supervision by the human.

Downing a Robot

In the course of a mission, a robot may be damaged or incapacitated to the extent where it can no longer contribute further to the mission. Alternatively, a robot may venture to an area where wireless communication cannot be sustained. Therefore, such a robot should be declared "down" by the human supervisor. The rationale for having to down a robot is to due to the need for the Apex robot to be informed about the number of as well as the identity of robots participating future coordination episodes[7]. Without this information, the Apex robot cannot successfully allocate new zones for surviving robots.

Mission End

The end of the mission is determined by the human supervisor and is independent of the number of waypoints visited because the objective of the mission is not to visit waypoints but rather to explore the area and search for victims. If the human supervisor ascertains that the mission area has been effectively explored and that all victims have been located, then it would be appropriate to end the mission.

At the end of the robots' mission, responders to the USAR scenario will be equipped with knowledge about the extent of damage to the disaster area, the location, as well as type of hazards within the area and most importantly, the location of human victims. The key advantage to deploying the system is that such information can be procured quickly without the need for human rescuers to enter the disaster area, requiring only a single human to supervise the robots while they are deployed.

A video of the developed GUI during a deployment of the robots in the mock USAR scenario is available at http://www.youtube.com/watch?v=NldotyktW1E.

FUTURE RESEARCH DIRECTIONS

While the proposed SHMRS has been successfully implemented and put through a series experimental runs in a mock USAR scenario, improvements on the system are definitely possible. Firstly, the current system has not addressed the need for robust communication between robots and the control station. If reliable wireless communication is to be possible, the system may have to consider the use of robots not only to search for victims in the mission area but also to act as relays for other robots to promote reliable communication between the control station and all members of the robot collective.

Secondly, the plans resulting from the designed autonomous coordination technique, while helpful, can be further enhanced. This may be achieved by imbuing robots with the capability to perform waypoint allocation not only by considering clusters of waypoints but also by considering the distance each robot has to travel to reach those waypoints. Currently, the robots are allocated waypoint clusters based only on straight-line distances from a robot's position to the centre of the cluster (generated via the K-means clustering algorithm) and not the actual distance a robot has to travel while avoiding walls and other known obstacles.

Apart from improving the SHMRS described here, further research may also be conducted to derive other creative methods of using organizational methods to address the challenges of multiple-robot deployment. For instance, robot collectives may be designed such that they encompass elements of both the hierarchical and horizontal structures. In this case, a robot collective may be populated with multiple hierarchies so that more than one Apex robot exists. Rather than being the Apex robot for the whole collective, a robot which is not a Subordinate would be a local Apex, responsible for a fraction of all the Subordinate robots in the collective. Local Apex robots would then become peers of other local Apex robots. In this manner, it may be possible to deploy collectives of larger sizes while not only avoiding a high level of susceptibility to single-point-failure (by having multiple Apex robots) for the collective but also deriving benefits of the hierarchical structure such as being able to field computationally simple Subordinate robots. The use of such a structure is akin to that of application of the organizational structure described by Horling and Lesser (2005) as a federation.

CONCLUSION

In this chapter, the key aspects of designing and implementing a system of mobile robots capable of wirelessly and autonomously coordinating their actions have been introduced. Such a system of robots would be suited for extending human presence in environments, which are typically highly dynamic and hazardous. Therefore, the system must be designed with the means to address the challenges of deploying robots in these sorts of environments. These challenges pertain to human-robot interaction challenges as well as to technical challenges.

Specifically, the human-robot interaction challenges refer to the proper management of human supervisor cognitive workload and the facilitation of adequate situation awareness. In the presented multiple-robot system, only a single human is used to supervise the deployed robots but the human-robot interaction challenges are still relevant, even when multiple human supervisors are available. Often, such HRI challenges are addressed by automating some portions of the robots' tasks. Here, the focus is on automating the process of coordinating robot actions.

However, care must be taken to ensure that the technical challenges of properly managing limited robot computational power and communication bandwidth are not overlooked when automating robot tasks. This is particularly true when automating coordination for large numbers of robots. Efficient use of these precious computational and communication resources is therefore necessary such that robots would ideally require only small, simple computers, and rely sparingly on wireless communication with one another. The proposed method of autonomous coordination using a hierarchically structured collective attempts to address these requirements by enabling the deployment of robots which perform only simple computations and establish communication links with just one robot.

Research into employing mobile robots for application on USAR has been found to be gaining in popularity and momentum because of advancements in sensing, locomotive, computational and communication technologies the past few decades. With continued endeavors such as the research described in this chapter, to further technologies in USAR robotics, it is hoped that future efforts to tackle the tragedy of real life USAR scenarios can be made less dangerous for rescuers while increasing the number of victims saved.

REFERENCES

Anderson, M., & Papanikolopoulos, N. (2008). Implicit cooperation strategies for multi-robot search of unknown areas. *Journal of Intelligent and Robotic Systems: Theory and Applications*, *53*(4), 381–397. doi:10.1007/s10846-008-9242-5

Barnes, M., Everett, H. R., & Rudakevych, P. (2005). ThrowBot: Design considerations for a man-portable throwable robot. In *Proceedings of SPIE-The International Society for Optical Engineering*, (pp. 511 - 520). SPIE.

Barrett, A., Rabideau, G., Estlin, T., & Chien, S. (2007). Coordinated continual planning methods for cooperating rovers. *IEEE Aerospace and Electronic Systems Magazine*, *22*(2), 27–33. doi:10.1109/MAES.2007.323296

Bellingham, J. G., & Rajan, K. (2007). Robotics in remote and hostile environments. *Science*, *318*(5853), 1098–1102. doi:10.1126/science.1146230

Burgard, W., Moors, M., Stachniss, C., & Schneider, F. E. (2005). Coordinated multi-robot exploration. *IEEE Transactions on Robotics*, *21*(3), 376–386. doi:10.1109/TRO.2004.839232

Burke, J. L., & Murphy, R. R. (2004). Human-robot interaction in USAR technical search: Two heads are better than one. In *Proceedings of the IEEE International Workshop on Robot and Human Interactive Communication*, (pp. 307-312). Kurashiki, Japan: IEEE Press.

Burmitt, B. L., & Stentz, A. (1998). GRAMMPS: A generalized mission planner for multiple mobile robots in unstructured environments. In *Proceedings of the IEEE International Conference on Robotics and Automation*, (pp. 1564 - 1571). IEEE Press.

Casper, J., Micire, M., & Murphy, R. R. (2000). Issues in intelligent robots for search and rescue. In *Proceedings of SPIE - The International Society for Optical Engineering*, (pp. 292 - 302). SPIE.

Casper, J., & Murphy, R. R. (2003). Human-robot interactions during the robot-assisted urban search and rescue response at the world trade center. *IEEE Transactions on Systems, Man, and Cybernetics. Part B, Cybernetics*, *33*(3), 367–385. doi:10.1109/TSMCB.2003.811794

Doroodgar, B., Ficocelli, M., Mobedi, B., & Negat, G. (2010). The search for survivors: Cooperative human-robot itneraction in search and rescue environments using semi-autonomous robots. In *Proceedings of the IEEE International Conference on Robotics and Automation*, (pp. 2858-5863). IEEE Press.

Durrant-White, H., & Bailey, T. (2006). Simultaneous localization and mapping: Part 1. *IEEE Robotics & Automation Magazine*, *13*(2), 99–110. doi:10.1109/MRA.2006.1638022

Fox, M. S. (1979). *Organization structuring: Designing large complex software*. Pittsburgh, PA: Carnegie Mellon University Press.

Fox, M. S. (1981). An organizational view of distributed systems. *IEEE Transactions on Systems, Man, and Cybernetics, 11*(1), 70–80. doi:10.1109/TSMC.1981.4308580

Haight, J. M., & Kecojevic, V. (2005). Automation vs. human intervention: What is the best fit for the best performance? *Process Safety Progress, 24*(1), 45–51. doi:10.1002/prs.10050

Halberstam, E., Navarro-Serment, L., Conescu, R., Mau, S., Podnar, G., Guisewite, A. D., et al. (2006). A robot supervision architecture for safe and efficient space exploration and operation. In *Proceedings of the 10th Biennial International Conference on Engineering, Construction, and Operations in Challenging Environments*, (pp. 92-102). IEEE.

Horling, B., & Lesser, V. (2005). A survey of multi-agent organizational paradigms. *The Knowledge Engineering Review, 19*(4), 281–386. doi:10.1017/S0269888905000317

Kadous, M. W., Sheh, R. K., & Sammut, C. (2006). Controlling heterogeneous semi-autonomous rescue robot teams. In *Proceedings of the IEEE International Conference on Systems, Man, and Cybernetics*, (pp. 3204-3209). Taipei, Taiwan: IEEE Press.

Khoshnevis, B., & Bekey, G. (1998). Centralized sensing and control of multilpe mobile robots. *Computers & Industrial Engineering, 35*(3-4), 503–506. doi:10.1016/S0360-8352(98)00144-2

Landauer, T. (1995). *The trouble with computers*. Cambridge, MA: MIT Press.

MacQueen, J. B. (1967). Some methods for classification and analysis of multivariate observations. In *Proceedings of the 5th Berkeley Symposium on Mathematical Statistics and Probability*, (pp. 281-297). IEEE.

Maturana, F., Shen, W., & Norrie, D. (1999). MetaMorph: An adaptive agent-based architecture for intelligent manufacturing. *International Journal of Production Research, 37*(10), 2159–2173. doi:10.1080/002075499190699

Montemerlo, M., Roy, N., & Thrun, S. (2003). Perspectives on standardization in mobile robot programming: The Carnegie Mellon navigation (CARMEN) toolkit. In *Proceedings of the International Conference on Intelligent Robots and Systems*. Las Vegas, Nevada: IEEE.

Murphy, R., Blitch, J., & Casper, J. (2002). AAAI/RoboCup-2001 urban search and rescue events: Reality and competitions. *AI Magazine, 23*(1), 37–42.

Murphy, R. R. (2004). Human-robot interactions in rescue robotics. *IEEE Transactions on Systems, Man and Cybernetics. Part C, Applications and Reviews, 34*(2), 138–153. doi:10.1109/TSMCC.2004.826267

Parker, L. E. (1998). ALLIANCE: An architecture for fault tolerant multirobot cooperation. *IEEE Transactions on Robotics and Automation, 14*(2), 220–240. doi:10.1109/70.681242

Parker, L. E. (2008). Multiple mobile robot systems. In Siciliano, B., & Khatib, O. (Eds.), *Springer Handbook of Robotics* (pp. 921–941). Berlin, Germany: Springer. doi:10.1007/978-3-540-30301-5_41

Scerri, P., Velagapudi, P., Sycara, K., Wang, H., Chien, S. Y., & Lewis, M. (2010). Towards an understanding of the impact of autonomous path planning on victim search in USAR. In *Proceedings of the IEEE/RSJ International Conference on Intelligent Robots and Systems*. Taipei, Taiwan: IEEE Press.

Schreckenghost, D., Fong, T. W., & Milam, T. (2008). Human supervision of robotic site survey. In *Proceedings of the AIP Conference*, (pp. 776-783). AIP.

Seib, V., Gossow, D., Vetter, S., & Paulus, D. (2011). Hierarchical multi-robot coordination. *Lecture Notes in Computer Science, 6556*, 314–323. doi:10.1007/978-3-642-20217-9_27

Shiroma, N., Chiu, Y. H., Sato, N., & Matsuno, F. (2005). Cooperative task execution of a search and rescue mission by a multi-robot team. *Advanced Robotics, 19*(3), 311–329. doi:10.1163/1568553053583670

Sugihara, M., & Suzuki, I. (1990). Distributed motion coordination of multiple mobile robots. In *Proceedings of the IEEE International Symposium on Intelligent Control*, (pp. 138-143). Philadelphia, PA: IEEE Press.

Sweeney, J. D., Li, H., Grupen, R. A., & Ramamritham, K. (2003). Scalability and schedulability in large, coordinated distributed robot systems. In *Proceedings of the IEEE International Conference on Robotics and Automation*, (pp. 4074-4079). IEEE Press.

Tadokoro, S. (2009). Earthquake disaster and expectation for robotics. In Tadokoro, S. (Ed.), *Rescue Robotics: DDT Project on Robots and Systems for Urban Search and Rescue* (pp. 1–16). London, UK: Springer.

Thrun, S., Fox, D., Burgard, W., & Dellaert, F. (2001). Robust Monte Carlo localization for mobile robots. *Artificial Intelligence, 128*(1-2), 99–141. doi:10.1016/S0004-3702(01)00069-8

Vincent, R., Fox, D., Ko, J., Konolige, K., Limketkai, B., & Morisset, B. (2008). Distributed multirobot exploration, mapping and task allocation. *Annals of Mathematics and Artificial Intelligence, 52*(2-4), 229–255. doi:10.1007/s10472-009-9124-y

Werger, B. B., & Matarić, M. J. (2000). Broadcast of local eligibility: Behavior-based control for strongly cooperative robot teams. In *Proceedings of the International Conference on Autonomous Agents*, (pp. 21 - 22). IEEE.

Wickens, C. D., & Hollands, G. J. (2000). *Engineering psychology and human performance* (3rd ed.). Upper Saddle River, NJ: Prentice Hall.

Wong, C. Y., Seet, G., & Sim, S. K. (2011). Multiple-robot systems for USAR: Key design attributes and deployment issues. *International Journal of Advanced Robotic Systems, 2*(3), 85–101.

Yamauchi, B. (2004). PackBot: A versatile platform for military robotics. *The International Society for Optical Engineering, 5422*, 228–237.

Yanco, H. A., Drury, J. L., & Scholtz, J. (2004). Beyond usability evaluation: Analysis of human-robot interaction at a major robotics competition. *Human-Computer Interaction, 19*, 117–149. doi:10.1207/s15327051hci1901&2_6

ENDNOTES

[1] A detailed review of the motivating factors for multiple robot deployment has been performed by Wong, Seet, and Sim (2011).

[2] It is foreseeable that there would be a limit to the number of Subordinate robots that can be included under the jurisdiction of an Apex robot. Beyond that limit, the Apex robot would run out of bandwidth for communication with its subordinates. In addition, the plans for a large collective may be too complex for the Apex robot's limited computing power to cope with.

[3] If the size disparity between Apex and Subordinate Robots is big enough, then the concept of marsupialism can be applied to the collective such that the Apex not only performs cumbersome computation for the collective but also acts as a mothership to deploy the Subordinate robots.

4   Each robot has a unique electronic identification tag.

5   During an experiment, a helper roams the mission environment with the robots. This helper acts as a referee and immobilizes a robot when the robot has been judged to have perished due to a crash or a "fall" into the hole. The robot will stop communicating with the control station and at the same time, the video transmitter on board the robot will be switched off. These measures simulate the loss of a robot.

6   If the collective is horizontally structured, then each robot would be provided with the map.

7   In a horizontally structured collective, all the robots would have to be provided with this information.

Chapter 10
A Mechatronic Description of an Autonomous Underwater Vehicle for Dam Inspection

Ítalo Jáder Loiola Batista
Federal Institute of Education, Science, and Technology of Ceará, Brazil

Tiago Lessa Garcia
Federal Institute of Education, Science, and Technology of Ceará, Brazil

Antonio Themoteo Varela
Federal Institute of Education, Science, and Technology of Ceará, Brazil

Daniel Henrique da Silva
Federal Institute of Education, Science, and Technology of Ceará, Brazil

Edicarla Pereira Andrade
Federal Institute of Education, Science, and Technology of Ceará, Brazil

Epitácio Kleber Franco Neto
Federal Institute of Education, Science, and Technology of Ceará, Brazil

José Victor Cavalcante Azevedo
Federal Institute of Education, Science, and Technology of Ceará, Brazil

Auzuir Ripardo Alexandria
Federal Institute of Education, Science, and Technology of Ceará, Brazil

André Luiz Carneiro Araújo
Federal Institute of Education, Science, and Technology of Ceará, Brazil

ABSTRACT

Driven by the rising demand for underwater operations concerning dam structure monitoring, Hydropower Plant (HPP), reservoir, and lake ecosystem inspection, and mining and oil exploration, underwater robotics applications are increasing rapidly. The increase in exploration, prospecting, monitoring, and security in lakes, rivers, and the sea in commercial applications has led large companies and research centers to invest underwater vehicle development. The purpose of this work is to present the design of an Autonomous Underwater Vehicle (AUV), focusing efforts on dimensioning structural elements and machinery and elaborating the sensory part, which includes navigation sensors and environmental

DOI: 10.4018/978-1-4666-2658-4.ch010

Copyright © 2013, IGI Global. Copying or distributing in print or electronic forms without written permission of IGI Global is prohibited.

conditions sensors. The integration of these sensors in an intelligent platform provides a satisfactory control of the vehicle, allowing the movement of the submarine on the three spatial axes. Because of the satisfactory fast response of the sensors, one can determine the acceleration and inclination as well as the attitude in relation to the trajectory instantaneously taken. This vehicle will be able to monitor the physical integrity of dams, making acquisition and storage of environmental parameters such as temperature, dissolved oxygen, pH, and conductivity, as well as document images of the biota from reservoir lake HPPs, with minimal cost, high availability, and low dependence on a skilled workforce to operate it.

1. INTRODUCTION

Unmanned Underwater Vehicles (UUVs) are mobile robots used to perform a wide range of activities in aquatic environments and used in some military, industrial (oil exploration and related activities), and scientific areas such as marine biology.

The Remotely Operated Vehicles (ROVs) and Autonomous Underwater Vehicles (AUVs) are the two main subgroups of UUVs (Yuh, 2000). The first is characterized by receiving energy and exchanging information with the control panel on the surface via an umbilical cord. From the control panel, the operator can plan tasks or use a joystick to directly maneuver the vehicle, features which are absent in AUVs as they do not require human intervention during their missions and have no umbilical cord. The power supply is loaded onto the vehicle, as well as the central processing unit. Because they use no cable, autonomous vehicles have greater movement freedom and their use is increasing because of advances in processors and energy storage means, allowing these vehicles greater autonomy.

There are few field studies of underwater robotics in Brazil. Dominguez (1989) realized a study on modeling and developed a program to dynamically simulate underwater vehicles. Cunha (1992) proposed an adaptive control system for tracking trajectories. Hsu et al. (2000) presented a procedure to identify the dynamic model of thrusters. Barros and Soares (1991) presented a proposal for a low cost vehicle that can operate as ROV or AUV. Souza e Maruyama (2012) investigated different control techniques for dynamic positioning.

Worldwide, there is a large number of published works. There are several research areas in the general context of underwater robotics, such as modeling the interaction between fluid and structure (Ridao & Batlle, 2001), modeling of the actuators (Blanke, et al., 1991), control techniques for the vehicle (Antonelli, et al., 1991), simulation environments (Conte & Serrani, 1991), and vehicle design (Miller, 1999).

There are currently 110 hydroelectric plants in Brazil, including small, medium, and large-size installations. Conducting an assessment of the physical condition of each dam is a task that requires time, money, and skilled workers. Considering 74% of Brazil's power is generated by hydroelectric plants, guaranteeing that these reservoirs dams are in good condition ensures the efficient production of energy for the entire population. An inspection task performed by a team of divers can take about a month, depending on the size of the reservoir. With the inclusion of an autonomous underwater vehicle, this task would take considerably less time. The use of an unmanned and autonomous vehicle to maintain dams makes inspection safer and more effective.

Currently, the integrity evaluation of submerged structures is performed with the aid of piezometers, level references, surface landmarks, containment devices, and downstream flow measurements to determine the degree of landfill. Although such devices represent valuable arguments for the maintenance of submerged structures, they do not allow inferences about the existence or characteristics of fractures and the proper localization of them. This assessment procedure is carried out by divers, who face the difficulty of water turbidity, which reduces visibility, interfering with the accurate evaluation of the structure and increasing the risk of diving.

Other important elements of the study is to optimize the operation and maintenance of HEP's, such as monitoring acidity, silting, control of shellfish, fish and fauna, macrophytes, and water quality, require additional mechanisms used without association with the studies of maintenance and this affects negatively strategic decisions.

Although research in the field of underwater vehicles is large, after a thorough review of the literature, as viewed previously, few studies were found dealing with the application of the AUV for the inspection of dams.

This chapter presents a project developed at the Federal Institute of Education, Science, and Technology of Ceará (IFCE/Brazil) in partnership with the Hydroelectric Company of São Francisco (CHESF/Brazil), which aims to design, implement, and test an Autonomous Underwater Vehicle (AUV) with autonomous locomotion, to monitor dams' physical structures and capture information about the local biota. Details related to the wireless AUV communication system will be further discussed.

This chapter is organized as follows: Section 2 presents a vehicle physical structure overview. Electronic systems details, instrumentation, and hardware systems are shown in Section 3. The experiments performed and results are discussed and analyzed in Section 4. Finally, conclusions and future proposals are presented in Section 5.

2. PHYSICAL STRUCTURE

The AUV developed in this study was designed and built following the paradigms of Mechatronics. According to Borenstein (1996), a mechatronic system is not only a combination of electrical and mechanical systems and is more than just a control system; it is a complete integration of all of them. For this reason, this project has been supported by students and engineers from different areas (Me-

chanics, Robotics, Control, Telecommunications, Computer Science, and Electronics) and divided into the following steps:

A. **Mechanical Systems:** CAD /CAE (Computer-Aided Design / Computer-Aided Engineering) have been employed to design the prototype. The project took several requirements into account, including environmental conditions, the appropriate position of sensors on the platform, the control system position, speed and maximum depth desired, etc. Then the prototype was built and assembled using CAD in the first step.

B. **Electronic Systems:** Some sensors were evaluated and the most suitable were acquired. In addition, a computer was selected to run the vehicle's control and navigation programs vehicle in an autonomous way. Along with the computer, some tasks are performed by distributed systems that are micro-controlled and connected via an appropriate communication bus.

C. **Information Technology:** Simultaneously with the previous phases, strategies for autonomous navigation and trajectory control were analyzed and studied.

Composition

Figure 1 shows the AUV's physical structure developed in this work. It consists of three parts: (a) a tube at the top center, where the sensors and electronic systems responsible for processing and controlling the vehicle are located; (b) two tubes at the lower base, where the floating, submersion, and emergence systems are located; (c) two propellers, fixed above the lower tubes, responsible for the submarine's rotation and translation.

It assumes an AUV form as described in the block diagram of Figure 1. The AUV is composed of multiple peripherals, including acceleration sensors, guidance, and pressure to power the

Figure 1. Submarine physical structure

robot's navigation system, and sensors to analyze water quality. The sensors and actuators are connected to different microcontroller processing units, interconnected to each other forming a distributed architecture in which all the systems function at the same hierarchical level, making the system robust, reliable, flexible, extensible, and fault tolerant. The failure of one subsystem does not necessarily imply the global system failure. The navigation and control system is controlled by a PC.

A system of radio frequency communication was incorporated between the PC and microcontroller processing units of the ballast system and drivers, because they use different lines. The units of instrumentation that are on the same tube as the PC are linked by means of a Control Area Network (CAN). The vehicle also has an image acquisition system to inspect dam walls (see Figure 2).

Floatation

The submarine has an initial volume of approximately 70.685 dm³ on each side and 35.342 dm³ in the central body; therefore, to maintain balance, the

robot can support a mass of up to approximately 175kg, making the thrust slightly larger than the weight which makes it float.

Initially, the submarine has to be on the water's surface. In this case, the thrust is greater than the weight for it to float on the surface. This does not generate disorder, however, because it is possible to compensate the imbalance with buoys or weights.

Flotation Control

Two individually coupled pistons on the side bases control the robot's floatation, emergence, and submergence. Today, some submarines use a pod system to carry out this task. For this project, a plunger-ballast system was chosen because, in addition to saving energy to perform the tasks of immersion and emersion, it does not affect the capture of images that is made to verify the dam's structure.

According to the laws of conservation of mass, the mass of a body cannot vary, but it is possible to change its volume. This variation is what makes the submarine perform submersions and emersions.

A slight variation in the difference between mass and volume changes the body's balance, causing it to sink, balance, or come to the surface in a liquid. The entire length of each side of the bases is used to control this difference, which are called compression tubes.

Each compression tube consists of a plunger, a stepper motor, a rack, a rail, and a support base for all these components illustrated in Figure 3.

The engine makes the rack moves horizontally on the rail by pushing the plunger back and forth, according to the need of the submarine to stay submerged or afloat.

There is an opening at the end of each tube through which the water fills the tube. When the piston is at its starting point (3/4 of total length), the thrust is equal to the weight and the submarine emerges in equilibrium. When the plunger is

Figure 2. AUV block diagram

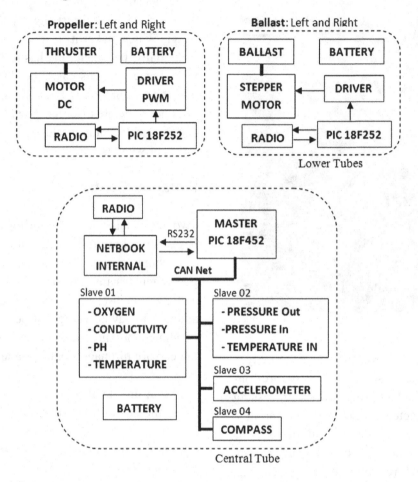

Figure 3. Lower tube and rack

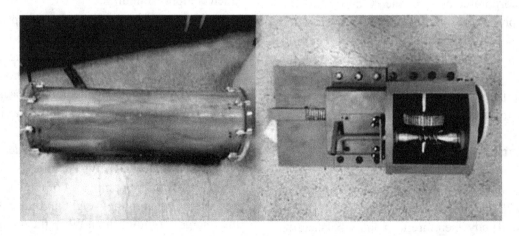

moved to occupy 2/4 of the tube, the volume will decrease, because the rest is occupied by water, and the submarine sinks. By moving along the entire length of the tube, the volume increases and the submarine comes to the surface, remembering that to keep it in balance it is necessary to move the tube to its starting point (3/4 of total length).

Within each compression tube, there is also a system of plates (control, communication, RF, and driver). The battery used will be detailed in the following sections.

Engine

This is a delicate point of the project, because the engine indicates the submarine's maximum effective working depth. A submersion depth of 10m was adopted for this work and the engine was chosen with this depth in mind.

According to the calculations, an engine with a torque of 300kg was regarded as optimal for successfully pushing the well-lubricated plunger to a depth of 10m with security.

Reduction

The method used in the reduction is the "Crown and worm" because of the submarine's size and the need for stability. Importantly, the engine also had to be in the same tube axis so that the force on the output shaft of the reduction could be applied to 90° of the input shaft.

The "crown and worm" pair is easily made and provides great power at a low cost while offering durability and high torque. Furthermore, the "crown and worm" occupies a small space when compared to other options and has the option of self-locking.

Notwithstanding its many advantages, it also has important disadvantages. In the case of the submarine, the ideal is a reduction is only 3:1, but for reductions in the range of 10:1 to 2:1 it needs

a large number of entries on the worm and a very high helix angle make its manufacture extremely complicated and expensive.

To get the mentioned advantages, only one entry was used. It was also necessary to scale the helix angle to ensure the self-locking; consequently, the cosine of the friction angle must be greater than the cosine of the helix angle and the calculations of scaling show that the reduction has a ratio of 14 (5:1) and a helix angle of 20.

The crown of this reduction could not be used to transfer motion to the rack because of its shape coupled with the worm, so in the same axis it was placed a helical gear teeth to transmit the movement to the rack, so the gear box is divided into two parts, facilitating the scaling:

- **Transfer of Power:** The "crown and worm" pair that increases the engine torque to the value specified for the project.
- **Transmission of Motion:** The "Crown and rack" pair that uses the power gained from the "Crown and the worm" to move the plunger.

Gasket

The gasket of the submarine is made from properly machined aluminum parts with brackets for screws and o-rings at the end of each 150 mm tube. The plunger gasket is comprised of two O-rings that, depending on the thickness, can stand pressures from depths of up to 100m.

To facilitate the movement, as well as increase the gasket and the lifetime of the plunger, silicone grease is used on the O-rings (better known as vacuum grease) because it is not soluble in water and prevents corrosion. In other words, a light film of a lubricant made with silicone grease was used that both sealed and acted as a permanent lubricant to reduce friction.

Propulsion

For the horizontal motion of the submarine-based thrusters, two drivers manufactured by Sea-Doo were used as illustrated in Figure 4.

These propellants are widely used by divers and are capable of moving a weight of approximately 150 kg, which, according to the scaling of the submarine, efficiently drive the vehicle.

Inside each propeller, there is a system of plates (control, communication, and driver) a nd drums.

3. EMBEDDED INSTRUMENTATION

Sensors are devices that change their behavior when subjected to an action or form of energy. Many are assembled in the form of Integrated Circuit (ICs).

They are used in many applications in the field of electronics, especially in control and regulation systems, instrumentation, processing, and signal generation, which may provide a sign that indicates this measurement. Robots that work in real, static, or dynamic environments have sensors that give them the ability to acquire information about the way they interact with the environment in which they operate and about their own internal state.

The submarine's sensory system is composed of two parts: one concerning the navigation of the vehicle and another referring to the capture of environmental data. The integration of these sensors

Figure 4. Thrusters

in an intelligent platform provides satisfactory control of the vehicle, allowing the movement of the submarine in the three spatial axes.

Navigation Sensors

Navigation is one of the most important aspects of underwater vehicles, to the extent that any use of a vehicle of this type largely depends on its ability to trace a designated path, or know where it is when there is some important event. A great challenge with navigation comes down to being able to determine, at every moment, the point in space where the vehicle is. For this purpose, we used two different sensors: acceleration and guidance.

Accelerometer

Accelerometers are inertial sensors used in inertial navigation systems to determine the accelerations of the vehicle. Additionally, an inclinometer is essential to maintain the stability and linearity of the vehicle.

The accelerometer used was a MEMS Freescale Semiconductor, model MMA7260QT shown in Figure 5. It has a triaxial inertial sensor, which has a selectable range from ±1.5g, ±2g, ±4g, or ±6g and operates within a temperature range from -20° to 85°.

This sensor has a sensitive element, able to feel acceleration, and an ASIC (Application-Specific Integrated Circuit) interface, capable of transforming the information from the sensitive element to an analog signal, allowing it to capture information. It should be noted that the detection of the signal direction is immediate in any of the three axes, which helps users determine the direction of movement, without the need for any calculations.

The use of accelerometers deployed on electronic chips to make inertial systems is achievable for two reasons: first, because it allows building a low-cost system and secondly, because the entire inertial system, using a microcontroller to process

Figure 5. Accelerometer MMA7260QT

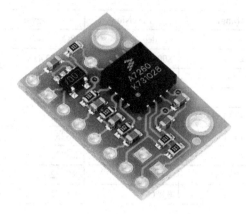

the electrical signals sent by the accelerometer, can be mounted in a single printed circuit board small.

Digital Compass

The compass provides the direction in space that the vehicle is following. Figure 6 shows the Honeywell HMC6352 electronic compass used. The HMC6352 combines two axes two magneto-resistive sensors, fully integrated, micro-processed and all the algorithms required to determine the orientation.

The module provides the steering angle, including temperature compensation algorithms and protection against magnetic parasite-fields. The signal of the compass is acquired through a I2C interface. Because the sensors used have a timely response time, it is possible to determine the acceleration and the inclination of the vehicle, and their attitude to the trajectory can be taken instantaneously.

Environmental Sensors

The sensory part for the monitoring of environmental conditions of the dam consists of: pH sensor (EPC-70 model), temperature (SC-200), level of oxygenation (SO-400), and electrical conductivity (SC-100), all of them manufactured by Instrutherm.

The sensors are located outside the structure of the submarine, attached to the frame with fixed structures. The sensor cables enter the structure through an opening, which is properly sealed to prevent water from entering the internal structure.

All data from the sensors are captured by a communication network and stored in an embedded microcomputer. Subsequently, this data will be sent to a specialist who will make a study of the environmental conditions.

The main characteristics of the sensors used are summarized in Table 1.

Internal Communication

Autonomous vehicles are typically controlled from a control system that is composed of sensor units, actuators, and a processing center. Sensors and actuators can be connected to a single processing system forming a centralized architecture or related to different processing units interconnected to each other forming a distributed architecture.

The processing systems of modern robotic systems are usually distributed, which requires a shared memory or a data communication network between the processing units to ensure adequate time constraints for message transmission. A data communication network that has this ability to communicate in real time is a CAN (Control Area Network).

Figure 6. Compass HMC-6352

Table 1. Digital compass characteristics

Sensor	Manufacturer	Range	Precision	Signal
Digital Compass	Honeywell/HMC-6532	± 20e	0,5 deg	I2C
Accelerometer	Freescale/MMA 6270QT	± 1,5g-±6g	800mV/g	0,85V-2,45V
External Pressure	NUOVA F./TPI-PRESS	0-2 bar	-	4-20mA
Internal Pressure	Novus/NP-430D	0-2 bar	0,002 bar	4-20mA
Temperature In	National/LM-35	+2°C - +150°C	0,5°C	10mV/°C
Dissolved Oxygen	Instrutherm/SO-400	0 - 20.0mg/L	±0,4mg/l	RS232
PH	Instrutherm/EPC-70	0 a 14pH	-	RS232
Conductivity	Instrutherm/SC-100	0,2 a 2,000mS	± (3% e.c.)	RS232
Water Temp.	Instrutherm/ST-200	0 a 65°C	0,8°C	RS232

The CAN network is a multi-master serial bus for distributed control systems that have the following main features: prioritized bus access, reconfiguration flexibility, high reliability in noisy environments, and a sophisticated and robust system of error detection.

Dealing specifically with the submarine, the information that travels over the CAN network includes: acceleration in three axes (x, y, and z), orientation of a digital compass, sensor data of water quality, oxygen levels, conductivity, water temperature, pH, internal temperature, and internal and external pressure. The CAN network of the vehicle is formed by 5 nodes as shown in Figure 7.

The submarine's navigation computer is in direct communication with the "master" board or "Node 0," which is responsible for receiving data from all sensors via the CAN network and sending commands to activate the propellers and weights. Nodes 1, 2, 3, and 4 perform the measurement and send the data to "Node 0" through the CAN network, as shown in Figure 7.

Figure 7. CAN network bus

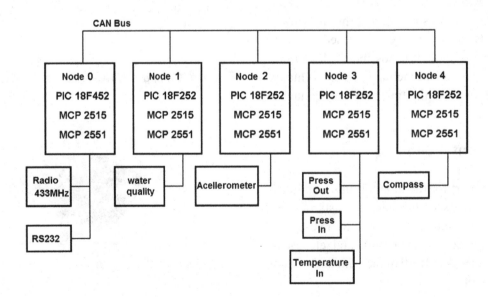

The standard CAN 2A is utilized due to the small number of messages received in the AUV (less than 20) and because of the software used and codes already provided by the manufacturer of the CAN controller, the baud rate of the network is 125Kbits/s.

This system is located in the central tube, along with the electronics for sensing and control. Figure 8 shows a real image of the plates connected to the CAN network.

External Communication

Distributed systems of the submarine must act together. That means that the elements: ballast, propellant and the CAN network must work in synchronism to move the vehicle. For this combined work, it is necessary to send messages and commands to each part of the structure of the submarine. However, as these elements are independent and placed in different locations of the vehicle, there is the need for wireless communication.

When it comes to underwater communication, there are serious difficulties, given that, in theory, radio waves do not propagate in water. However, after some testing, a radio module was found. We selected the Easy-Radio ER400TRS transceiver, which offers a limit of wave propagation in water

of up to 2 meters. This feature made it possible to integrate the CAN network, ballast, and thrusters.

According to the literature, there are several approaches to solving problems related to mobile robot communication. The ability to form ad-hoc (self-forming, self-maintaining) networks from an assorted collection of platforms, ranging from simple sensors to unmanned autonomous vehicles to manned submersibles, would allow a richly interactive environment for data collection, surveillance, data distribution, collaborative planning, and processing. Traditional existing architectures for distributed control of robotic and/or sensor systems have assumed a fixed, high bandwidth environment. This has enabled adaptive communication architectures to evolve to where they currently support powerful distributed applications such as distributed simulations, supercomputer based analysis of distributed databases, and so forth. While current network-based distributed computing techniques exist, they require high bandwidth. In this regard, these techniques are not suited to the underwater domain. The environment in which autonomous undersea systems must operate is fundamentally different due to the low bandwidth inherent to undersea communications systems. Benton (2004, 2005) and Mupparapu (2005) propose architectures for distributed control in Underwater Vehicles and ad hoc routing protocols.

Figure 8. CAN network platform

This chapter will discuss only the internal communication between the propulsion systems, ballast, and control system, based on a modular RF (Radio Frequency).

Radio Module

Description

The ER400TRS consists of a low-power RF transceiver, a microcontroller, and a voltage regulator, as shown in Figure 9. The microcontroller is responsible for the serial interface to the host system and RF transceiver functions. It also stores configuration data for the various operation modes of the transceiver module. The module has one output (Received Signal Strength Indicator – RSSI) that is used to measure the levels of the received signal.

Characteristics

The transceiver module operates at a 433MHz frequency of the Pan-European band and it is fed by a 3.6V power source, transmitting input and output data serially. A crystal is used to obtain better accuracy in the RF module synthesizer frequency. The receiver module is highly sensitive (105dBm@19.2 Kbps). Additionally, the transmitter output power uses up to 10mW, and can be programmed by the user; the operating frequency and data rate are also programmable.

Operation

Figure 10 shows how the communication takes place between two transceptor modules and how the ER400TRS is connected to the microcontroller. Communication between the module and the microcontroller is carried out with two serial control signals; the first signing when the module is busy performing some internal task and the other for receiving data stored in the buffer module (Host Ready).

The voltage levels used by the radio module levels are , while the netbook uses the standard serial RS-232. To match the voltage levels, a TTL/RS232 converter is used. Figure 11 shows the electronic integration layout.

Figure 9. Transceiver module block diagram

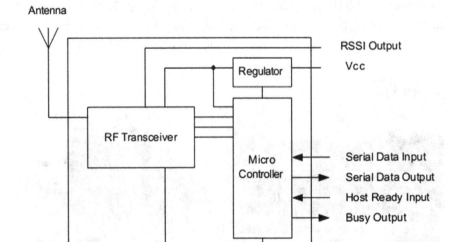

Figure 10. Connecting the module to the host

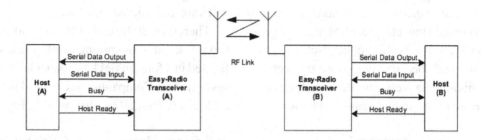

Figure 11. Connecting the RF module to the netbook

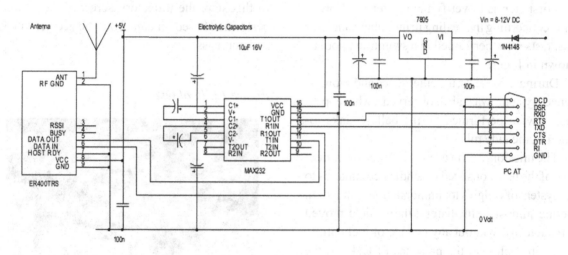

Protocol

As shown in Figure 2, underwater wireless communication occurs between the two drivers, internal ballast and netbooks, which also receives data from all sensors connected to the CAN network and performs the system control.

The data sent by the netbook to the propeller includes:

- Reference speed of the propellers;
- Constant control (Kp, Ti, and Td);
- Request the current speed;
- Request the battery voltage.

The data sent by the netbook to the ballasts includes:

- Depth reference;
- Constant control (Kp, Ti);
- Request the current position;
- Request the battery voltage.

4. EXPERIMENTAL TESTS AND DISCUSSIONS

After the physical structure and peripherals are properly shown, severe experiments will be conducted for calibration and testing of the sensory system, floatation control, navigation, and inspection, in order to evaluate performance and validate the developed platform.

Although some experimental results of navigation are presented (trajectory control and image processing), these issues will not be discussed in detail in this work because the necessary techniques for this are relatively complex and deserve a unique space to properly discuss this topic.

1st Step: Gasket, Immersion, and Emersion

The first step is to verify the existence of problems concerning the sealing in the submarine. For this, tests were performed in a swimming pool as shown in Figure 12.

During these tests, the emergence and submergence system was also observed, whose processes were performed by the ballast tanks located in the side tubes.

The mission, a term used to indicate the trajectory of the test, consisted in sending commands to the system of weights for immersion and emersion. Acting almost immediately, the vehicle moved satisfactorily, without any change of inclination. Later, in each test, the presence of leaks in the structure of the material was checked.

2nd Step: Sensors and Communication

The second step is related to data acquisition from the navigation sensors and environmental sensors through the CAN network, in addition to activities already undertaken previously. At first, these data were used to identify the dynamic model of the robot, a topic not discussed in this work.

The implemented CAN network provided a good position to evaluate performance. It was possible to see the advantages that are obtained by using a distributed architecture such as: robustness, reliability, speed, and expandability to new control points. As observed, the implementation of

a CAN bus involves concepts of hardware, which is strongly dependent on the strategy of using the software and microcontroller.

Therefore, at the end of the second step, after some tests in a swimming pool, a test was performed in a foreign field, the dam Gavião, in possession of the Companhia de Gestão dos Recursos Hídricos (COGERH), illustrated in Figure 13.

3rd Step: Navigation, Control, and Validation

In this step, the trajectory screening tests were performed based on data from the accelerometer and compass.

Figure 12. Pool test

Figure 13. Gavião dam test

The control system must not only compensate the nonlinearities of the vehicle dynamics, but also the dynamics that were not modeled, or unstructured uncertainties, as well as external disturbances. These uncertainties include river currents, the actuator system hydrodynamics, navigation, and control subsystem delays.

The goal stated in this step consists of carrying out some preliminary trajectory control tests in the horizontal plane to evaluate vehicle performance and validate the developed platform. The study of control position was conducted with the use of the linear PID control technique, due to the ease of implementation and easy tuning of controller parameters. The vehicle performed well following the planned trajectory, reaching the final desired position and orientation.

4th Step: Inspection

The inspection module uses images from the registration module and simultaneously performs a treatment on the object to highlight cracks, if any. The techniques used to detect cracks on the images are the application of adaptive threshold, mean detection, and crack by the size of the area.

The application of an adaptive threshold consists of applying a 3x3 mask on the pixel matrix to increase the points of interest. The mean detection technique consists on calculating the average value obtained from the pixels and based on this value a scan on the pixel matrix is performed and it is marked in a different color of the image to illustrate the location of the crack. The indication and decision on the existence of a crack in the image is performed by calculating the area of the crack

Figure 14. Pool inspection

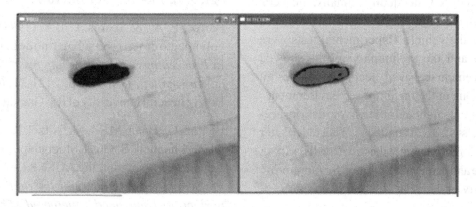

Figure 15. Dam inspection

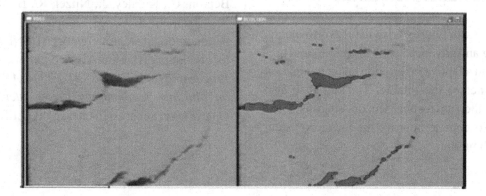

in the image, that is, having the value of the area, the interface reports whether the image has a crack. Initially, a test was performed to detect cracks in a pool as seen in Figure 14. Finally, an evaluation to detect cracks is performed in a underwater dam inspection video, shown in Figure 15.

5. CONCLUSION

This work presents a mechatronic description of an AUV designed to inspect dams. The AUV we developed is based on the synergistic integration of mechanical engineering, electronics, and control.

The mechanical design of the mobile robot was performed using CAD/CAE technologies in which the main features of the dam and its peripherals were considered. Then, some sensors were installed on the platform; some of the sensors were used by the control and navigation system and others to analyze water quality. Finally, the AUV was subjected to several tests in a real environment to validate the vehicle. Experimental tests were successful and the preliminary results showed that the vehicle behaved according to design specifications in the the proposed environment.

Combining safety and efficiency to dam inspection projects is a great challenge. The underwater robot developed integrates these two skills, with an immediate and achievable solution for monitoring the integrity of dams.

FUTURE RESEARCH DIRECTIONS

Perspectives for future work will focus on dynamic modeling and advanced control techniques (predictive and adaptive) to provide better navigation performance of the vehicle. The aim is to also include other navigation sensors such as sonar and a gyroscope to improve the location system of the submarine.

ACKNOWLEDGMENT

This work was supported by the Hydroelectric Company of San Francisco (CHESF/Brazil) and the National Council for Scientific and Technological Development (CNPq/Brazil). The authors thank the Federal Institute of Education, Science, and Technology of Ceará (IFCE/Brazil) for its support, lending us their installations for the submarine tests, and the Water Resources Management Company (COGERH) for permitting us to field test the submarine at one of their facilities.

REFERENCES

Antonelli, G., Chiaverini, S., Sarkar, N., & West, M. (1991). Adaptative control of an autonomous underwater vehicle: Experimental results on Odin. *IEEE Transactions on Control Systems Technology, 9*(5), 756–765. doi:10.1109/87.944470

Barros, E. A., & Soares, F. J. A. (1991). Desenvolvimento de um robô submarino de baixo custo. In *Proceedings of the XIV Congresso Brasileiro de Automática,* (vol. 4, pp. 2121-2126). Natal, Brazil: Federal University of Rio Grande do Norte.

Benton, C., Haag, M. M. M., Nitzel, R., Blidberg, D. R., Chappell, S. G., & Mupparapu, S. (2005). Advancements within the AUSNET protocol. In *Proceedings of the Fourteenth International Symposium on Unmanned Untethered Submersible Technology (UUST 2005)*. Durham, NH: UUST.

Benton, C., Kenney, J., Nitzel, R., Blidberg, D. R., Chappell, S. G., & Mupparapu, S. (2004). Autonomous systems network (AUSNET) – Protocols to support ad-hoc AUV communications. In *Proceedings of IEEE/OES AUV 2004: A Workshop on Multiple Autonomous Underwater Vehicle Operations*. Sebasco Estates, ME: IEEE Press.

Blanke, M., Lindegaard, K. P., & Fossen, T. I. (1991). Dynamic model for thrust generation of marine propellers. In *Proceedings of the IFAC Conference on Maneuvering and Control of Marine Craft*, (vol. 9). Aalborg, Denmark: IFAC.

Borenstein, J. H. R. (1996). *Navigation móbile robots: Systems and techniques*. New York, NY: AK Peters.

Conte, G., & Serrani, A. (1991). Modeling and simulation of underwater vehicles. In *Proceedings of the IEEE International Symposium on Computer-Aided Control System Design*, (pp. 62-67). Dearborn, Ml: IEEE Press.

Cunha, J. P. V. S. (1992). *Projeto e estudo de simulação de um sistema de controle a estrutura variável de um veículo submarino de operação remota*. (Unpublished Master Dissertation). Federal University of Rio de Janeiro. Rio de Janeiro, Brazil.

Dominguez, R. B. (1989). *Simulação e controle de um veículo submarino de operação remota*. (Unpublished Master Dissertation). Federal University of Rio de Janeiro. Rio de Janeiro, Brazil.

Hsu, L., Cunha, J. P. V. S., Lizarralde, F., & Costa, R. R. (2000). Avaliação experimental e simulação da dinâmica de um veículo submarino de operação remota. *Controle & Automação, 11*(2), 82–93.

Miller, D. P. (1996). Design of a small cheap UUV for under-ship inspection and salvage. In *Proceedings of the IEEE Symposium on Autonomous Underwater Vehicle Technology*, (pp. 18-20). Monterey, CA: IEEE Press.

Mupparapu, S. S., Bartos, R., & Haag, M. M. (2005). Performance evaluation of ad hoc protocols for underwater networks. In *Fourteenth International Symposium on Unmanned Untethered Submersible Technology (UUST 2005)*. Durham, NH: UUST.

Ridao, P., Batlle, J., & Carreras, M. (2001). *Dynamic model of an underwater robotic vehicle. Research Report*. Girona, Spain: University of Girona.

Souza, E., & Maruyama, N. (2002). An investigation of dynamic positioning strategies for unmanned underwater vehicles. In *Proceedings of the XIV Congresso Brasileiro de Automática, XIV Congresso Brasileiro de Automática*, (pp. 1273-1278). Natal, Brazil: Federal University of Rio Grande do Norte.

Yuh, J. (2000). Design and control of autonomous underwater robots: A survey. *International Journal of Autonomous Robots, 1*(3), 1–14.

KEY TERMS AND DEFINITIONS

AUV: Is a robot which travels underwater without requiring input from an operator.

Dams: Is a barrier that impounds water or underground streams.

Inspecion: Is, most generally, an organized examination or formal evaluation exercise.

Mechatronics: Is the combination of Mechanical engineering, Electronic engineering, Computer engineering, Software engineering, Control engineering, and Systems Design engineering in order to design, and manufacture useful products.

Microcontroller: Is a small computer on a single integrated circuit containing a processor core, memory, and programmable input/output peripherals.

Navigation: Is the process of monitoring and controlling the movement of a craft or vehicle from one place to another.

Thrusters: Is a small propulsive device used by spacecraft and watercraft for station keeping, attitude control, in the reaction control system, or long duration low thrust acceleration.

Chapter 11
A Virtual Simulator for the Embedded Control System Design for Navigation of Mobile Robots applied in Wheelchairs

Leonimer Flávio de Melo
State University of Londrina, Brazil

Silvia Galvão de Souza Cervantes
State University of Londrina, Brazil

João Maurício Rosário
State University of Campinas, Brazil

ABSTRACT

This chapter presents a virtual environment implementation for embedded design, simulation, and conception of supervision and control systems for mobile robots, which are capable of operating and adapting to different environments and conditions. The purpose of this virtual system is to facilitate the development of embedded architecture systems, emphasizing the implementation of tools that allow the simulation of the kinematic, dynamic, and control conditions, in real time monitoring of all important system points. To achieve this, an open control architecture is proposed, integrating the two main techniques of robotic control implementation at the hardware level: systems microprocessors and reconfigurable hardware devices. The utilization of a hierarchic and open architecture, distributing the diverse actions of control in increasing levels of complexity and the use of resources of reconfigurable computation are made in a virtual simulator for mobile robots. The validation of this environment is made in a nonholonomic mobile robot and in a wheelchair; both of them used an embedded control rapid prototyping technique for the best navigation strategy implementation. After being tested and validated in the simulator, the control system is programmed in the control board memory of the mobile robot or wheelchair. Thus, the use of time and material is optimized, first validating the entire model virtually and then operating the physical implementation of the navigation system.

DOI: 10.4018/978-1-4666-2658-4.ch011

Copyright © 2013, IGI Global. Copying or distributing in print or electronic forms without written permission of IGI Global is prohibited.

INTRODUCTION

This work's main objective is to present an implementation of a virtual environment for rapid design and simulation prototyping. The conception of supervision and control systems for mobile robots are also discussed. Mobile robot platforms are capable of operating and adapting themselves to different environments and conditions. The purpose of this virtual system is to facilitate the development of embedded architecture systems, emphasizing the implementation of tools that allow the simulation of the kinematic, dynamic, and control conditions, with real-time monitoring of all important system points. To accomplish these tasks, this work proposes an open control architecture that integrates the two main techniques of robotic control implementation at the hardware level: microprocessor systems and reconfigurable hardware devices, like a CPLD (Complex Programmable Logic Device). The utilization of a hierarchic and open architecture, distributing the diverse actions of control on increasing levels of complexity, the use of resources of reconfigurable computation, and the validation of this environment are made in a virtual simulator for mobile robots (Normey-Ricoa, et al., 1999).

Locomotion planning, under some types of restrictions, is a very vast field of research in the field of the mobile robotics (Siegwart & Nourbakhsh, 2004). The basic planning of trajectory for mobile robots implies the determination of a way in the space-C (configuration space), between an initial configuration and the final configuration, in such way that the robot does not collide with any obstacle in the environment, and that the planned movement is consistent with the kinematic restrictions of the vehicle (Graf, 2001). In this context, one of the points boarded in this work was the development of a trajectory calculator for mobile robots.

The implemented simulator system is composed of a module of a trajectory generator, a kinematic and dynamic simulator module, and an analysis module for results and errors. The simulator was implemented from the kinematic and dynamic model of mechanical drive systems of the robotic axles and can be used to simulate different control techniques in the field of mobile robotics, allowing the simulator system to deepen the concepts of navigation systems, trajectory planning, and embedded control systems. All the kinematic and dynamic results obtained during the simulation can be evaluated and visualized in graphs and table formats in the results analysis module, allowing the system to minimize errors using necessary adjustments and optimizations (Antonelli, et al., 2007).

The controller implementation in the embedded system uses a rapid prototyping technique, which allows, along with the virtual simulation environment, the development of a controller design for mobile robots. After being tested and validated in the simulator, the control system is programmed into the memory of the embedded mobile robot board. In this way, time and material are economized by first virtually validating the entire model and then operating the physical implementation of the system.

The hardware and mechanical validation and the tests were accomplished with a nonholonomic prototype of the mobile robot model with a differential transmission.

THE MOBILE ROBOT PLATFORM

This chapter presents the implementation of a virtual environment for simulation and conception of supervision and control systems for mobile robots and is focuses on the study of the mobile robot platform, with differential driving wheels mounted on the same axis and a free castor front wheel, whose prototype is used to validate the proposal system, as depicted in Figure 1.

Figure 2 illustrates the principal elements and components of the mobile robot platform.

Figure 1. Mobile robot prototyping used for testing and validations

In this study, an embedded processor, with dedicated control software was used on the mobile robot platform. Additionally, another platform, a commercial one, coupled to a communication net, is analyzed. The set of platforms, whose objective is to make use of the existing communication interfaces and to provide an alternative embedded user interface for the mobile robot, allows us to create a powerful link with the external world. The objective of this platform is to make use of the existing communication interfaces and to

provide an embedded interface alternative for mobile robot users. Another important aspect is the flexibility of the hardware design, which needs to facilitate the expansion of the mobile robot control system. To do this, new combinations of sensors should be used. Different models of supervision and control should equally be used to carry out the mobile robot tasks.

DIRECT KINEMATICS FOR DIFFERENTIAL TRACTION

It is necessary to mathematically model the mobile robot to extract the equation and algorithms that will be used in the simulator's blocks. This section is going to provide this model. The mobile robot has differential driving wheels with nonholonomic restrictions as well as a wheelchair. Both have the same restrictions and can be modeled by the same equations.

Suppose that a robot is in a position (x, y) and is "facing" along a line making an angle θ with the x axis (Figure 3). Through manipulation of the control parameters v_e and v_d, the robot can move to different positions (Dudek & Jenkin, 2000).

Figure 2. Elements and components of the mobile robot platform

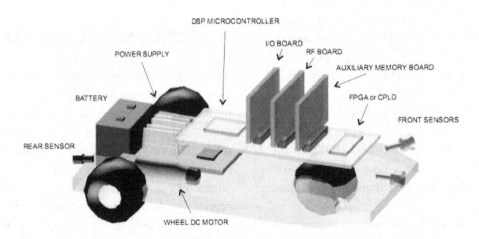

Figure 3. Forward kinematics geometry

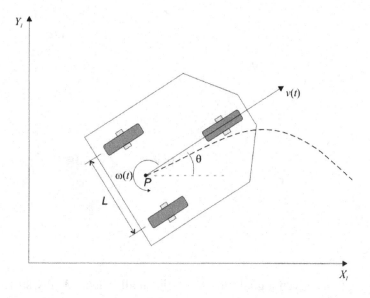

Figure 4 depicts the nonholonomic restrictions for this situation.

Determining the positions that are reachable at given control parameters is known as the forward kinematics problem for the robot. Because of v_e, v_d and hence R and θ are functions of time, it is straightforward to show that, if the robot has position (x, y, θ) at some time t, and if the left and right wheels have ground-contact velocities v_e and v_d, during the period $t{\rightarrow}t{+}dt$, the ICC (Instantaneous Center of Curvature) is given by Equation 1:

$$ICC = \left[x - R\sin\left(\theta\right), y + R\cos\left(\theta\right)\right]. \qquad (1)$$

For better comprehension of the equation, it is possible to simplify ICC = I, cos(θdt) = C and sin(θdt) = S, then, at time $t{\rightarrow}t{+}dt$, the position of the robot is given by Equation 2:

$$\begin{bmatrix} x' \\ y' \\ \theta' \end{bmatrix} = \begin{bmatrix} C & -S & 0 \\ S & C & 0 \\ 0 & 0 & 1 \end{bmatrix} \begin{bmatrix} x - I_x \\ y - I_y \\ \theta \end{bmatrix} + \begin{bmatrix} I_x \\ I_y \\ \omega dt \end{bmatrix}. \qquad (2)$$

Equation 2 describes the motion of a robot rotating a distance R about its ICC with an angular velocity given by ω.

The forward kinematics problem is solved by integrating Equation 2 from some initial condition (x_0, y_0, θ_0). It is then possible to compute where the robot will be at any given time t based on the control parameters $v_e(t)$ and $v_d(t)$. For the special case of a differential drive vehicle, it is provided by Equation 3:

$$x\left(t\right) = \frac{L}{2}\frac{v_d + v_e}{v_d - v_e}\sin\left[\frac{t}{L}\left(v_d - v_e\right)\right],$$

$$y\left(t\right) = -\frac{L}{2}\frac{v_d + v_e}{v_d - v_e}\cos\left[\frac{t}{L}\left(v_d - v_e\right)\right] + \frac{L}{2}\frac{v_d + v_e}{v_d - v_e},$$

$$\theta\left(t\right) = \frac{t}{L}\left(v_d - v_e\right), \qquad (3)$$

Figure 4. Forward kinematics for nonholonomic restrictions

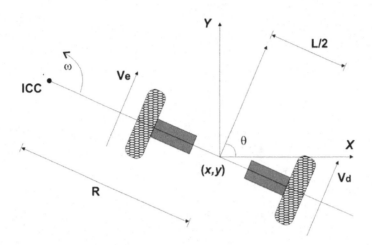

where $(x, y, \theta)_{t=0} = (0, 0, 0)$, given a goal time t and goal position (x, y). Equation 3 solves for v_e and v_d, but does not provide a solution for independent control of θ. There are, in fact, infinite solutions for v_e and v_d from the equation, but all correspond to the robot moving about the same circle that passes through $(0, 0)$ at $t = 0$ and (x, y) at $t = t+dt$; however, the robot goes around the circle different numbers of times and in different directions (Zhao & BeMent, 1992).

SUPERVISION AND CONTROL: EMBEDDED OPEN ARCHITECTURE

Hardware and software technologies have experienced significant growth in fields related to embedded system projects. This is particularly true for the evolution of engines, sensors, microcontrollers, and microprocessors, as well as communication and power interfaces. Starting with this fact, the idea is to elaborate opened structures, which can adapt easily to the development of these technologies.

In this section, we propose an embedded open control and supervisory architecture to be used in mobile robot systems. The project is described in small independently structured modules with communication interfaces inserted in an open architecture hierarchy.

Reconfigurable Systems

Reconfigurable systems are systems that can replace part of their software or hardware to better adapt to specific tasks. The objective is get high performance at a low production cost as an alternative to Von Neumman machines, implemented by systems with microprocessors. Reconfigurable software systems are quite common, with examples of embedded systems in automobiles, household appliances or even in video games. In these systems, changing a ROM memory or a CD-ROM reconfigures the functions responsible for the operation of the system (Miyazaki, 1998). Reconfigurable systems are newer and are associated with the development of FPGA (Field Programmable Gate Array) in the 90s. The advent of devices such as FPGAs has changed the point of balance between flexibility and performance. The goal of FPGA is to maintain performance goals while increasing flexibility.

Compared to reconfigurable software, reconfigurable hardware has greater potential in terms of performance and adaptability. Other names are associated with reconfigurable systems: CCM (Custom Computing Machines) or FCCM (FPGA-based Custom Computing Machines). The terms "system reconfigurable computing," "reconfigurable logic," and "reconfigurable systems" are also associated with reconfigurable hardware.

Traditionally, the execution of an algorithm in conventional computing can follow two methods: using a hardware technology, such as ASICs (Application Specific Integrated Circuits) or printed circuit boards, or by using programmable microprocessors. The first method has the advantage of high speed of execution, but is not flexible to change later. The second method, while providing high flexibility to modifications, does not run with the same speed of the algorithms implemented by the hardware (Compton, 2002). The fact that the microprocessors perform their tasks sequentially (Von Neumann machine) increases processing time that is not acceptable for many applications (Pimentel, 2000). The reconfigurable computing aims to link the gap between the software solution and hardware solution for reaching performance far superior to the performances achieved by software, but getting a greater flexibility than that obtained by a flexible hardware solution (Coric, 2002).

Otherwise, embedded systems have great technological dynamics, i.e., subject to major technological changes in a short space of time, be they by new tasks and demands for performances or by the availability of new technologies for sensors and actuators. Restricted resources in embedded systems accentuate the need for new design ideas. It is common to use blocks of functionally tested software to accelerate the design time of a system (API – Application Program Interface). Hardware functional blocks may undertake the same function. These blocks can be combined with other circuits to implement a particular algorithm. This modular design enables easy maintenance and modernization of the project (Renner, 2002). The use of IP-CORE (Intellectual Property Core) allows the integration of solutions already de-

veloped by several suppliers to minimize design time. PCI bus interfaces, communication and signal processing functions, as FFT (Fast Fourier Transform) coding and decoding digital images and even microprocessors and DSPs are examples of available IP-CORE (Ito, 2000).

The PLDs (Programmable Logic Devices), also known as EPLD (Erasable Programmable Logic Devices) or CPLD (Complex Programmable Logic Devices) and FPGA or DFGA (Dynamically Field Programmable Gate Array) are devices that execute algorithms directly in hardware. With these programmable devices, we can run algorithms exploiting the parallelism inherent in the solution of hardware, running them much faster than if the same algorithm was run by microcontrollers or sequentially by DSP, subject to the Von Neumann model. The PLDs and FPGAs have differences in application and in the internal structure. The FPGA are usually applied on systems with less cost sensitivity and greater complexity (Diniz, 2002).

Parallel to devices that allow computer reconfiguration, design tools, debugging, simulation, and testing have been developed by various manufacturers. These environments permit users to create modules developed with high-level languages of abstraction, called hardware description languages, like VHDL and AHDL (Altera, 2011). Otherwise, it is possible to implement blocks with representations of lower levels of abstraction, like schematic. The integration of blocks created with different languages permits persons to create projects in a very flexible way, allowing easy interaction with teamwork.

Open architecture systems have been studied and applied in recent years by various engineering research institutions used in machine tools engineering production among others, in order to aspects of modularity and its effect on system performance (Bhasker, 2010). The aim of a reconfigurable architecture is to allow for the easy and rapid adaptation of the mobile robot command structure to provide better portability and interchangeability of the final system. By dividing the structure in small functional blocks, with specific and dedicated interfaces, the modularization of the project becomes more effective. This allows persons to better specify the first tasks to be developed by a multidisciplinary team of researchers and subsequently the adaptation of a particular block to a new technological development. All of these characteristics cannot be found up to now in most common commercial mobile robots, which allow such manipulation. Mobile robots that consolidate knowledge in various areas of teaching and research, such as modeling, control, automation, power systems, sensors, actuators, data transmission, embedded electronic systems, and software engineering, are being increasingly used in teaching and research institutes that working in this field (Chang, 2009).

The aim of this concept applied to a reconfigurable architecture is to allow quick and easy adaptation of embedded devices to new technological developments, for better portability, exchange capacity, and final system prototyping. The division of the structure into small functional blocks and specified interfaces permits users to better specify the tasks of a multidisciplinary project-team. This type of solution cannot be achieved in many presently available commercial systems because they do not provide information or do not have resources to allow for this type of handling (Melo, 2007).

Digital Signal Processing

Digital Signal Processors (DSPs) are designed to handle most common digital processing operations such as addition, multiplication and the consecutive transfer to memory. To perform these tasks, there are instructions that precede the abovementioned instructions, making it possible to implement these often by using only one memory cycle. DSPs have a Harvard architecture (bus and memory separate program) and some can operate with floating point numbers (Hayes, 2008).

Using a DSP as the main controller of the embedded system offers the advantage of offering a much higher processing speed, which better meets the needs of project control since the embedded system has to autonomously navigate both in indoor and outdoor environments. Thus, the chosen processor must be fast enough to handle all data received by sensors through the reconfigurable hardware devices, sending commands to actuators, performing navigation algorithms, triangulating with markers, generating trajectories, etc. It must also manage all communication with the supervisory system via Wi-Fi links, so, we chose to use a processor from the Texas Instruments C6400 family (model TMS320C6416) which operates at a frequency of 1 GHz running up to 8000 million instructions per second (MIPS) at peak performance. The TMS320C6416 processor is based on the C6400 family platform designed for applications requiring high performance and intensive use of memory. Some examples of typical applications using TI's C6400 family include network software implementations, video processing images, embedded controllers, and multi-channels, among others. Implementation can be made using the Texas Instruments DSP TMDSDSK6416 development kit (TI, 2009) containing the processor and all necessary peripherals for system integration.

Embedded System Proposal

According to Alami (2009), an autonomous robot is preceded by an integrated architecture that enables the robot to plan its tasks. In such way, we present a proposal for an open architecture for mobile robotic systems illustrated in Figure 4. The different blocks are implemented into both software and hardware. The architecture, from the point of view of mobile robot systems, is organized into several independent blocks, connected through a bus data, with address and dedicated control. One Master-Block controls several slave-blocks. The blocks related to the sensors and actuators interfaces, communication, and memory are under direct control of the Master-Block.

The advantage of using the common bus is the facility to proceed if there is occasional need to expand the system. For example, when resource limitations are overcome, new blocks can be added, allowing reconfiguration of the robot for each task (Melo, 2006).

The proposed environment is a set of hardware and software modules, implemented with emphasis on the latest generation DSP processors in terms of trajectory control, reconfigurable logic, local control, and integration to support prototypes of embedded systems. Structuring of environments using DSP technologies associated with reconfigurable logic uses the best features of real-time processing that is obtained through DSP processors, along with the inherent qualities of reconfigurable hardware systems.

The reconfigurable logic used in this context is justified as follows (Melo, 2005):

- **Facility and Speed in Project Execution:** By using a series of tools provided by the market, it is possible to execute, test, simulate, and debug projects very efficiently. Dividing the project into smaller functional blocks allows a better distribution of tasks. Blocks can be previously developed leveraged into new projects, minimizing the time required to implement them.
- **Minimizing Costs:** The development of new hardware to meet requirements of a new project or the demands of an old project involves costs that can be avoided by using reconfigurable logic.
- **Easy Expansion:** New modules can be added gradually to the original system permitting new features to be implemented without requiring new design of hardware.
- **Speed of Operation:** To operate tasks in parallel, algorithms implemented with reconfigurable logic are typically faster than the algorithms implemented with conventional logic (execution of one line of code at a time by a processor).

- **Technological Domain:** Reconfigurable logic is presented as a solution for many practical engineering problems. This is justified by the effort to increase processing speed in the area of embedded electronics systems.

The use of DSP processors in the proposed system structure has the following justifications in terms of trajectory control:

- **Processing Speed:** DSP processors have a processing speed that is superior to others that are frequently used for real-time signal analysis.
- **Fast Internal Communication:** They are fast enough to handle all data received by the sensors through the reconfigurable hardware devices.
- **Processing Capacity:** DSP processors have enough processing power to perform navigation algorithms, triangulation with markers, path generation, artificial intelligence, and so on.

The objectives of the proposed environment include:

- **Flexibility:** The environment should be able to adapt easily to the desired application. A new set of requirements must be easily met, without creating new hardware projects.
- **Expandability:** The environment should be easy to expand, allowing new features to be added to the existing system.
- **Transparency:** The environment must be fully open, allowing new users complete access. This is necessary to meet the different academic needs of undergraduate and postgraduate environments it must be able to assist.
- **Task Suitability:** The environment must be reconfigurable to solve major problems with the least complexity possible. This al-

lows persons to use adequate resources to solve problems without wasting resources.

- **Budget:** The optimization of costs required to develop prototypes of embedded systems is considered an important consideration. Design tools can be used at little or no cost.
- **Command Capacity and Remote Monitoring:** In many embedded systems applications, monitoring and modification of internal system parameters can be very useful, reducing design time, and the required to meet project goals.
- **A Prototyping Framework:** The environment should be structured and be able to communicate with simulation systems (Matlab, Simulink, Labview, etc.) to quickly perform prototyping and Hardware In-the-Loop (HIL).

Control Architecture System

The control architecture system can be visualized at a logical level in the block diagram depicted in Figure 5.

The system was divided into three control levels, organized according to different control strategies (Shim & Koh, 1995). The levels can be described as:

- **Supervisory Control Level:** This consists in a high-level control. At this level, it is possible to manage the supervision of one or more mobile robots through the execution of global control strategies.
- **Local Control Level:** At this level, the control is processed by embedded software for the mobile robot, which is implemented by a 16-bit microcontroller. Local control strategies allow decision making to be done at a local level with occasional corrections from the supervisory control level. Without communication with the supervisory control level, the mobile robot

just carries out actions based on obtained sensor data and on information previously stored in its memory.

- **Interface Control Level:** This control level is restricted to strategies of control associated with the interfaces of the sensor and actuators. The strategies at this level are implemented by the hardware through PLD (Programmable Logic Devices).

The hardware architecture, from the point of view of the mobile robot, is organized into several independent blocks, connected through the local bus, which is comprised of data, addresses, and the control bus (Figure 6).

The Master-block manages several slave-blocks. Blocks associated with the sensors and

Figure 5. Different control levels of the proposed system

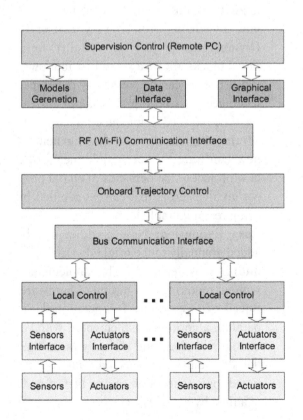

Figure 6. Hardware architecture: block diagram of the proposed system

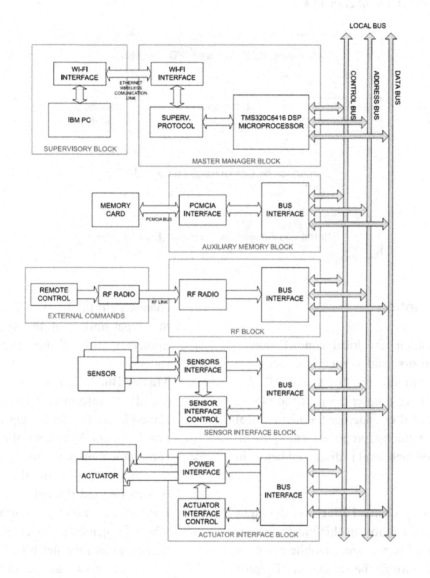

actuators interfaces, communication, and auxiliary memories were subjected to direct control of the master-block. The advantage of using a common bus was the facility to expand the system. Despite the limited resources, it was possible add new blocks, allowing an adapted configuration of the robot for each task (Melo & Rosario, 2006).

Remote Communication

One of the aspects necessary to implement the proposed architecture is remote communication or remote operation between the level of supervisor or remote control and level of embedded control. In other words, the ability to perform remote operations on embedded considered systems. This functionality can be implemented by Radio Frequency (RF) with Wi-Fi, Infrared (IR), fiber optic, or wire.

Remote operation function is a bi-directional communication between the embedded system (level of embedded control) and a remote computer (control level supervisor). This wireless data link and the RF link to the manual control of the mobile robot are illustrated in Figure 7.

Figure 7. Block diagram showing the interconnection supervisory control layer and the embedded control layer through a radio frequency link

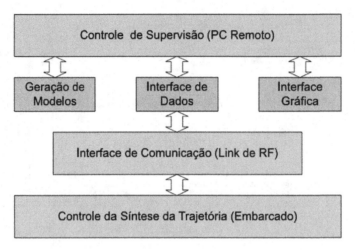

Block Description

In this application, the local control blocks, interfaces, sensors, and power interfaces are implemented through by PLDs and FPGAs. Figure 6 illustrates a description of the blocks to be implemented considering the hierarchical architecture and structure proposal for this project. The structure of each level is discussed below in more detail.

- **Supervisory Control Block:** Is the highest level of control. In this block, the supervision of one or more mobile robots is managed through the execution of global control strategies. It is implemented in an IBM PC platform and is connected with the local control level in the mobile robot through an Ethernet wireless WI-FI link. This protocol uses the IEEE 802.11a standard for wireless TCP/IP LAN communication. It guarantees up to 11 Mbps in the 2.4 GHz band and requires fewer access points to cover large areas. It also offers high-speed access to data at up to 100 meters from a base station. Fourteen available

channels (in the band of 2.4 GHz) assure the expansibility of the system and the implementation of strategies for multiple robots.

- **Master-Block:** Is responsible for processing all the information received from the slave-blocks for generating the trajectory profile for the local control blocks and for communicating with the external world. Communication with the master-block was carried out through a serial interface. For this, a commercial platform was used which implemented external communication using an Ethernet WI-FI wireless protocol. The robot was visualized as a TCP/IP LAN point in a communication net, allowing remote supervision at supervisory level. It was implemented with a Texas Instruments TMDSDSK6416 DSP board Kit that uses the TMS320C6416 DSP, a 1 GHz device that can deliver up to 8000 million instructions per second (MIPs).

- **Interface Sensor Block:** Is responsible for sensor acquisition and for processing data to be sent to the master-block. The implementation of the interface through the PLD permitted the integration of information from sensors locally, thus reducing the processing demand of the master-block.

- **Actuator Interface Block:** This block carried out speed control or position control of the motors responsible for the traction of the mobile robot. The reference signals were supplied through bus communication in the form of digital words. Derived information from the sensor was also used in the controller implemented in PLD. Because large hardware volume integration capacity PLD was appropriate to implement state machines, reducing the need for block manager processing. Besides integrating hardware resources, PLD facilitated implementation and debugging. The possibility of modifying PLD programming allowed, for example, changing control strategies of the actuators, adapting them to the required tasks.

- **Auxiliary Memory Block:** It stores sensor information and operates as a library for possible control strategies of sensors and actuators. Apart from this, it comes with an option for operation registration, allowing a register of errors. The best option for our work was a PCMCIA interface because it is easily available and it is well adapted for mobile robot applications because of its low power consumption, low weight, small dimensions, high storage capacity, and good immunity to mechanical vibrations.

- **RF Communication Block:** Allows users to establish a bi-directional radio link for data communication. It was operated in parallel with a commercial platform with a WI-FI link. The objective of these communication links was to allow the use of remote control. The remote control has a high trajectory priority, much like a supervisory control block, and can take control of the mobile robot to execute, for example, emergency or other necessary movements,

or even stop. To implement this block, we used a low-power BiM-433-40 UHF data transceiver module.

DATA FUSION

The question of how to combine data from different sources has generated a great amount of research in the academic realm. In the context of mobile robot systems, the fusion of data must be effective in at least three distinct ways: arranging measurements of different sensors, of different positions, and of different times.

Kalman Filter is a data fusion methodology used to combine position sensor data to obtain the best next possible pose estimation and correct real-localization of the robot. The next section will show how we used the Kalman Filter in our work.

Kalman Filter

To control a mobile robot, it is frequently necessary to combine information from multiple sources. The information derived from trustworthy sources must be given greater importance than information collected by less trustworthy sensors.

A general way to compute the trustworthiness of sensors and the relative weights which must be given to the data from each source is provided by the Kalman Filter (Kalman, 1960). The Kalman Filter is one of the most widely used methods used for sensorial fusing in mobile robotics applications (Leonard & Durrant-Whyte, 1991). This filter is frequently used to combine data obtained from different sensors in a statistically optimal estimate. If a system can be described with a linear model and the uncertainties of the sensors and the system can be modeled as white Gaussian noises, then the Kalman Filter gives an optimal statistical estimate for the casting data. This means that, under certain conditions, the Kalman Filter is able to find the best estimate based on corrections of each individual measure

Figure 8. Schematic for Kalman filter mobile robot localization

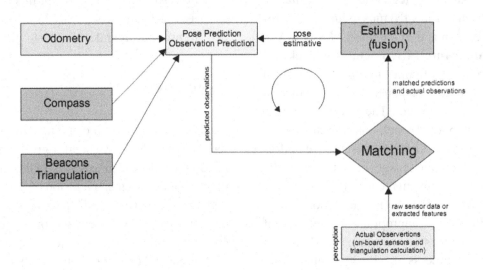

(Dudek, et al., 1996). Figure 8 depicts the particular schematic for Kalman Filter localization (Siegwart & Nourbakhsh, 2004).

If the measurement process satisfies certain properties, such as zero mean error, the Kalman Filter provides an optimal method for the fusion of data, especially when is used the method of least squares. The Kalman Filter, particularly in mobile robotics applications, is used to keep an advance estimate of the position and orientation of the vehicle, or parameters that describe objects of interest in the environment, such as another mobile robot that is traveling in the same environment. The Kalman Filter allows an estimate of the existing position of the robot, for example, which can be combined with information from the position from one or more sensors. An attribute of the Kalman Filter is it provides an advance estimate of not only as a variety of parameters, but also of relative confidence in these estimates in the form of covariance matrix. In certain circumstances, the Kalman Filter performs these updates in an optimal manner and effectively minimizes the expected error estimate.

The use of the Kalman Filter consists of the following presented stages. It is assumed, for model simplification, that the state transition matrix Ψ and the observation function Λ_E remain constant in function of time. Using the plant model and computing a system state estimate in the time $(k + 1)$ based on the known robot position at the instant of time k, the following formula shows how the system evolves in the time with the input control u(k):

$$x(k + 1) = \Psi x(k) + Yu(k). \tag{4}$$

In some practical equations the input u(k) is not used. It can also, to actualize the state certainty as expressed for the state covariance matrix P() through the displacement in the time, as:

$$P(k + 1|k) = \Psi P(k)\Psi^T + Q(k). \tag{5}$$

Equation 5 expresses the ways system state knowledge gradually decays with the passing of time in the absence of external corrections. The Kalman gain can be express as Equation 6.

$$K(k + 1) = P(k + 1) \rangle\, {}^{T}_{E}R_i^{-1}(k + 1). \tag{6}$$

However, as it does not compute P(k + 1), this can be computed by Equation 7:

$$K\left(k+1\right) = P\left(k+1\mid k\right) \rangle\, _E^T[\rangle\, _E P\left(k+1\mid k\right)\rangle\, _E^T$$
$$+\, R_i^{-1}\left(k+1\right)]^{-1}.$$

$$(7)$$

Using this matrix, an estimate of revised state can be calculated that includes the additional information obtained by the measurement. This involves the comparison of the current sensors data $z(k+1)$ with the data of the foreseen sensors using it as a state estimate. The difference between the two terms of Equation 8, or at the linear case presented by Equation 9, is known as the innovation matrix.

$$r(k+1) = z(k+1) - h_{st}(x(k+1\mid k),\ p_t)$$
$$(8)$$

$$r(k+1) = z(k+1) - \rangle\, _E x(k+1\mid k)$$
$$(9)$$

If the state estimate is perfect, the innovation must not be zero. Then, the updated state estimate is given by Equation 10:

$$x(k+1) = \ \hat{}\,x(k+1\mid k) + K(k+1)r(k+1),$$
$$(10)$$

and the up-to-date state covariance matrix is given by Equation 11:

$$P(k+1) = [I - K(k+1)\rangle\, _E]P(k+1\mid k),$$
$$(11)$$

where I is the identity matrix.

Some Experimental Results

Figure 9 shows the results of the Kalman-Filter-based method for the pose estimate applied in the robot's trajectory which has triangulation measurements marked by yellow crosses. In this experimental result, we allocated four beacons,

Figure 9. The result of KF pose estimative applied at irregular curvilinear trajectory

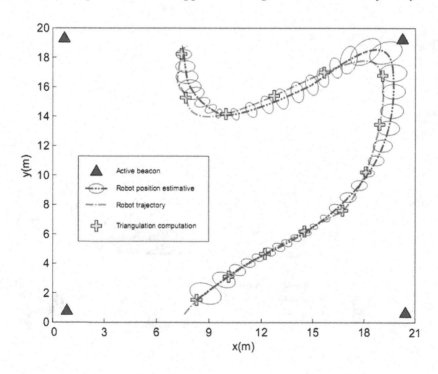

one in each corner, about two meters high. In the middle of the room, there was a camera for registering the robot's true trajectory. The ellipses delimit the area of uncertainty in the estimates. It can be observed that these ellipses are bigger in the curve's trajectory extremities, because at these points the calculus of position by triangulation suffers some loses. The average quadratic error varies depending on the chosen trajectory. It can be noticed that the pose estimative improves for more linear trajectories and with high frequency of on-board RF beacon triangulation measurements.

MOBILE ROBOT AND WHEELCHAIR SIMULATOR

The use of the system starts with collecting the main points for mobile robot trajectory generation. The idea is to use a system of photographic video cameras that catch the image of the environment where the mobile robot navigates. This initial

system must be able to identify the obstacles of the environment and to generate a matrix with some strategic points that will input data for the trajectory generation system. Figure 10 presents a general vision of the considered simulator system.

The Figure 11 illustrates an example of an environment with some obstacles where the robot must navigate. In this environment, the robot is located initially at point P1 and the objective is to reach the point P4. The trajectory generating system must be supplied by Cartesian points P1, P2, P3, and P4, which are the main points of the traced route.

This system is particularly interesting and can be used, for example, in robotic soccer games, where the navigation strategies are made from images of the environment (soccer field), and the obstacles (robot players). With this system, the best trajectory can be defined and traced, respecting always the kinematic, holonomic, or nonholonomic constraints of the robotic systems in question. The system is able to carry out all the

Figure 10. General vision of the simulator system

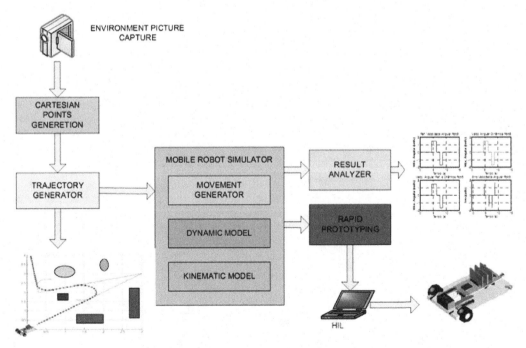

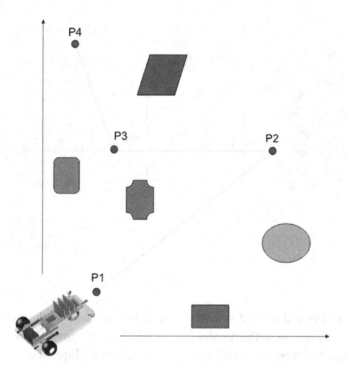

Figure 11. Example of an environment with some obstacles where the robot must navigate

simulation of the robot displacement, showing trajectory or movement imperfections and analyzing results before the final implementation of the control strategy at mobile robot.

Mobile Robot Control Structure

The tasks carried out for the mobile robots are based on the independent movement of each degree of freedom, coordinated from a trajectory plan based on a kinematic model. In the most of the cases, task programming is planned with anticipation and a map of the environment is loaded into the robot memory board. The mobile robot accomplishes the trajectory with a sequence of independent movements carried out by each axle until reaching the desired final position. From the knowledge of these articulated positions, a reference generator (speed profile) can be easily implemented based on the kinematic characteristics of each joint (Melo, 2005).

For the robot's task accomplishment and for reference signal generation to the position controller of each robotics joint of the mechatronics system in study, the establishment of mathematical model based in the kinematics of the system becomes necessary. Therefore, the control of a robot needs to transform the positioning data such as the linear speed and the bending radius into Cartesian coordinates, when one wants to realize robot control through a Cartesian referential. Figure 12 illustrates the mobile robot and the wheelchair control structure with the representative blocks of the system trajectory generation and the dynamic and kinematic models.

The trajectory generator receives reference data including the positioning vector $X_{ref} = \left[x_{ref}, y_{ref}, \theta_{ref} \right]$, the robot reference linear speed V_{ref}, and the robot instantaneous trajectory radius $R_{curv,}$ which are converted into VE_{ref} (linear speed of the left wheel) and VD_{ref} (linear speed of the right wheel). These differentiated speeds

Figure 12. The mobile robotic and wheelchair control structure

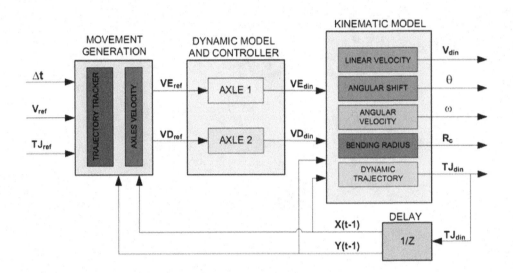

are received by the controller and in the dynamic model of the system, they are sent to the respective robot wheels through its actuators. Then they are generated by the controller the vectors VE_{din} (dynamic linear speed of the left wheel) and VD_{din} (dynamic linear speed of the right wheel). At the kinematic model's block, these data are converted into the vector final positioning of the robot $X = [x, y, \theta]$.

PID Controller

A PID controller was used to control the angular speed of DC motors and, consequently, the speed and linear movement of each wheel. If a mathematical model of the plant can be obtained, then it can be applied to several design techniques to determine the parameters of the controller that will enforce the specifications of the transitional and steady state of a closed loop system. The process of selecting the controller parameters that guarantee a given performance specification is known as controller tuning. The tuning of the PID controller was made based on the experimental step response which results in a marginal stability, when only a proportional action is used. The PID

controller tuned by the first method of Ziegler-Nichols rules provides a transference function G(s) given by Equation 12.

$$G(s) = 0,6T \frac{\left(s + \dfrac{1}{L}\right)^2}{2} \tag{12}$$

The Controller Module and Dynamic Model

The controller module and dynamic model show the components for the electrical and mechanical modeling of the drive system (in this case DC motors), and also the implementation of the controller used. The input parameters of this block are the reference linear velocity of the right wheel (V_d) and reference linear velocity of the left wheel (V_e). The outputs are the dynamic speeds of each wheel. Figure 13 illustrates the components of the controller block on the right wheel. The left wheel has an identical block. All data on the dynamics of the system are available for review. The angular velocity of the wheel, the angular velocity of the motor, the dynamic angular displacement of the wheel, the wheel dynamic linear displacement, and linear velocity of the wheel dynamics are presented. All dynamic analysis of these variables are available in the parser module results.

Figure 13 illustrates the implementation of the PID controller. Each axis of the mobile robot has a PID controller that acts independently. The drives of the two axes are DC motors whose representative blocks can be seen in Figure 14.

The kinematic model of the mechatronic system under study was implemented using S-function features within the built-in MATLAB© Simulink© blocks. Figure 15 shows the implementation of the function of the kinematic model of the mobile robotic system.

Trajectory Embedded Control

Figure 11 illustrates an example of an environment with some obstacles where the robot must navigate. In this environment, the robot is located initially at point P1, and the objective is to reach point P4. The supervisor generates initial Cartesian points and then must supply to the module of embedded trajectory generation the Cartesian

points P1, P2, P3, and P4, which are the main points of the traced route.

Figure 10 presents a general vision of the trajectory generating system. The use of the system begins with the capture of the main points to generate the mobile robot trajectory. The idea is to use a system of photographic video camera that catches the image of the environment where the mobile robot will navigate. This initial system must be capable of identifying obstacles and generating a matrix with some strategic points that will serve as input for the embedded system for trajectory generation. The embedded control system for a mobile robot initially receives the trajectory to be executed through the supervisory system. These data are loaded in the robot memory that is sent to the trajectory generation module. At a specified time, the robot starts to execute the trajectory and the dynamic data are returned to the embedded controller, which makes the comparisons and necessary trajectory corrections using the information provided by the sensors.

The trajectory embedded control system of the mobile robot is formed by three main blocks. The

Figure 13. Controller module and dynamic model of the right wheel

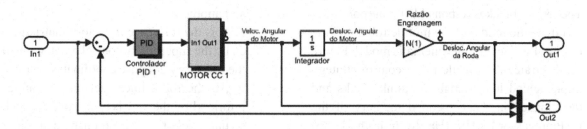

Figure 14. PID controller of the mobile robot actuators

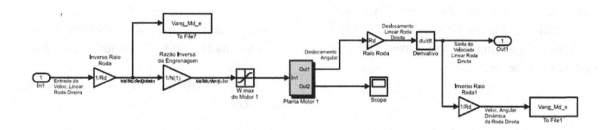

Figure 15. Implementation of the kinematic model of mobile robot

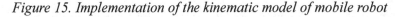

```
1       function [X,Y,V_robo,R_c,Teta,W] = trajetoria(V_d,V_e,L,t,Teta_ant,X_ant,Y_ant)
2
3       % Monta os vetores X, Y e o Teta da trajetória do robo móvel
4 -     Delta_t = .01;
5
6 -     if V_d == V_e         %trajetória reta
7 -         V_robo = V_d;
8 -         Teta = Teta_ant;
9           %X = posição do robo no eixo X
10 -        X = X_ant + V_robo * cos(Teta) * Delta_t;
11          %Y = posição do robo no eixo Y
12 -        Y = Y_ant + V_robo * sin(Teta) * Delta_t;
13 -        R_c = inf;
14 -        W=0;
15      else                  %trajetória circular
16          %W = velocidade angular do robô em rad/s
17 -        W = (V_d - V_e) / L;
18          %R_c = raio de curvatura da trajetória do robô
19 -        R_c = (V_d + V_e) / (2 * W);
20          %Teta = deslocamento angular do robo no tempo em rad
21 -        Delta_Teta = W * Delta_t;
22 -        Teta = Teta_ant + Delta_Teta;
23          %Centro de curvatura instantânea
24 -        CCIx = X_ant - R_c * sin(Teta);
25 -        CCIy = Y_ant + R_c * cos(Teta);
26          %Velociade linear do robô
27 -        V_robo = ((V_d - V_e) * R_c) / L;
28          %Posição X e Y do robô
29 -        X = (X_ant - CCIx) * cos(Delta_Teta) - (Y_ant - CCIy) * sin(Delta_Teta) + CCIx;
30 -        Y = (X_ant - CCIx) * sin(Delta_Teta) + (Y_ant - CCIy) * cos(Delta_Teta) + CCIy;
31      end
32
```

first one is responsible for generating movements. The second block is responsible for the controller and dynamic model of the mobile robot and the third is the block for the kinematic model. Figure 19 illustrates the mobile robot control strategy implemented into Matlab Simulink blocks and then loaded in the embedded memory of the DSP processor by HIL (Hardware-in-the-Loop) technique.

The embedded control for the mobile robot is implemented using a kinematic and dynamic model for axle control and the movement generator module. Figure 12 illustrates the block diagram representing those modules. The input system variables are:

- Δt is the period between one pose point and another.
- TJ_{ref} is the reference trajectory matrix given by the supervisory control block with all the trajectory dot pose coordinates (x, y, θ).
- V_{ref} is the robot linear velocity dynamics informed by the supervisory control block so that a robot can accomplish a particular trajectory.

The embedded system output variables are:

- TJ_{din}, that is the robot dynamic trajectory matrix, given in Cartesian coordinate format.
- V_{din}, is the dynamic linear velocity of the robot.

- R_c, is the mobile robot ICC.
- θ, is the orientation angle.
- ω, is the angular velocity vector.

Trajectory Generation Block

The Trajectory Generation Block receives some important points from the camera system so that the trajectory can be traced to then be realized by the mobile robot. These points form a Cartesian matrix containing greater or fewer points, depending on the complexity of the environment. For tests purpose and for system validation, the number of points to be fed in the system was fixed at four. Nevertheless, the number of points can be increased depending on the complexity of the environment where the mobile robot navigates. Important data to be used by the system are related to the holonomic constraints of the modeled mobile robot. The bending radius must be informed to the system to be used at trajectory calculus. It is also necessary to know the radius of the two curves to be executed for the robot. The information of distinct radii makes the system more flexible, enabling the trajectory to be traced with different radii bending which are dependent on the displacement direction angle and on the robot restrictions. The graphic of Figure 16 illustrates

the initial points for the trajectory generator. In this example, it was captured from the camera system and transmitted for the simulator through vector x = [.13 .3 .1] and the vector y = [.13 2 4], with the bending radius for the first and the second semicircle given by the vector r = [.4 .3]. All the measures are in meters.

Figure 16 shows that it is possible to see the tracing of the straight lines, the semicircles, and an intermediate segment of straight line, indicating the beginning and end of the trace of each circular movement executed by the mobile robot. Figure 17 shows the final tracing of the mobile robot trajectory, which is represented by red spots.

The Virtual Simulator Implementation

This section presents the main characteristics of a mobile robotic system simulator. It is implemented from the kinematic and dynamic models of the mechanical drive systems of the robotic axles to simulate different control techniques in the field of mobile robotics, allowing researchers to deepen the concepts of navigation systems, trajectory planning, and embedded control systems. This simulator, designed in a modular and open architecture, allows persons to directly apply concepts

Figure 16. Initial points given by camera system and assay to the delineated trajectory

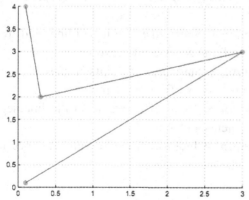
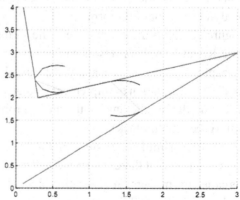

Figure 17. The end tracing of the mobile robot trajectory

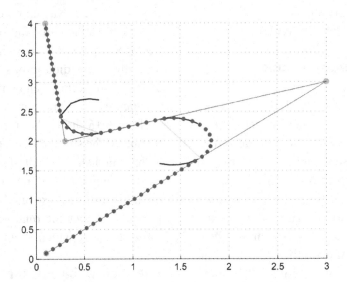

related to the area of mobile robotics, being used for validation. The main objective of this study is the model implementation of a mobile robot with nonholonomic kinematic constraints and differential drive with two degrees of freedom (movement of linear displacement and rotation), based on the actual prototype. For the simulation, constructive aspects of the mobile robot prototype were considered, including the kinematics and dynamics modeling of drive and control systems. The simulator presents the trajectory-generating module that is the first block of the system and was implemented to generate a trajectory for the mobile robot from a matrix of supplied points. Another presented block is the controller, implemented in traditional PID form.

Figure 18 shows the initial page of the Virtual Mobile Robot Simulator. Users can choose one of the captured trajectories to analyze the graphic results of the simulation or implement changes in the robot model by clicking on the mobile robot system (green main block).

The virtual simulator system of the mobile robot is composed of three main blocks. The first one is called the movement generation block. The second one is the mobile robot controller and dynamic model block and the third one is the block of the kinematic model. Figure 19 illustrates the mobile robot simulator implemented using Simulink functions.

The Motion Generation Module

The motion generation module is implemented internally by a trajectory tracking system and a reference linear speed of the two-wheeled mobile robot. The trajectory tracking system contains a reference trajectory generated by the trajectory generator block as one of the data inputs of the array. Figure 20 depicts this module.

The trajectory matrix (traj1) is composed of:

Column 1: Start and end points of lines and arcs ref. the X offset of the robot,

Column 2: Start and end points of lines and arcs ref. the Y offset of the robot,

Column 3: Radii of curvature in the line segments and arcs,

Column 4: Inclination angle with reference to the line segments.

Figure 18. The virtual simulator initial page

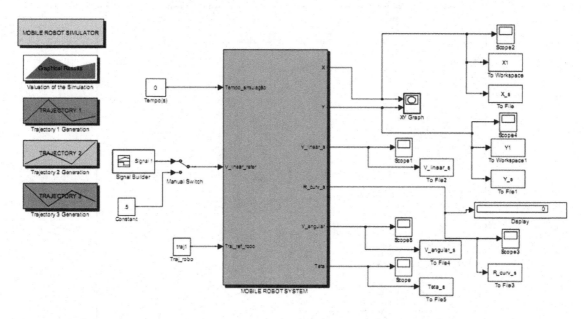

Figure 19. Mobile robot simulator implemented into Simulink

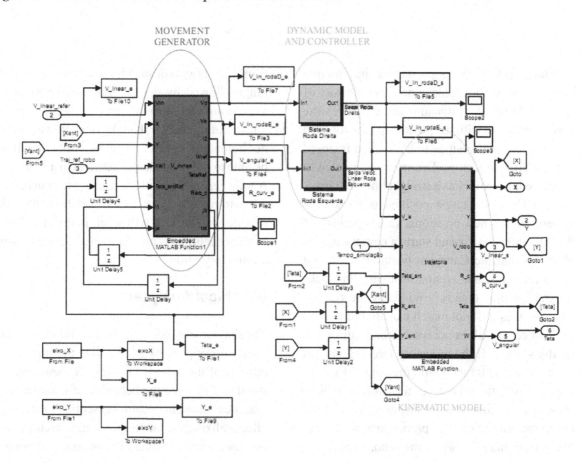

Figure 20. The motion generation module

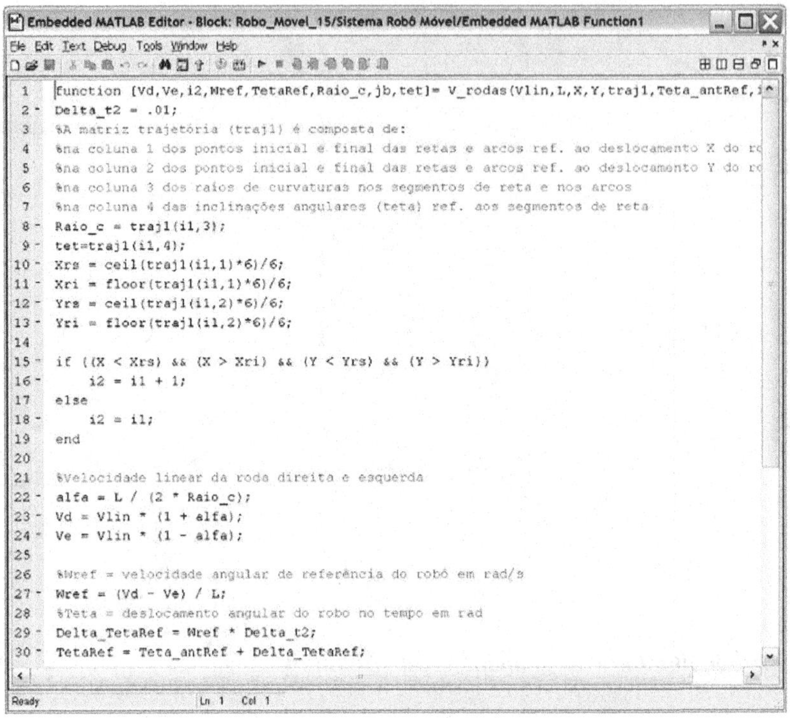

The trajectory tracking system has its principle of operation in tracing the dynamics trajectory of the mobile robot. It dynamically compares the current position with the position of the robot to be achieved. The position to be reached is stored in the matrix trajectory. This matrix is related to the current trajectory as a straight line or a curve. Once a Cartesian target position is found, the system reloads new positional values (x, y, θ) of the trajectory matrix and starts a new trace. It is observed that the Cartesian position is a target value reference and comparison data for tracking the dynamic data is provided by the system. These values will not match exactly but will be in a range of values acceptable and adjustable by the system. If the trajectory tracking system cannot find the target within the range of values programmed, the robot loses the path and will be running an infinite straight line or a circle by the end of the simulation. This particularly interesting occurrence happens when the robot tries to execute movements or speeds outside the maximum range allowed by the programmed restrictions. For example, if a person tries to execute a curve with a radius smaller than the one allowed by the nonholonomic restrictions of the robot, it will not be able to run the path and gets lost in a loop ring. Another example would be to execute a curve with a linear velocity greater than the allowed by the robotic system. This will result in linear velocity of the wheels that is greater than allowed, causing a trajectory error.

Graphical Analyzer

The simulator implemented in a Simulink environment allows persons to visualize the inputs and outputs of the system in study. To better understand and analyze the behavior of the system, a graphical analyzer becomes essential. In this way, after realizing the simulations, data archives are obtained corresponding to better study important variables (angular and Cartesian position, linear and angular speed, and control signals). These data, after treatment, makes it possible to verify important results for better analysis of the system behavior. Figure 21 illustrates a menu of the graphical analyzer of the mobile robotic system of this study with an example of generated graphics.

One kind of analysis that is made is in relation to the linear displacement of the robot in axis X and Y. In Figure 22, the dynamic behavior of the robot with regard to these parameters, as well as the presented errors is illustrated.

Another important graphic generated for the system in the Cartesian trajectory sub-menu is the graphic of the Cartesian trajectory kinematics and dynamics of the mobile robotic system on the XY plane. Figure 23 shows the dynamic tracing of reference and of the trajectory of the mobile robot.

Figure 24 illustrates the graphic of the trajectory error.

Figure 25 illustrates an example of the output graphic of the displacement angle of the robot (θ). It presents the results of the kinematic and dynamic system as the error over time. Note that the error in θ has values in a module just above of 0.05 rad in when executing curves, where the graph of θ describes a straight tilted line. Where $d\theta/dt \rightarrow 0$ the error also tends to be zero, thus stabilizing the trajectory of the robot. In this way, Figure 25 illustrates the graphic of the error of the kinematic and dynamic angular displacement of the mobile robot trajectory.

MOBILE ROBOT RAPID PROTOTYPING

This section shows the advantages of using the technique of RCP and HIL simulation in the rapid prototyping of mobile robotic systems. This approach to development of embedded controllers

Figure 21. Submenu of the mobile robot graphical analyzer with an example of generated graphic

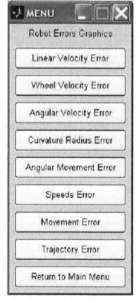

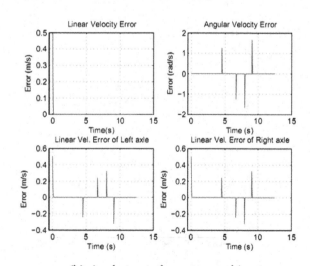

(a) Robot errors analyzer submenu.

(b) A robot speeds errors graphics.

Figure 22. Dynamic behavior graphics of the robot in the X- and Y-axis with their errors

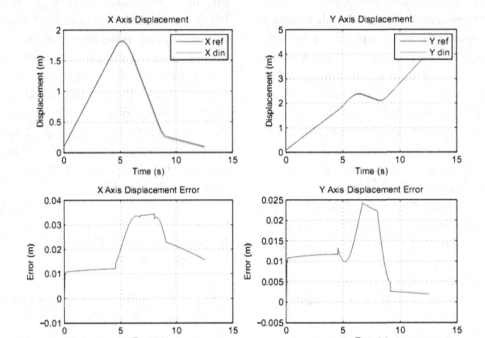

Figure 23. Cartesian trajectory kinematics and dynamics of the mobile robotic

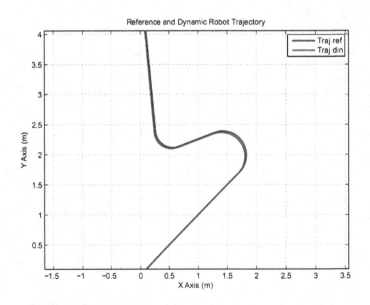

Figure 24. Mobile robot kinematics and dynamics trajectory errors

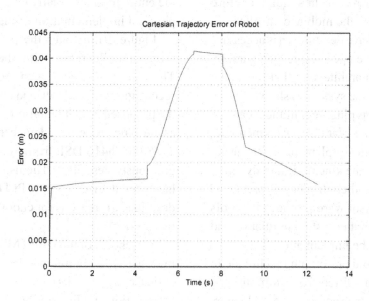

Figure 25. Error of the kinematics and dynamics angular displacement of the mobile robot trajectory

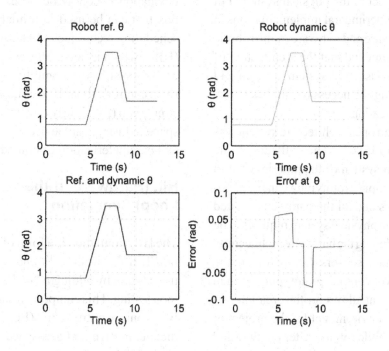

is included in this work so that the entire system can be tested and approved in the simulator before being implemented by the mobile controller.

For the system to have flexibility, it is necessary that its architecture be open and reconfigurable. In this sense, it is an architecture that meets the operational needs of the system as shown above.

The rapid prototyping techniques can be performed only if the plant has its kinematic, dynamic, and controller implemented in a virtual simulator. This time, the kinematic and dynamic model of the plant and controller are presented above. The Matlab© software provides the tools necessary for the computer system simulation and rapid prototyping to be implemented.

The use of the rapid prototyping technique in mobile robotic systems differs from the traditional target used in mechanics engineering and enters into new field of research and development for projects of mobile robot mechatronics systems (Kamiriski, et al., 2004). In this way, the rapid prototyping of these systems is associated not only with the project of the physical system, but mainly with the experimental implementations in the fields of hardware and software of the robotic system. It is fundamental that the architecture of hardware of the considered system be open and flexible in effecting the necessary modifications for system optimization.

A proposal of an open architecture system was presented in Melo (2003). The software of the embedded control system of the mobile robot in the context of the rapid prototyping can be elaborated in simulators and all the parameters tested for adjustments the physical system requires to be implemented, including the hardware architecture, the actuators, and the sensors. In this way, in the context of this work, rapid prototyping is then the methodology that allows for the creation of a virtual simulation environment for the project of a controller for mobile robots. After being tested and validated in the simulator, the control system is programmed in the control board memory of the mobile robot. In this way, an economy of time

and material are obtained sooner by validating the entire model virtually and later operating the physical implementation of the system.

Figure 2 illustrates the platform of the prototype mobile robot with the main elements. The idea is to use an open architecture system, reconfigurable, as described above. In this sense, the proposed implementation of the hardware has as the main processing device the microprocessor TMS320C6416 DSP from Texas Instruments as previously described. The use of reconfigurable hardware devices such as PLDs and FPGAs has described his reasons for choosing them in previous sections.

The choice of the DSP TMS320C6416 device manager as the main embedded system rests on two main factors. First, its large capacity for processing information and instructions, on operating clock frequency of 1 GHz, and running up to 8 billion information per second at peak performance, making the system efficient for the complex implementations required for robotics systems navigation. Secondly, because it has a specific tool box that can be used in Matlab for its real-time scheduling techniques of Hardware-in-the-Loop (HIL), which is one of the simulation techniques used in systems for rapidly prototyping mobile robot controller boards. This new technique of HIL simulation (previously only available in the aerospace industry) can be used for the development and parameterization of embedded controllers.

HIL (Hardware-in-the-Loop) Simulation

The HIL simulation is a form of real-time simulation. It differs from traditionally real-time simulations found by adding a "real" component in the loop testing. This component can be an embedded Electronic Control Unit (ECU) system used in the automotive and aerospace industries (Bona, et al., 2006).

The HIL simulation technique is used to develop and tests real time embedded systems.

HIL simulations provide a platform to add greater complexity to test platforms. The control system is enclosed in the tests and developments through its mathematical model representations and the entire respective dynamic model (Melo, 2006). Figure 26 illustrates the use of the HIL simulation technique for real-time simulation of the considered mobile robotic system.

The HIL simulation also includes emulation of electrical sensors and actuators that act as interface between the plant simulation and embedded system being tested. The value of each sensor emulated by the simulator is controlled electrically from the plant and is used by embedded system. Furthermore, the embedded system implements the control algorithms acting on the actuators system. Changes in the control signals result in variations in the variables values of simulator plant.

In many cases, the most effective way of developing embedded systems is connecting the embedded system with the real plant. In these cases, the HIL simulation is more efficient. The development and efficiency of tests is typically included in the formula that includes the following factors: cost, duration, and security. The cost approach is a measure of the cost of all tools and efforts. The duration of development and testing affects the market in time for planning or trading a certain product. The safety factor and duration are typically equated with a cost measure. Specific conditions that guarantee the use of HIL simulation include:

- **Tight Deadlines for Development:** The terms of new product development in the automotive, aerospace, and defense sectors are increasingly short and do not allow embedded systems tests to wait until the evaluation of the final prototype. In fact, most of the deadlines for development of new products assume that the HIL simulation is

Figure 26. HIL simulation for mobile robot and wheelchair system

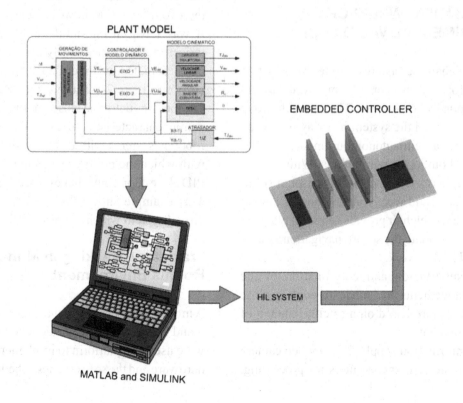

used in parallel with the misdirection of the plant. For example, while a new prototype automobile engine becomes available for testing its control system, 95% of the controller tests are being completed by HIL simulation.

- **High Cost of the Plant:** In most cases, the physical implementation of the system is much more expensive than a real-time simulator. It is therefore more economically viable to develop and test embedded systems connected to a HIL simulator than to build a complete prototype.

- **Development of Systems with Human Factors (Ergonomic):** The HIL simulation is very important in the development of ergonomic systems. It is a method that can guarantee usability and consistency of ergonomic systems. For real-time technology, test data for components that have an interface with humans interacting with it are vital.

CASE STUDY: WHEELCHAIR EXPERIMENTAL VALIDATION

The wheelchair used as an experimental validation of the proposed system has improved navigation for quadriplegic users. The research group has implemented the system in a way that it can navigate in a semiautonomous mode, in both indoor and outdoor environments. With a set of ultrasonic sensors, a laser beam sensor, a solid-state compass, and accelerometer sensors, among others, the wheelchair provides an ideal platform to test and validate the secure navigation system proposed in this work.

The supervision and coordination of the chair's movements are made through a close loop architecture based on a satellite camera of the environment.

The information supplied for a video camera is sent to one or two computers for processing.

The data obtained from this process are used to generate a sequence of instructions that are sent for the wheelchair. The chair receives the instructions and carries through the actions obeying the predetermined tasks. The instructions are the result of programmer strategies to execute the tasks and carry out the chair navigation in the environment. Figure 27 depicts this platform.

The trajectory to be executed by the robot, obtained from trajectory generation software, is illustrated in Figure 28. This trajectory allows a comparison between the real system, represented by the robot prototype, and the virtual system.

A dynamic and kinematic strategy was implemented on the onboard control of the mobile robot, in order to find best trajectory and avoid obstacles, as can be seen in Figure 11.

The graphic presented in Figure 29 illustrates the difference of the reference trajectory, shown by the slim line with square blocs (violet) and the dynamic trajectory executed by the wheelchair, shown by the thick line (blue).

Even so, the wheelchair executes the proposed trajectory with success. We can see errors between the two trajectories, however, mainly on curves. It was foreseen on simulator of the Cartesian trajectory kinematics and dynamics of the wheelchair, depicted on Figure 23 with the trajectory error on Figure 24. Those errors are predictable for the simulator because of the intrinsic dynamic characteristic of nonholonomic mobile robotics systems with differential drive wheels, with which the prototype was made, and for the PID close-loop controller of wheel velocity axles. This example shows the need for an efficient virtual system which we propose in this work.

Trajectory Tracking in Multiple Rooms Environment

A mobile robot prototype that is shown in Figures 1 and 2, and a wheelchair depicted in Figure 27 were used as a platform to implement the control hardware and the software described in this work.

Figure 27. Wheelchair platform used for experimental system validation

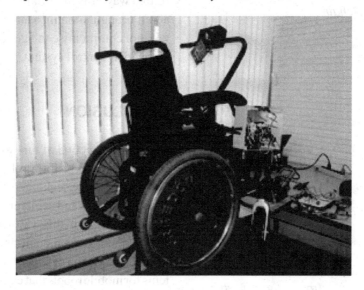

Figure 28. Best trajectory to be executed by the wheelchair

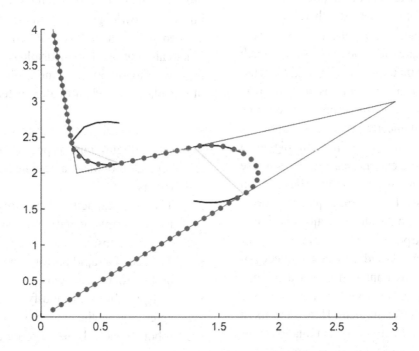

Figure 29. Reference and dynamic trajectory executed by the wheelchair

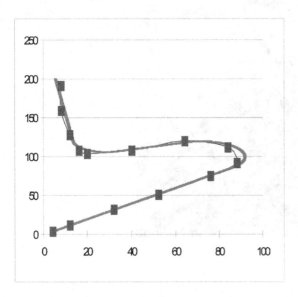

Figure 9 shows the result of EKF pose estimative applied in the robot's trajectory with triangulation measurements marked by yellow crosses. This experiment allocates four beacons, one in each corner, about two meters high. In the middle of the room, there was a camera for registering the true robot trajectory as a way to compare. The ellipses delimit the area of uncertainty in the estimates. It can be observed that these ellipses are bigger in the trajectory curve extremities because at these points the calculus of position by triangulation suffers some loss. The average quadratic error varies depending on the chosen trajectory. It can be noticed that the pose estimative is improved for more linear trajectories and with high frequency of on-board RF beacons triangulation measurements.

Figure 30 depicts the robot navigation inside environment with two ambient. Here were fixed six actives beacons for massive triangulation computation. The trouble with massive triangulation is that the on-board embedded processor keeps busy computing it and has not enough time for other important tasks like control management of wheels, supervisory communication, EKF calculation, etc.

In this way, it is important to find a middle term about triangulation measurement frequency that will not compromise the good functioning of the embedded control system.

CONCLUSION

The main objective of this work was to propose a generic platform for a robotic mobile system, like a wheelchair, seeking to obtain a support tool for autonomous and semiautonomous navigation design systems. In this way, it presents the virtual environment implementation for simulation and design conception of supervision and control systems for mobile robots that can operate and adapt in different environments and conditions. This came from encountering the growing need to propose research that integrates the knowledge acquired in several domains that stimulates teamwork in order to reach a result. Another objective was to improve knowledge in the mobile robotic area, applied with wheelchairs, aimed at presenting practical solutions for quadriplegic user's problems with locomotion. Some promising aspects of this platform and simulator system are:

- The flexibility of a great variety of possible configurations to provide solutions for several problems associated with mobile robots.
- The great capacity of memory storage allowing implementation of sailing strategies for maps.
- The use of a rapid prototyping technique in mobile robotic systems.
- The possibility of modifying control strategies during the operation of the mobile robot in special mechatronic applications, like wheelchairs.

The use of the rapid prototyping technique in mobile robot systems was validated here. Even though we are collecting the initial positive results,

Figure 30. Mobile robot and wheelchair navigation in multiple rooms environment

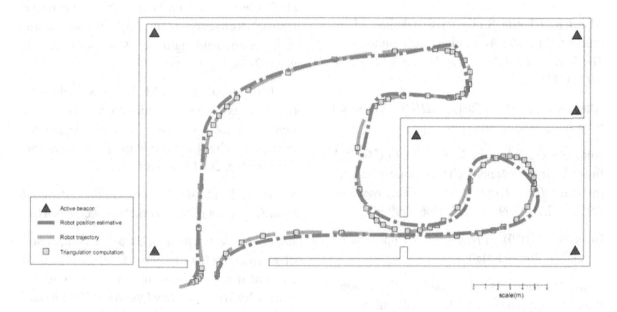

these results encourage us to keep researching and improving the system. This is a relatively low-cost implementation system that provides great results for mobile robot for indoors navigation. This is now possible because of the high performance of the embedded multicore DSP processors for mobile robots.

Sensor data obtained by an odometer, compass, and the result of Cartesian triangulation estimative, was used with the Kalman Filter to perform an optimal estimation and correct robot positioning. Once given the nonlinearity of the system in question, the use of the EKF became necessary. It does not have, therefore, in the beginning, theoretical guarantees of optimality nor of convergence. Therefore, a kinematic and dynamic model was implemented with the embedded robot control system, which allows us to verify the performance of this technique under real conditions. Among other parameters that were seen in this realistic model, we witnessed an increased error in measuring the beacon position when the distance between the robot and it increases. This effect also was introduced in the

estimate of the observation covariance matrix to allow a more coherent performance of the filter. An extremely important factor is the characterization of the covariance matrices of the present signals in the system.

The results of the simulations associated with experimental validations confirm that this technique is valid and promising so that the mobile robots, in an autonomous way, can correct their own trajectories. The consistency of the data fusing relative to the odometry and compass sensing of the mobile robot is obtained even after inserting disturbances in the system. The presented method does not make an instantaneous absolute localization, but successive measurements show that the estimative state converges to the real state of the robot or wheelchair.

REFERENCES

Alami, R., Chatila, R., Fleury, S., Ghallab, M., & Ingrand, F. (2009). An architecture for autonomy. *The International Journal of Robotics Research, 31*(11), 315–337.

Altera Corporation. (2011). *AHDL*. Retrieved from http://www.altera.com

Antonelli, G., Chiaverini, S., & Fusco, G. (2007). A fuzzy-logic-based approach for mobile robot path tracking. *IEEE Transactions on Fuzzy Systems, 15*(2). doi:10.1109/TFUZZ.2006.879998

Bhasker, J. (2010). *A VHDL primer*. Upper Saddle River, NJ: Prentice Hall.

Bona, B., Indri, M., & Smaldone, N. (2006). Rapid prototyping of a model-based control with friction compensation for a direct-drive robot. *IEEE/ASME Transactions on Mechatronics, 11*(5), 576–584. doi:10.1109/TMECH.2006.882989

Chang, K. C. (2009). *Digital design and modeling with VHDL and synthesis*. Washington, DC: IEEE Computer Society Press.

Compton, K. (2009). Reconfigurable computing: A survey of systems and software. *ACM Computing Surveys, 34*, 171–210. doi:10.1145/508352.508353

Coric, S., Leeser, M., & Miller, E. (2002). Parallel-bean backprojection: An FPGA implementation optimized for medical imaging. [Monterey, CA: FPGA.]. *Proceedings of FPGA, 2002*, 217–226.

Diniz, P. C., & Park, J. (2002). Data reorganization engines for the next generation of system-on-a-chip FPGAs. In *Proceedings of FPGA*, (pp. 237-244). Monterey, CA: FPGA.

Dudek, G., & Jenkin, M. (2000). *Computational principles of mobile robotics*. Cambridge, UK: Cambridge University Press.

Dudek, G., Jenkin, M., Milios, E., & Wilkes, D. (1996). *Reflections on modeling a sonar range sensor*. Retrieved from http://www.cs.cmu.edu/~motionplanning/papers/sbp_papers/integrated2/dudek_snr_mdl.pdf

Graf, B., Wandosell, J. M. H., & Schaeffer, C. (2001). Flexible path planning for nonholonomic mobile robots. [Stuttgart, Germany: Angenommenzur Eurobot.]. *Proceedings of Angenommenzur Eurobot, 2001*, 456–565.

Hayes, M. H. (2008). *Schaum's outline of digital signal processing*. New York, NY: McGraw-Hill.

Ito, S. A., & Carro, L. (2000). A comparison of microcontrollers targeted to {FPGA}-based embedded applications. In *Proceedings of the IEEE 13th Symposium on Integrated Circuits and Systems Design*, (pp. 397-402). Manaus, Brazil: IEEE Press.

Kalman, R. E. (1960). A new approach to linear filtering and prediction problems. *Transactions of the ASME – Journal of Basic Engineering, 82*, 35–45.

Kamiriski, B., Wejrzanowski, K., & Koczara, W. (2004). An application of PSIM simulation software for rapid prototyping of DSP based power electronics control systems. In *Proceedings of the 2004 IEEE 35th Annual Power Electronics Specialists Conference PESC 2004*, (Vol. 1, pp. 336-341). IEEE Press.

Leonard, J. J., & Durrant-Whyte, H. F. (1991). Mobile robot localization by tracking geometric beacons. *IEEE Transactions on Robotics and Automation, 7*(3), 376–382. doi:10.1109/70.88147

Melo, L. F., Lima, C. R. E., & Rosario, J. M. (2005). A reconfigurable control architecture for mobile robots. In *Proceedings of 2nd International Symposium on Mutibody and Mechatronics -MuSMe 2005*, (vol. 1, pp. 1-8). MuSMe.

Melo, L. F., & Rosario, J. M. (2006). A Proposal for ahybrid opened architecture with hardware reconfigurable control applied in mobile robots. Proceedings of IEEE International Conference on Robotic and Bionemetics - ROBIO 2006,v. 1.pp. 1101-1106, China.

Melo, L. F., & Rosario, J. M. (2007). A hybrid opened architecture with hardware reconfigurable control applied in nonholonomic mobile robots. In *Proceedings of the 19th International Congress of Mechanical Engineering - COBEM 2007*, (pp. 100-108). COBEM.

Miyazaki, T. (1998). Reconfigurable systems. In *Proceedings of the Design Automation Conference, ASP-DAC 1998*, (pp. 447-452). IEEE Press.

Normey-Ricoa, J. E., Gómez-Ortegab, J., & Camachob, E. F. (1999). A Smith-predictor-based generalized predictive controller for mobile robot path-tracking. *Control Engineering Practice, 7*(6), 729–740. doi:10.1016/S0967-0661(99)00025-8

Pimentel, J. C., & Le-Huy, H. (2000). A VHDL-based methodology to develop high performance servo drivers. In *Proceedings of the Industry Applications Conference*, (pp. 1505-1512). IEEE Press.

Renner, M., Hoffmann, K. J., Markert, K., & Glesner, M. (2002). Desing methodology of application specific integrated circuit for mechatronic systems. In *Microprocessors and Microsystems* (pp. 95–102). London, UK: Elselvier Science Ltd.

Shim, H.-S., Kim, J.-H., & Koh, K. (1995). Variable structure control of nonholonomic wheeled mobile robot. In *Proceedings of IEEE International Conference on Robotics and Automation*, (vol. 2, pp. 1694-1699). IEEE Press.

Siegwart, R., & Nourbakhsh, I. R. (2004). *Introduction to autonomous mobile robots*. Cambridge, MA: The MIT Press.

TI. (2009). *DSP*. Retrieved from http://www.dsp.ti.com

Zhao, Y., & BeMent, S. L. (1992). Kinematics, dynamics and control of wheeled mobile robots. In *Proceedings of IEEE International Conference on Robotics and Automation*, (vol. 1, pp. 91-96). IEEE Press.

ADDITIONAL READING

Bobrow, J. E. (1985). Solid modelers improve NC machine path generation techniques. *Computers in Engineering, 1*, 439–444.

Dudek, G. L. (1996). Environment representation using multiple abstraction levels. *Proceedings of the IEEE, 8*(11), 1684–1704. doi:10.1109/5.542415

Ellepola, R., & Kovesi, P. (1997). Mobile robot navigation using recursive motion control. In *Proceedings of the Second EUROMICRO Workshop on Advanced Mobile Robots*, (pp. 168-174). IEEE Press.

Evans, W. R. (1948). Graphical analysis of control systems. *Transactions of IEEE, 69*(2), 547-551.

Everett, H. R. (1995). *Sensors for mobile robots: Theory and application*. New York, NY: AK Peters.

Fierro, R., & Lewis, F. L. (1998). Control of a nonholonomic mobile robot using neural networks. *IEEE Transactions on Neural Networks, 9*(4). doi:10.1109/72.701173

Figueroa, J. F., & Barbieri, E. (1991). Increased measurement range via frequency division in ultrasonic phase detection methods. *Acoustic, 73*, 47–49.

Gorinevsky, D. M., & Formalsky, A. M. (1997). *Force control of robotics systems*. Boca Raton, FL: CRC Press.

Hsu, L., & Guenther, R. (1993). Variable structure adaptive cascade control of multi-link robot manipulators with flexible joints: The case of arbitraty uncertain flexibilities. In *Proceedings of the IEEE Conference on Robotics and Automation*, (pp. 340-345). Atlanta, GA: IEEE Press.

Melo, L. F., & Rosario, J. M. (2005). Hardware reconfigurable control with opened architecture for mobile robots. In *Proceedings of the 18th International Congress of Mechanical Engineering – COBEM 2005*, (vol. 1, pp. 1-8). COBEM.

Melo, L. F., & Rosario, J. M. (2006). A proposal for a hybrid opened architecture with hardware reconfigurable control applied in mobile robots. In *Proceedings of the IEEE International Conference on Robotic and Bionemetics - ROBIO 2006*, (Vol. 1, pp. 1101-1106). Kunming, China: IEEE Press.

Rosario, J. M., Melo, L. F., Lima, C. R. E., & Saramago, M. A. P. (2005). A reconfigurable control architecture for mobile robots. In *Proceedings of MuSMe 2005 – 2nd International Symposium on Mutibody and Mechatronics*, (vol. 1, pp. 1-8). IEEE Press.

Chapter 12
Design of a Mobile Robot to Clean the External Walls of Oil Tanks

Hernán González Acuña
*Universidad Autónoma de Bucaramanga,
Colombia*

Jairo de Jesús Montes Alvarez
*Universidad Autónoma de Bucaramanga,
Colombia*

Alfonso René Quintero Lara
*Universidad Autónoma de Bucaramanga,
Colombia*

Hernando González Acevedo
*Universidad Autónoma de Bucaramanga,
Colombia*

Ricardo Ortiz Guerrero
*Universidad Autónoma de Bucaramanga,
Colombia*

Elkin Yesid Veslin Diaz
Universidad de Boyacá, Colombia

ABSTRACT

This chapter describes a Mechatronics Design methodology applied to the design of a mobile robot to climb vertical surfaces. The first part of this chapter reviews different ways to adhere to vertical surfaces and shows some examples developed by different research groups. The second part presents the stages of Mechatronics design methodology used in the design, including mechanical design, electronics design, and control design. These stages describe the most important topics for optimally successful design. The final part provides results that were obtained in the design process and construction of the robot. Finally, the conclusions of this research work are presented.

DOI: 10.4018/978-1-4666-2658-4.ch012

Copyright © 2013, IGI Global. Copying or distributing in print or electronic forms without written permission of IGI Global is prohibited.

1. INTRODUCTION

Robotic systems have been applied to tasks where work by humans has drawbacks like the possibility of making mistakes during repetitive tasks, low levels of precision in tasks calling for rapidity, risks of exposure to hazardous settings, or the completion of work that calls for the application of force.

One specific application of robots is the maintenance of tanks in the industry of oil and oil derivatives. These tanks:

- Avoid shortages of products necessary in the refining process.
- Ensure the continuous flow of products in the refining process.
- Measure the barrels of product processed per day.
- Allow the sedimentation of water, sludge, and other elements that come with oil that are to be removed during the refining process.

Because of the harsh sun, water, wind, and salt conditions these tanks are exposed to, damage to paint often results, thus reducing usable life. Consequently, these tanks constantly require both internal and external cleaning, wall inspection, and repairs.

This chapter presents the design of a mobile robot for externally cleaning tanks of hydrocarbons and hydrocarbon derivatives by removing paint and corrosion from external tank walls by means of high pressure techniques like sandblasting, hydroblasting and hydrosandblasting.

The vortex method, one existing vacuum generation technique, allows mobile robots to maintain a solid grip by means of a constant airflow that creates a suction that in turn produces negative pressure in this closed area, resulting in the necessary force of attraction between the robots and the wall.

Traction in the four wheels enable this umbilical-cable powered mobile robot to climb walls from the ground, controlled through wireless systems like Xbee.

2. STATE OF THE ART

2.1 Adhesion Techniques

Literature contains many examples of adherence techniques, divided into 5 principal groups (Marques, et al., 2008; Longo & Muscato, 2008):

1. **Magnetic Adhesion:** Magnetic force exists between two electrically charged moving particles. Magnetic adhesion can use either coils or permanent magnets, where the latter is likely to be an easier way to obtain adhesion. Ferromagnetic materials will yield the best results. Some applications use this technique in mobile robot that climb walls on magnetic wheels that are designed for inspecting interior surfaces of gas tanks made out of thin metal sheets (Fischer, et al., 2008). Other works developed analyze permanent magnetic systems for wall-climbing robots with permanent magnetic tracks, used for rust removal in vessels (Wang, et al., 2009) (see Figure 1).

2. **Chemical Adhesion:** Chemical adhesion occurs because of a chemical reaction. Generally, chemical substances are used that generate adhesive forces between the robot and the surface. In some cases these, chemical forces originate at the molecular level. Some robots employ adhesive tape to gain grip (Daltorio, et al., 2006), while others use elastomer adhesives (Unver, et al., 2006) (see Figure 2).

3. **Electrostatic Adhesion:** Electrostatic force is generated by conductive materials, which allow electrons to form a difference in electrical charge between the robot and the

Figure 1. Robots that work with magnetic adhesion: (a) magnetic wheels (Fischer, et al., 2008); and (b) magnetic tracks (Wang, et al., 2009)

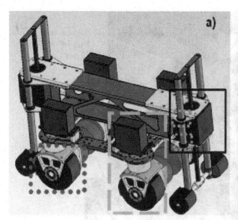 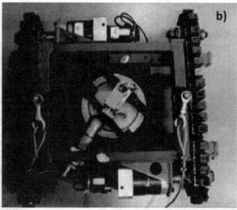

Figure 2. Robots that work with chemical adhesion: (a) (Daltorio, et al., 2006) adhesion stick; and (b) (Unver, et al., 2006) elastomer adhesives

surface. Some robots use electro-adhesion, a force controlled by inducing electrostatic charges on a wall substrate using a power supply connected to compliant pads placed on the moving robot (Prahlad, et al., 2008). Other robots developed use several design principles adapted from the gecko with a control of tangential contact forces to achieve control of adhesion (see Figure 3).

4. **Pneumatic Adhesion:** This adhesion force is generated from a pressure difference between two points. Pneumatic adhesion is obtained through techniques like vortex adhesion, vacuum adhesion and the Bernoulli Effect. Robot Alicia, described in Longo et al. (2005) is used to explore vertical walls in search of potential damage or problems in oil tanks or dams. The robot described

Figure 3. (a) Robot with electro-adhesive flaps (Prahlad, et al., 2008), adhesion stick and (b) gecko robot (Kim, et al., 2008; Unver, et al., 2006)

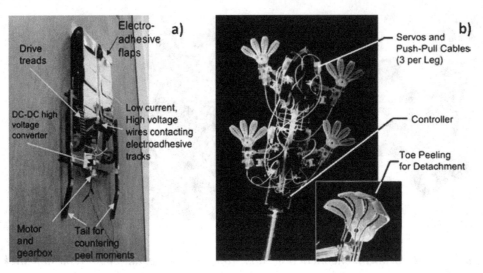

in Wagner et al. (2008) uses the Bernoulli Effect to produce adhesion, independent of wall material and surface conditions; the force/weight ratio with this technique can be as high as 5. In Zhang et al. (2007), the design of the Sky Cleaner robots, used to clean glass in buildings, is shown (see Figure 4).

5. **Mechanical Adhesion:** Robots can attach to surfaces by using mechanical elements like hinges, hooks, or mechanical cups. The adhesion system reported in Miyake et al. (2007) uses suction cups to adhere to wet surfaces or present adherence problems due to rough surfaces. A pole-climbing robot that uses claws to adhere to poles is presented in Haynes et al. (2009) (see Figure 5).

Figure 4. (a) Robot Alicia (Longo, et al., 2005); (b) robot that uses the Bernoulli effect (Wagner, et al., 2008); (c) sky cleaners robot (Zhang, et al., 2007)

 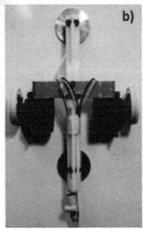 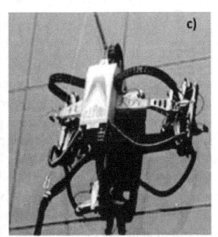

Figure 5. (a) Robot with suction cups (Miyake, et al., 2007); (b) robot RiSE V3 with claws (Haynes, et al., 2009)

2.2 Cleaning Robots

Some companies use robots industrially to clean vessel walls. Because these robots were developed by private companies, they are not offered commercially. Figure 6 presents the VB400, a robot developed by Urakami R&D, which uses a vacuum system to adhere to walls while recovering both the water used and the rust removed in the process.

Another company with important achievements is *Flow International Corporation,* developer of the "Hydrocat," shown in Figure 7, that employs adhesion and rust-collection features similar to those of the Urakami VD400. The Hydrocat's approximate weight and maximum flow are 240 kg and 5, 6 gpm, respectively.

3. METHODOLOGY

The methodology design used for the robot, divided into four stages Figure 8 is presented.

- **Problem Analysis:** This first step analyzes the problem and proposes a solution.

- **Design Proposal:** The proposal design studies different topics like adhesion and cleaning methods.

- **Design:** This stage divided into three parts: mechanical, electronic, and control. These three parts, combined, create field solutions.

- **Analysis of Results:** In this final stage analyzes the results to define if the goal was achieved. These results may call for redesign and changes.

4. VORTEX ADHESION SYSTEM

Our selection of a vortex system, a pneumatic method, was decided on after a comparison with the other methods:

- **Vortex-Magnetic:** The magnetic method uses many smaller magnets or a few or one large magnet to create the necessary invariable adhesion force.

- **Vortex-Chemical:** Its incipient development phase does not allow its use a degree of force level is called for.

Figure 6. Urakami wall-cleaning robot

Figure 7. Wall-cleaning robot Hydrocat

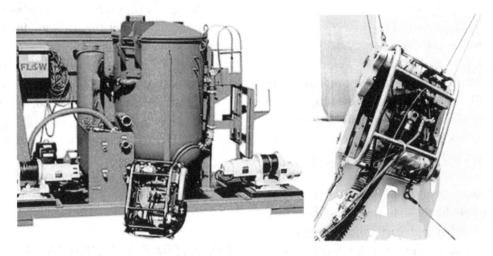

- **Vortex-Electrostatic:** Just like the chemical method, this new method is not usable when high degrees of force are called for.
- **Vortex-Mechanical:** This method uses multiple mechanical cups or hooks, but similar to the magnetic method, it does not offer the high adhesion of the pneumatic method does.
- **Vortex-Vacuum:** These two methods provide high adhesion forces, but sustained

adhesion on irregular surfaces is better reached with the vortex system because the vacuum system's poor sealing on uneven surfaces cause a loss of adhesion.

A vortex is a turbulent flow with a spiral motion around the center. The speed and rotation rate of the fluid in a vortex are greatest at the center and decrease progressively with distance from the center. Importantly, the speed of a forced

Figure 8. Methodology design for the cleaning robot

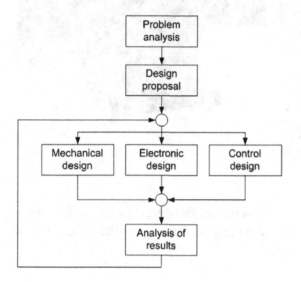

$$v_\theta = \frac{\Gamma}{2\pi r} \qquad (2)$$

A system called vortex attractor is shown in Kensington et al. (2003). This system presents the use of a centrifugal impeller into a cup to produce the air flow that creates vortex adhesion. Figure 9 shows how the air flow entering through the center of the centrifugal impeller is expelled in a radial direction. This flow of air goes to the walls of the cup, creating a continuous flow with the air expelled and the air entering.

5. MECHANICAL SYSTEM

The main objective of the mechanical system is to define the robot's physical characteristics like shape, size, and drive motors. Figure 10 shows the CAD design and configuration of the climbing robot. There are three main parts: drive motors, suction cups, and vortex mechanism. The robot has four drive motors, one for each wheel, a configuration that allows the robot to move from the floor to the wall thanks to the climbing force of the two frontal motors, aided by the thrust of the two back motors. The vortex motor starts any time the robot inclination reaches 45°, creating adhesion to the wall.

vortex is zero at the center and increases proportionally to the distance from the center. Mathematically, vorticity $\vec{w}$ is defined as the curl of the fluid velocity $\vec{u}$).

$$\vec{w} = \nabla \times \vec{u} \qquad (1)$$

The tangential velocity is given by equation Equation (2), where Γ is the circulation and r is the radial distance from the center of the vortex.

Figure 9. (a) Airflow in the impeller; (b) centrifugal impeller used in the vortex attractor

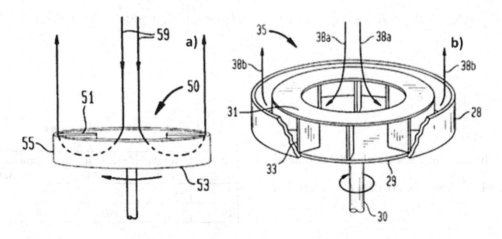

Figure 10. Components of the climbing robot

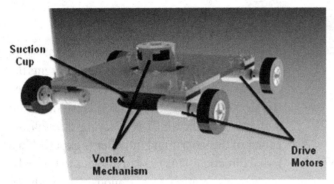 

5.1 Static Analysis

Static analysis calculates the minimum adhesion force yielded by the vortex system. This calculation is necessary for the robot to keep its balance (Li, et al., 2009). Static analysis only covers the position of the robot on the wall (Figure 11).

Nomenclature:

Ff_n = Friction force in the wheel, for n=1...4.
Fr_{sand} = Reaction force of the sand
Rw_n = Reaction force in the wheel, for n=1...4.
F_{adh} = Adhesion force.
W_R = Weight of the robot and the hoses.

Equilibrium force in axes X, Equation (3) and Y, Equation (4):

$$\sum F_x = 0$$
$$F_{adh} = R_{w1} + R_{w2} + R_{w3} + R_{w4} + Fr_{sand} \cos \theta \tag{3}$$

$$\sum F_y = 0$$
$$W_R + Fr_{sand} \sin \theta = F_{f1} + F_{f2} + F_{f3} + F_{f4} \tag{4}$$

The adhesion force is located in the middle of the length L $\left(\dfrac{L}{2}\right)$ and the position of the reaction

force for the sandblasting flow is calculated with Equation (5), measured from the same point of the F_{adh} :

$$\Delta y = c \times \tan \theta \tag{5}$$

Symmetry of the robot results in equal forces in R_{w3} and R_{w4} and successful torque equilibrium around point F, Equation (6) was calculated at the center point of the robot's width:

$$\sum T_F = 0 + \circlearrowleft$$
$$Fr_{sand} \cos \theta \times \left(\frac{L}{2} - \Delta y\right) + R_{w3,4} \times L = F_{adh} \times \frac{L}{2} + W_R \times c \tag{6}$$

Adhesion force is obtained from Equation (6) when the robot is working on the wall. Adhesion force is a function of the Reaction force of sand, that can change with the variation of the angle.

$$F_{adh} = \frac{2}{L}\left(Fr_{sand} \cos \theta \left(\frac{L}{2} - \Delta y\right) + R_{w3,4} \times L - W_R \times C\right) \tag{7}$$

Table 1. Test robots specifications

Robot	Weight (kgf)	Motor speed (rpm)	Adhesion Force (Kgf)
1	0,6	9,000	1.1
2	3	25,000	10

Figure 11. Free body diagram of the climbing robot

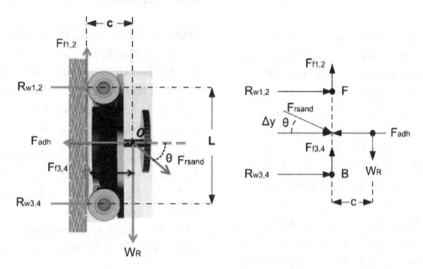

Figure 12. Climbing robot, RLT V1

Figure 13. Climbing robot, RLT V2

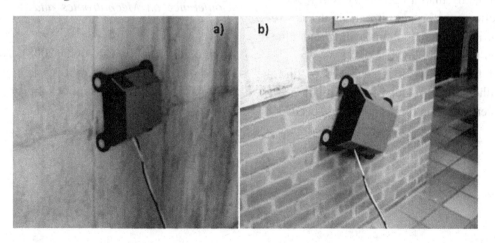

6. RESULTS

Two experimental robots were built for the test. Table 1 presents technical specifications like robot weight, motor speed along the test, and adhesion force levels proportional to speed.

Figure 12 shows the first test robot. Each picture shows the robot on a different surface type. In Figure 12(a), the robot is on a concrete wall, and in Figure 12(b) it is on a wooden door. The following picture shows the left wheels successfully tackling door surface level changes.

Figure 13 shows the second version of the climbing robot. In Figure 13(a), the robot is climbing a regular concrete wall surface, and in Figure 13(b), the robot is climbing an irregular brick surface without loss of adhesion force.

CONCLUSION

- Application of the vortex system in the cleaning robot allows a greater separation between the suction cup and the wall, which helps the system reduce the torque in the driver motors because friction is reduced when the robot is moving. Wheel skidding was also eliminated when the robot operated on an irregular surface.

- It was found that the vortex system on irregular surfaces maintains a vacuum more efficiently than a vacuum system made of suction cups because irregular surfaces result in poor suction cup sealing and adhesion loss.

- For future work, it is necessary to find a mathematical model with the relation between adhesion force and motor speed.

REFERENCES

Daltorio, K., Gorb, S., Peressadko, A., Horchler, A. D., Ritzmann, R. E., & Quinn, R. D. (2006). A robot that climbs walls using micro-structured polymer feet. In *Climbing and Walking Robots* (pp. 131–138). Berlin, Germany: Springer. doi:10.1007/3-540-26415-9_15

Fischer, W., Tâche, F., & Siegwart, R. (2008). Magnetic wall climbing robot for thin surfaces with specific obstacles. In Laugier, C., & Siegwart, R. (Eds.), *Field and Service Robotics* (pp. 551–561). Berlin, Germany: Springer-Verlag. doi:10.1007/978-3-540-75404-6_53

Haynes, G. C., Khripin, A., Lynch, G., Amory, J., Saunders, A., & Rizz, I. A., & Koditschek, D. E. (2009). Rapid pole climbing with a quadrupedal robot. In *Proceedings of the IEEE Conference on Robotics and Automation*. IEEE Press.

Kensington, L. I., & Englewood, D. R. (2003). *Vortex attractor*. United States Patent No 6.595.753. Washington, DC: US Patent Office.

Kim, S., Spenko, M., Trujillo, S., Heyneman, B., Santos, D., & Cutkosky, M. R. (2008). Smooth vertical surface climbing with directional adhesion. *IEEE Transactions on Robotics*, 24(1).

Li, J., Gao, X., Fan, N., Li, K., & Jiang, Z. (2009). BIT climber: A centrifugal impeller-based wall. In *Proceedings of the 2009 IEEE International Conference on Mechatronics and Automation*. IEEE Press.

Longo, D., & Muscato, G. (2008). Adhesion techniques for climbing robots: state of the art and experimental considerations. In *Proceedings of CLAWAR 2008*. CLAWAR.

Longo, D., Muscato, G., & Sessa, S. (2005). Simulation and locomotion control for the Alicia3 climbing robot. In *Proceedings of the 22nd International Symposium on Automation and Robotics in Construction, ISARC 2005*. ISARC.

Marques, L., Almeida, A., Tokhi, M. O., & Virk, G. S. (2008). *Advances in mobile robotics*. Singapore, Singapore: World Scientific.

Miyake, T., Ishihara, H., & Yoshimura, M. (2007). Application of wet vacuum-based adhesion system for wall climbing mechanism. In *Proceedings of Micro-Nano Mechatronics Symposium*, (pp. 532-537). IEEE Press.

Prahlad, H., Pelrine, R., Stanford, S., Marlow, J., & Kornbluh, R. (2008). Electroadhesive robots-wall climbing robots enabled by a novel, robust, and electrically controllable adhesion technology. In *Proceedings of the IEEE International Conference on Robotics and Automation*. IEEE Press.

Unver, O., Uneri, A., Aydemir, A., & Sitti, M. (2006). Geckobot: A gecko inspired climbing robot using elastomer adhesives. In *Proceedings of ICRA*, (pp. 2329–2335). IEEE Press.

Wagner, M., Chen, X., Nayyerloo, M., Wang, W., & Chasepp, J. G. (2008). A novel wall climbing robot based on Bernoulli effect. In *Proceedings of Mechatronics and Embedded Systems and Applications* (pp. 210–215). IEEE Press. doi:10.1109/MESA.2008.4735656

Wang, X., Yi, Z., Gong, Y., & Wang, Z. (2009). Optimum dynamic modeling of a wall climbing robot for ship rust removal. [Berlin, Germany: Springer-Verlag.]. *Proceedings of ICIRA, 2009*, 623–631.

Zhang, H., Zhang, J., Wei Wang, W., Liu, R., & Zong, G. (2007). A series of pneumatic glass-wall cleaning robots for high-rise buildings. *Industrial Robot: An International Journal, 34*(2), 150–160. doi:10.1108/01439910710727504

KEY TERMS AND DEFINITIONS

Cleaning: It is the process used to remove dirt, rust, etc.

Climbing: It is an activity where the goal is to ascend by walls or surfaces.

Hydroblasting: Cleaning method with a jet of water to remove elements for instance, rust or paint on surfaces.

Methodology: It is a systematic process followed to achieve some goal.

Robot: It is an electromechanical system controlled by software for perform tasks programmed.

Sandblasting: Cleaning method with a jet of sand to remove elements for instance, rust or paint on surfaces.

Storage Tanks: These are containers used like reservoirs of liquids or fluids.

Chapter 13
RobotBASIC:
Design, Simulate, and Deploy

John Blankenship
RobotBASIC, USA

Samuel Mishal
RobotBASIC, USA

ABSTRACT

Unlike most chapters in this book, this chapter does not introduce new methods or algorithms related to robotic navigation systems. Instead, it provides an overview of a simulation tool that, in some situations, can be useful for quickly evaluating the overall appropriateness of a wide variety of alternatives before focusing more advanced development activities on a chosen design. In addition, since the tool described herein is totally free, it can be used to help students and others new to robotics understand the value of utilizing a design-simulate-deploy approach to developing robotic behaviors. Robot Simulators can emulate nearly all aspects of a robot's functionality. Unfortunately, many programming environments that support simulation have steep learning curves and are difficult to use because of their ability to handle complex attributes such as 3D renderings and bearing friction. Fortunately, there are many situations where advanced attributes are unnecessary. When the primary goal is to quickly test the feasibility of a variety of algorithms for robotic behaviors, RobotBASIC provides an easy-to-use, economical alternative to more complex systems without sacrificing the features necessary to implement a complete design-simulate-deploy cycle. RobotBASIC's ability to simulate a variety of sensors makes it easy to quickly test the performance of various configurations in an assortment of environments. Once algorithm development is complete, the same programs used during the simulation phase of development can immediately control a real robot.

DOI: 10.4018/978-1-4666-2658-4.ch013

Copyright © 2013, IGI Global. Copying or distributing in print or electronic forms without written permission of IGI Global is prohibited.

INTRODUCTION

The use of simulators is an indispensable and effective strategy during an engineering design process. The process of designing behavioral-control algorithms for robotic platforms can be made more effective as well as faster and easier with the use of a simulator. Nevertheless, ultimately a real robot must be used. When the simulation language also provides access to I/O ports, then the simulated algorithms and programs can often be ported over to the physical robot without any changes or further programming. This allows the time and effort spent developing during simulation to become an effective and integral part of the whole design process.

Many computer languages and language libraries that support robot simulation are available. The most comprehensive offerings certainly include Matlab toolboxes and Microsoft's Robotic Studio and many others can be found with an Internet search. The ultimate choice of which system to use can be based on many factors including price, graphical rendering, available sensors, and ease-of-use. High-end programs can simulate nearly every aspect of a physical machine such as the rolling resistance of a wheel and 3D movement in realistic environments. Such detail is remarkable when needed, but such features can add tremendous complexity for users. If the primary goal is to quickly test the basic principles of a behavioral algorithm, then often a less complex tool can dramatically reduce the development time. A programming language that offers this uncomplicated approach is RobotBASIC.

RobotBASIC

RobotBASIC is a full-featured, powerful, yet easy-to-use programming language with an integrated robot simulator. RobotBASIC's inbuilt functions and commands enable full control over the serial and other I/O ports thus providing the ability to directly control external hardware. Additionally

RobotBASIC has a specialized protocol for seamlessly implementing wireless control over remote physical robots equipped with the appropriate firmware. Moreover, RobotBASIC may be freely used by students, universities, and researchers, even for end-user application development. It can be downloaded from www.RobotBASIC.com.

The 2D implementation of the RobotBASIC simulator is deceptively simple because the real power of the design rests in the simulator's sensory capabilities. The wide range of sensors available on the simulated robot allows it to be a realistic development tool for many applications. Some of the sensors available include contact-sensing bumpers, no-contact IR object proximity detection, distance-measuring ranging sensors, a compass, line and drop-off detectors, a beacon detector, an x-y positioning GPS, and a color detecting camera. The fact that these features are fully implemented in the language and are ready to use without any user supplied programming greatly decreases development time.

EASY-TO-USE

While complex programs and advanced algorithms can be implemented and tested with RobotBASIC, it is not the purpose of this chapter to explore such topics. Instead, we will utilize several unsophisticated examples to demonstrate ease-of use so that potential users can more easily determine if the system can be beneficial for them.

The example program in Figure 1 demonstrates how easily algorithms can be implemented and tested. In this example, the goal is to have the robot move to the object closest to it. The main module uses a subroutine to create an environment consisting of two randomly placed objects. The simulated robot is then initialized at the center of the environment.

In order to find the closest object the robot rotates through 360° and remembers the angle (obtained from the compass sensor) associated

Figure 1. This program allows the robot to find and approach the object closest to it

```
Main:
  gosub CreateEnvironment
  gosub InitializeRobot
  gosub FindClosestObject
  gosub MoveToObject
end

CreateEnvironment:
  CircleWH 100+Random(200),100+Random(400),50,50,RED,RED
  RectangleWH 500+Random(200),100+Random(400),50,50,RED,RED
return

InitializeRobot:
  rLocate 400,300 // locates the robot at the center of screen
  rInvisible Green // allows robot to leave an invisible trail
  rPen Down
return

FindClosestObject:
  distance=500
  for i=1 to 360
    if rRange()<distance
      angle=rCompass()
      distance=rRange()
    endif
    rturn 1
  next
return

MoveToObject:
  while rCompass()<> angle
    rTurn 1
  wend
  while not rBumper()
    rForward 1
  wend
return
```

with the shortest distance measured with the ranging sensor. The robot then rotates to the proper angle and moves forward until contact is made with a bumper sensor. Again, this example is trivial, but it shows how easily it is to read and utilize sensory data.

If you run the program in Figure 1, you will see an output similar to that in Figure 2. Remember, the objects will assume different random positions each time the program is run. Notice that this program has commanded the robot to leave a trail as it moves to make it easier to visualize the robot's action with a static figure.

This short program shows how easy it is to utilize the inbuilt sensors to test concepts and principles. It is also worthy to note that the BASIC language syntax allows even novice programmers to concentrate their attention on algorithm devel-

opment instead of cryptic syntax. More advanced programmers will appreciate that RobotBASIC also allows C-style syntax when writing programs. For example, while not rBumper() can be replaced by while (!rBumper()). This ability to freely mix syntaxes allows students to initially learn in a friendly environment while providing an easy transition to languages such as C and Java.

INTERFACING WITH HARDWARE

For many situations, RobotBASIC's simplicity and short learning curve makes its simulator a powerful tool for quickly developing algorithms and prototyping applications. Perhaps the real power though, is its ability to utilize the simulator programs to immediately control real-world

Figure 2. The program in Figure 1 produces this output

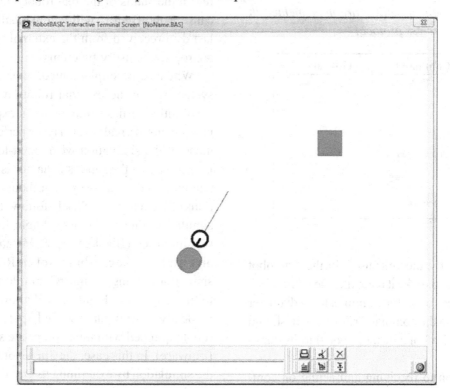

robots. This is accomplished through a unique built-in serial protocol for wireless adapters (such as Bluetooth or Zigbee) that allow programs developed with the simulator to immediately control a real robot equipped with appropriate firmware. Let us see how this works.

Normally, when you want to tell the simulated robot to move, you need a command like rForward 20 which tells the simulated robot to move forward on the screen twenty pixels (negative numbers move backward). A command such as rTurn 15 tells the robot to turn 15° to the right (negative numbers turn left).

If you want these same commands to control a real robot, you simply have to issue the command rCommPort N at the beginning of your program. This command indicates that a Bluetooth adapter (or other communication device) has been connected and is using serial port N (typically a Virtual Serial Port). Just replace N with the actual number of the port being used.

Once the rCommPort command has been issued, all robot commands (including rForward and rTurn) cease to control the simulation. Instead, they automatically send two bytes to the specified serial port. The first of these bytes is an operation code that identifies the command. The table in Figure 3 shows the codes used for some of the simulator commands. In addition to the standard commands, RobotBASIC also provides an rCommand function that allows the RobotBASIC programmer to implement custom commands needed for their particular implementation.

The second byte sent to the robot is either zero (if not needed) or a parameter related to the command. For example, when the command rForward 20 is issued, the PC will send out a 6 followed by a 20. The 6, of course, indicates a FORWARD operation is being requested and the 20 specifies how far forward to move. When a remote robot receives this example command, its firmware must activate its motors to move the robot 20 units

Figure 3. These examples of RobotBASIC commands (and their op-codes) are implemented in the internal wireless protocol

Command	Op-code
rLocate	3
rForward	6
(backward)	7
rTurn (right)	12
(left)	13
rCompass	24

forward. Ideally, the units used for the real robot are calibrated to make it react like the simulation. The simulated robot, for example, has a diameter of 40 pixels so the command rForward 40 should move the real robot a distance equal to its diameter.

In addition to receiving and acting on commands from the PC, the external robot also has the responsibility to return sensory data to Robot-BASIC. The protocol requires that the robot return five bytes of sensory data *every* time it receives a command. Three pieces of data (information from the bumper sensors, the infrared proximity sensors, and the line sensors) are generally particularly time sensitive and are nearly always returned in the first three of the five bytes. The remaining two bytes are usually zero because they are typically not needed. When commands such as rCompass, rBeacon, and rRange are executed though, these two bytes are used to return the requested information (along with the standard information in the first three bytes).

When RobotBASIC receives the five sensory bytes, it automatically extracts the individual pieces of information and uses them appropriately. For example, the rBumper() function used in Figure 1 normally provides the status of the simulated robot's bumpers. After an rCommPort command has been issued though, rBumper will

return the status of the real robot's bumpers because RobotBASIC will automatically map the last data received from the external robot to the appropriate sensory functions.

When fully implemented, this provides a system where the external robot has all of the capabilities of the simulation. As expected, the movements of a real robot may not perfectly mimic those of the simulation when open-loop control is used. When the robot's behavior is constantly altered based on sensory data though, the reactions of the real robot closely follows those of the simulation. Figure 4 shows a Pololu 3pi robot that was modified (Blankenship & Mishal, 2010) to give it most of the capabilities of the RobotBASIC simulator. Adding a single rCommPort command to the program in Figure 1 will allow it to automatically control our modified 3pi (or any other robot equipped with the appropriate sensors and firmware). In this case, the 3pi robot will mimic the simulation by examining its environment and moving to the closest object.

Figure 4. This modified 3pi robot has most of the capabilities of the RobotBASIC simulator

The RobotBASIC HELP file and website provides general information on how to implement the wireless protocol on numerous platforms such as the Arduino, Basic Stamp and the Parallax Propeller chip. The code for such implementation will be in a variety of languages native to their respective platforms (C, PBasic, and Spin). Books with specific details are also available on the website for those needing a more turnkey approach. A RobotBASIC Robot Operating System on a Chip (RROS) (Blankenship & Mishal, 2012) makes it easy to build RobotBASIC-compatible robots that can utilize a variety of standard sensors and various motors types without any low-level programming by the user. Visit www.RobotBASIC.com for more information on any of these topics.

INTERFACING USING LIBRARIES

In cases where it is not convenient or possible to alter the target robot's internal firmware, another approach can be used. Includable libraries can be easily implemented in RobotBASIC to allow both the simulation and nearly any real robot to be controlled by the same application programs. This concept has been demonstrated using a low-cost Lego NXT robot as pictured in Figure 5 and two custom libraries both of which are available as downloads from the RobotBASIC website. A LegoLibrary provides communication with and control of the Lego robot and a LegoSimulationLibrary allows the same functions to control the simulation. Applications can automatically control the simulation or the real robot based on which library is included. It is worth mentioning that no firmware of any kind was required for the Lego robot as all control is implemented in the library utilizing Lego's direct Bluetooth commands.

The internal protocol approach generally offers a more sophisticated and seamless transition from simulation to real-world control, but the library approach can sometimes be the easiest and fastest

Figure 5. This Lego robot is controlled from RobotBASIC using an includable library

method for interfacing RobotBASIC with a new robot platform.

The library approach may also be preferred for robots that have sensory types and configurations that differ dramatically from those on the RobotBASIC simulator because the library can be customized to accommodate the specific needs of the target robot. The downloadable source code for the Lego project serves as template for similar implementations on other platforms. Libraries have also been implemented for more advanced platforms such as robots built by the Robotics Group Inc that uses Wi-Fi (TCP/UDP) and wireless serial communications.

ALGORITHMIC DEVELOPMENT

Since the very same programs used to control the simulator can be used to control real-world robots, the efforts expended on algorithmic development can often be extremely productive. In order to demonstrate this principle we will examine another simple example. Remember, this example was chosen only to demonstrate how easy it is to implement and test an idea with RobotBASIC. More sophisticated robotic behaviors that utilize other sensors can certainly be explored. Examples of various behaviors are documented in our book *Robot Programmer's Bonanza* (Blankenship & Mishal, 2012).

In this example, we will consider the feasibility of a variety of options for making a robot follow the contour of an object and test one of those options with the simulator. Following the contour of an object can be very useful in obstacle-avoidance situations where the robot needs to find its way around an object blocking its path.

Assume for a moment you are faced with the task of developing a wall-following algorithm. Multiple options are open to you. You could develop a working program using multiple bumper or IR sensors, or even with one or more ranging sensors. An appropriate simulator can be the easiest and fastest way to determine which of these approaches is best able to handle the environment where your robot is expected to operate. Confusing and/or unpredictable operating environments may require more complex algorithms that involve utilizing two or even three types of sensors simultaneously in order to meet the desired performance criteria. The use of a simulator allows the evaluation of numerous potential solutions to evaluate the effectiveness of various sensor configurations in the specific environment where the robot is expected to operate. This allows the programmer to determine the best types of sensors to use, the minimum number of sensors that will be required, and even the placement of the chosen sensors, even before a physical prototype is built.

In order to demonstrate the viability of this approach, let us examine how easily a simple workable algorithm can be developed and tested using a simulator. For this example, we will only use one ranging sensor mounted on the front of the robot and angled to the left toward the wall to be followed.

The principle involved is straightforward. If the wall is too far away, the robot should turn toward the wall; when too close it should turn away. A

basic algorithm might determine the size of these turns experimentally while a more advanced algorithm might vary the turn amounts in real time based on the current nature of the environment being explored. Only a basic algorithm will be examined here.

The actual orientation of the ranging sensor itself is also an important consideration. If, for example, the sensor points 90° to the left, then the distance to the wall would increase if the robot turned either left or right. This would render the sensor's data useless in many situations.

The above discussion indicates that there are at least five parameters that have to be determined to ensure that the algorithm works properly. These are the amount to turn (both left and right), the mounting angle of the sensor itself, and the maximum and minimum distances from the wall that will be used to initiate the turns. The program in Figure 6 shows an implementation of these principles. Notice that all of the relative parameters are initialized at the beginning of the main program. This makes it easy to change these parameters and see the affect on the robot's performance.

If the program in Figure 6 is executed, it will produce the output shown in Figure 7. Since the robot has been commanded to leave a trail, it is easy to see how well the robot follows the contour of the object. Unfortunately, in this case, sometimes when an irregular obstruction is encountered, the robot collides with the wall. If we change the value of maximum to 60, the robot stays further from the wall and the algorithm is successful, as shown in Figure 8.

Alternatively, we could experimentally change the value of angle instead of changing the value of maximum. Using a value of -85 for angle, produces the output shown in Figure 9. Notice the robot avoids collision as in Figure 8, but stays closer to the wall as it did in Figure 7.

Figure 6. This program allows you to experiment with parameters affecting a wall-following algorithm

```
Main:
  minimum=30
  maximum=40
  angle = -80
  rightturn = 2
  leftturn = 2
  gosub Initialize
  gosub FollowWall
end

Initialize:
  Circle 200,200,500,350,RED,RED
  Circle 400,200,470,360,RED,RED
  Rectangle 230,180,300,300,RED,RED
  rLocate 300,380,90
  rInvisible GREEN
  rPen DOWN
return

FollowWall:
  while true
    d=rRange(angle)
    if d<minimum then rTurn rightturn
    if d>maximum then rTurn -leftturn
    rForward 1
  wend
return
```

Figure 7. With the original parameters, the robot follows the wall, but fails at certain types of obstructions

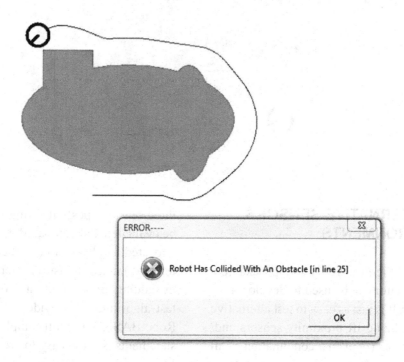

Figure 8. When the algorithm keeps the robot further from the wall, the algorithm is successful

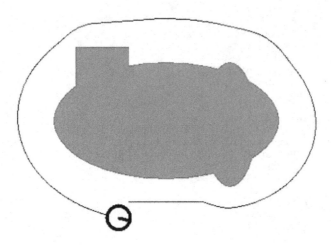

Figure 9. Altering the orientation of the ranging sensor also produces an acceptable behavior

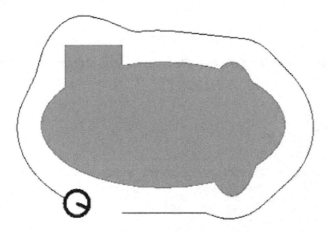

USING ALTERNATIVE SENSORS AND ENVIRONMENTS

The program in Figure 6 is only a simple example of how a simulator can be used to develop a robot's behavior. It is just as easy to test alternative approaches utilizing IR proximity sensors and/or bumper sensors (or even a combination of all three sensor types), allowing various options to be evaluated and compared before resources are allocated to physical construction. The simulator also makes it easy to alter the environment by adding protrusions and other obstacles to test algorithms in a wide variety of situations. RobotBASIC has a full complement of graphical statements, so it is easy to create anything from the simple environment used in this example to complex office or warehouse floor plans.

Remember, as long as the real robot's sensors and motors respond similarly to the simulator, then closed-loop algorithms developed for the simulator generally work with the real robot with minimal, if any, modification. Even if there are significant differences between the simulation and the real robot, an algorithm developed on the simulation is far easier to fine-tune with the real robot than developing the entire algorithm from scratch.

CONCLUSION

A simulator can dramatically reduce the time to develop robotic behaviors for both general and ad hoc situations. If there is no need to emulate physical characteristics such as friction or deal with detailed 3D environments, then RobotBASIC is a tool to consider. Its quick-learning curve and

the fact that it is both free and powerful makes it appropriate for many situations. Refer to www. RobotBASIC.com for more information about RobotBASIC.

REFERENCES

Blankenship, J., & Mishal, S. (2008). *Robot Programmer's Bonanza*. New York, NY: McGraw-Hill.

Blankenship, J., & Mishal, S. (2010). *Enhancing the Pololu 3pi with RobotBASIC*. Retrieved from http://www.amazon.com/Enhancing-Pololu-RobotBASIC-John-Blankenship/dp/1438276052/ref=sr_1_6?ie=UTF8&qid=1346469953&sr=8-6&keywords=robotbasic

Blankenship, J., & Mishal, S. (2012). *The RROS User's Manual: A RobotBASIC Robot Operating System on a Chip*. Retrieved from http://www.robotbasic.org/6.html

Chapter 14
Study and Design of an Autonomous Mobile Robot Applied to Underwater Cleaning

Lafaete Creomar Lima Junior
Federal University of Rio de Janeiro, Brazil

Armando Carlos de Pina Filho
Federal University of Rio de Janeiro, Brazil

Aloísio Carlos de Pina
Federal University of Rio de Janeiro, Brazil

ABSTRACT

The chapter describes the stages of an autonomous mobile robot project, in this case, an underwater cleaning robot. First, the authors analyze the products already available for costumers, mainly focusing on the tasks they can perform (instead of the systems they use), in order to define the requirements of their project. Then, they build some models, based in the literature available. Based on them, the authors dimension the parts and systems by evaluating the results of these models. Finally, the authors use all information gathered to create a prototype, modeled with a CAE system.

INTRODUCTION

The main objective of this chapter is to design a robot, which "does a task," specifically, cleans a pool. Therefore, we will design an autonomous robot that cleans a pool. This is the concept of domotics: the use of available technology to make some house chore or just to provide a friendly interface to control the home systems.

Thinking about autonomous mobile robots, we usually cite the vacuum cleaner, although there are many others.

Also concerning operations on the floor, besides the vacuum cleaner, some manufacturers have released floor-washing robots. Both are very similar, but the vacuum cleaner does not use water. Other autonomous mobile robots are window cleaners, which climb glass as they remove dirt. Since our work will be based on pool cleaning machines, we will briefly review their characteristics.

DOI: 10.4018/978-1-4666-2658-4.ch014

Copyright © 2013, IGI Global. Copying or distributing in print or electronic forms without written permission of IGI Global is prohibited.

Thus, we will be able to start the modeling process and later design the prototype.

BACKGROUND

Nowadays, there are many options available to the costumer, and the prices are intimately related with their complexity.

The simplest model available is human controlled and it can just clean the pool floor. This means that the owner must not only spend his money, but he has to spend his time controlling the machine. Obviously, this does not fulfill the autonomy requirements of our project. However, this is an exception because the majority of the available models are autonomous. Nevertheless, they are differentiated by their features.

The simpler of them just cleans the pool floor randomly. It just has to determine the time needed to go through the pool ensuring that almost the whole pool is cleaned. Following this philosophy, a new feature was added: cleaning the pool walls.

Lastly, there are the robots which are able to map the pool and ensure that the whole pool is cleaned after a certain time. They usually clean the floor and walls, but some of them can just clean the floor.

Based on these points, we will design our own equipment, focusing on simplicity in order to reduce costs.

MAIN FOCUS OF THE CHAPTER

Pre-Project

The first step is to determine what features the robot will offer based on others available and on the market and consumer needs.

As an essential feature, our equipment has to clean the pool floor because this characteristic is available in all pool-cleaning robots.

However, on second thought, if the owner has to clean the walls by himself, he will not be inclined to buy this equipment, so the pool wall cleaning is an important feature too. Besides, since the customer looks for convenience, the equipment must also be autonomous.

Thus, only these few requirements guide our project.

Pool floor cleaning will demand brushes and pool wall cleaning will need a floater. The robot will also need a pump and a filter. For autonomy, the robot will require some sensors and some control devices.

Floating

We can think of two ways of performing the pool wall-cleaning task, one is by using a sucker or a floating system. The use of a sucker, or an array of them, depends on the wall material, and we do not want to restrain this. Therefore, the floating feature is the more reliable way to allow pool wall cleaning, because it depends only of specific mass of water instead of the pool physical characteristics.

The floating system consists, basically, of an empty tube, which can be filled with water and submerged. A piston controls the input and the output of water. To get water inside the system, a steel cable pulls the sealing disc, reducing the chamber pressure. Since the opening is in contact with the pool water, the chamber is filled. When the cable is released, a spring pulls the seal, increasing the pressure and discharging the water into the swimming pool.

Actually, the input of water increases the equipment's specific mass, making it greater than the specific mass of the water, making the equipment submerge. When the floater expels the water, the specific mass of the equipment is less than the water's specific mass and it floats. Our calculation will assume a Newtonian fluid and the load loss is negligible. Let ρ_w be the water specific mass, ρ_e the specific mass of the equipment when the floater is empty, and ρ_f when it is full of water. V is the volume of displaced water, g is the gravity acceleration, T is the thrust, W is the weight, M is the mass of the equipment without the water of the floater and m is the floater water mass. We can see the forces acting in the equipment in Figure 1.

So, when the equipment is leaving the pool floor, the normal force tends to zero. Therefore, the wheels cause no force and there are no flow forces:

T-W<0

But:

$$W = \rho_f V g$$

$$T = \rho_w V g$$

$$\rho_w - \rho_f < 0$$

$$\rho_w < \frac{M + m}{V}$$

Figure 1. Control volume and free body diagram

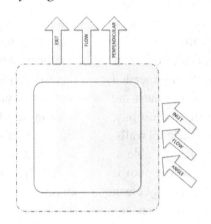

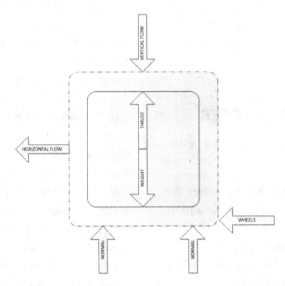

Analogously, when floating:

$$\rho_w > \frac{M}{V}$$

Thus:

$$m > \rho_w V - M$$

This is the minimum mass of water necessary to make the equipment submerge and we have to keep in mind that the specific mass of the empty equipment should be less than the water's specific mass, thus allowing flotation.

The wall cleaning will be performed when the robot is rising. But, since a precise control is very difficult to achieve, the cleaning process has to be done in a relatively short time. Therefore, it is necessary that the specific mass difference be small. This will ensure low acceleration and an increased time to operate (see Figure 2).

At this point, another difficulty appears: we have to ensure the contact between the brushes and the wall. To solve this, we will use a pump and some pipe. Consequently, we must simply orientate them in the correct direction.

To do this, we will need to create a control volume around the equipment as shown in Figure 1. We will then model a continuous flow caused by the pump. We know that the intake mass should leave the equipment because there is no water storage in this flow. Thus, according to Fox and McDonald (1981), the mass conservation:

$$\frac{dm}{dt} = \frac{\partial}{\partial t} \int_\forall \rho_w d\vec{\forall} + \int_S \rho_w \vec{V} \cdot d\vec{A}$$

Becomes:

$$0 = -\rho_w V_i A_i + \rho_w V_o A_o$$

$$V_i A_i = V_o A_o$$

Analyzing momentum conservation:

$$F = \frac{d\vec{P}}{dt} = \frac{\partial}{\partial t} \int_\forall \rho_w \vec{V} \cdot d\vec{\forall} + \int_S \vec{V} \rho_w \vec{V} \cdot d\vec{A}$$

$$F = [V_i(\cos(\theta)\hat{\imath} + \sin(\theta)\hat{\jmath})][\rho_w V_i(\cos(\theta)\hat{\imath} + \sin(\theta)\hat{\jmath})$$
$$\bullet A_i(-\cos(\theta)\hat{\imath} - \sin(\theta)\hat{\jmath})] + V_o\hat{\imath}\rho_w(V_o\hat{\imath} \bullet A_o\hat{\imath})$$

$$F = \rho_w[(V_o^2 A_o - V_i^2 A_i \cos(\theta))\hat{\imath} - V_i^2 A_i \sin(\theta)\hat{\jmath}]$$

Using mass conservation:

$$F = \frac{\rho_w V_o^2 A_o}{A_i}[(A_i - A_o \cos(\theta))\hat{\imath} - A_o \sin(\theta)\hat{\jmath}]$$

This relation allows us to calculate the force in the wall direction during the flotation. More than just this, it shows us the parameters that we could change to maximize this reaction. θ is the measured based on the normal intake surface. As we can notice, $\theta = 0°$ gives no force because the flow enters and leaves the equipment without any changes in its direction. $\theta = 180°$ gives the maximum force in the wall direction. Although this seems to be the best setup, we must remember a restraint: the intake area of this system must

Figure 2. The floating system

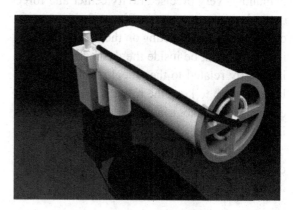

be under water at all time, but it will be hard to ensure this when $90° < \theta < 180°$. So we will get better results using $0° < \theta < 90°$, because this will insure water intake. Then, the best choice is $\theta = 90°$, because this will get the best results:

$$F = \frac{\rho_w V_o^2 A_o}{A_i} [A_i \hat{\imath} - A_o \hat{\jmath}]$$

By analyzing this result, the area influence became more evident. We know the product area versus velocity gives us the volume of water per unit of time, which will be determined by the pump capacity. Therefore, if we take a careful look into the component $\hat{\imath}$, which presses our robot against the wall and substitute the area versus volume by the pump capacity C we will have:

That is, with the same pump, fixing the value for C, we improve the force, thus improving the velocity. In other words, by mass conservation, we improve the force by reducing the exit area. At the first estimation, we will use a 45° angle, because we still need the suction of the suspended particles.

Locomotion

Now we will choose the locomotion settings. Basically, we must choose the type and the number of wheels or rolls (see Figure 4).

Theoretically, only two wheels are needed to maintain the robot work position. Clearly, this demands a very precise gravity center and force control. As we can see in Figure 3, if there are only vertical forces acting on the robot, the gravity center must be inside the dashed box, which is directly related to the wheel width. Rolls are large enough to keep the robot working, but our environment makes it impossible the use of wheels in this case.

Since the free space under the robot is a requirement, we will not use rolls, because they obstruct a large area.

Figure 3. The pump and the filter

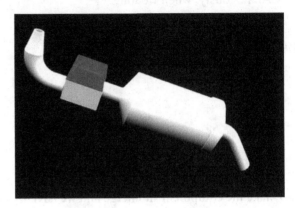

The solution is, then, to add another wheel. Then our gravity center has to be inside the triangle determined between the wheels. Again, this disposition has some problems; it does not allow us to take advantage of the robot base space very well.

Therefore, we decide to use four wheels. This option gives us a bigger useful area as the wheels can be placed on the equipment border. It also demands a smaller gravity center control, as we can see in Figure 3, in the large dashed area. Besides, it resists more adverse conditions when the gravity center is placed near the dashed square center.

As we know, there are many types of wheels (Siegwart & Nourbakhsh, 2004), but the environment requires simpler ones because the wheels will be in direct contact with water and, therefore, do not need lubrication. Besides, more complex wheels may deadlock the robot due to a small particles.

Although there are some types of wheels, which make omnidirectional movement possible, this is not necessary because the environment has generally simple shapes. Therefore, two-directional simple wheels are enough.

In order to allow movement, we will have another question to answer: will we transmit torque to wheels or will we use a water flow to provide propulsion? Even if the second option was adopted, we still would need wheels and, even more importantly, we would need a bigger pump. The biggest problem is that the equipment will not run if the

Figure 4. Wheel location

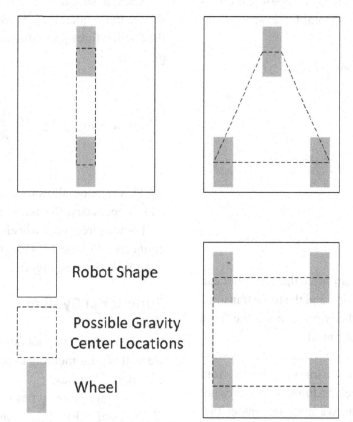

Robot Shape

Possible Gravity
Center Locations

Wheel

pool has areas that are too shallow. Thus, we will opt for an engine and torque-transmitted to-wheel model, which allows the equipment movement, even if it is not in the water.

Since the floor force will be transmitted by friction, this implies that we have to analyze the normal and the friction forces.

The first approach is to determine the dry friction, although the word dry does not seem adequate. It is a simple model in which the friction coefficient has to be adapted to the environment. Our starting point to this analysis is to suppose only a roll movement, that is, there is no relative movement between the contact point of the wheel and the floor. So we have a friction force proportional to the normal force. However, how do we know the real normal force?

First, we must remember that in aquatic environments we need to consider the weight and the thrust. Furthermore, there is the pump force, which contributes to the normal forces and to the motion forces. Thus, by using the previous results we have:

$$N = W - T + \frac{\rho_w V_o^2 A_o}{A_i}[A_i - A_o \cos(\theta)]$$

$$N = M + m - \rho_w V g + \frac{\rho_w V_o^2 A_o}{A_i}[A_i - A_o \cos(\theta)]$$

And the dry friction theory says:

$$F = \mu N$$

But we have to add the flow force in this direction in order to know the total force, so:

$$F = \mu N + \frac{\rho_w V_o^2 A_o^2}{A_i} \sin(\theta)$$

Then:

$$F = \mu \left[M + m - \rho_w V g + \frac{\rho_w V_o^2 A_o}{A_i} [A_i - A_o \cos(\theta)] \right] + \frac{\rho_w V_o^2 A_o^2}{A_i} \sin(\theta)$$

This is the maximum force that the robot can use to transmit by friction and the force transmitted by the flow, in other words, this is the force available for the robot's motion.

At this point, there is an interesting question to approach: how do we transmit torque from the engine to the wheels? Although this does not seem to be a problem, we must remember that each wheel will go a different distance when the equipment changes its direction. Therefore, there are two common ways to do this.

The first way is to just use two engines, where each engine can operate in a different rotation, allowing the curve movement. This choice has two problems for our project. The first problem is the fact that buying and placing two engines is harder than one. The other trouble is how to control the difference of rotation between them.

Therefore, we will adopt another solution, which requires only one engine. This solution is the differential gear, which allows different wheel rotations. This consists of an input gear or shaft, a crown, which receives the torque from the gear and maintains the robot in line. However, when it is turning, the planetary and satellite gears move, and let the external wheel rotate faster than the internal wheel.

Then, to calculate the necessary torque to the differential available by the engine or a gearbox is the friction force part influence in the differential gears, so:

$$T_d = \frac{r_p}{r_c} r \mu \left[M + m - \rho_w V g + \frac{\rho_w V_o^2 A_o}{A_i} [A_i - A_o \cos(\theta)] \right]$$

This relation allows us to specify the engine and, if necessary, the gearbox.

For the directional wheels, we will use a servo connected to wheel forks, providing simplicity and reliability (see Figure 5).

Functional Systems

In order to clean the pool floor and walls properly, we will need something to remove the dirt, most commonly brushes.

There are many types of brushes available, like rolls of bristles. This one, besides cleaning by its rotation, also implies a force, adding or subtracting from the motion forces. That is, this type of brush can interfere with the motion devices by adding friction and by adding new resulting forces in the system.

Figure 5. The powertrain

So, we immediately think about the rotating brushes. With this type of brushes, due to their rotation axis, imply a friction force. Importantly, however, the forces caused by the rotation cancel themselves out. In other words, this type of brush does not interfere significantly with the motion of robot.

In order to cover a wider range per pass by the robot, we will use two brushes, preferentially rotating in opposite directions. They will be preferentially placed in front of the filter elements. This configuration allows the suspended particles to enter the filter system and be retained on the filtering element (see Figure 6).

The filter, due its characteristics, retains dirt, which can restrain the passage. To avoid this, we adopt more than one filter, each one to retain a part of the particles and this arrangement will work for more time.

However, a smarter solution is to adopt a cup filter. Besides retaining particles, this type of filter has lateral surfaces to keep filtering even if the filter is full of dirt. Therefore, this type of filter makes the project simpler and functional. With it, the robot will run a complete cleaning cycle before it needs a filter cleaning.

For this reason, the filter compartment must be easily accessed. This implies that the pump and the pipes should be installed inside the robot, but the filter lid, with the intake curve, must be removable from the outside. Thus, a quick coupling lid would be a good choice. Although a gasket is necessary, this arrangement prevents any damage caused by a costumer mistake while reassembling the filter lid after cleaning, because the joint is placed outside the seal compartment.

In order to force the water into the filter and to allow the floating control as described before, we will adopt a pump. There are many pump types, and although this makes our selection more complex, it also gives us a more adequate pump.

Considering the floating requirements, our equipment must have few mass changes, keeping the specific mass as stable as possible. Therefore, pumps, which employ a variable amount of water, do not meet our requirements.

Obviously, the more adequate ones are the centrifuge pumps whose water mass is constant because their circular movement apply a constant force into the equipment structure too.

A very common pump for underwater application is the bilge pump, largely used in artisanal ROVs (Remotely Operated Vehicles) as a thrust source. This is our choice, mainly because of its low cost and simple installation.

The pipes can be just the commercial ones assembled in the needed shapes. However, this option causes a very large load loss. The best option is then to manufacture a unique element with the required shapes.

Sensing and Controlling

As an autonomous robot, our equipment has to be able to decide the next action. Basically, every point where the robot must stop doing something to do another action is identified by a specific sensor. Siciliano and Khatib (2008) published a table with the available sensors and their main characteristics.

Many robots opt for mapping the environment, by using a group of sensors. This type of robot can achieve the best efficiency. Nevertheless, it

Figure 6. The brushes

requires very accurate sensing, and although it seems not to be a problem, it must remember the environment.

Wheel sensing is a problem because of the underwater environment which lowers friction, as well as the lower than normal force due to the presence of thrust, which reduces friction even more, inducing the robot to slip instead of roll. Since the main way to control distances is to monitor the wheel and relate its rotation to the translation movement of the equipment, when this relation is not valid anymore, we loss accuracy, making mapping useless.

However, there is another problem: when floating, the robot is subjected to flow forces, which can move the equipment laterally. Although the flow may seem to be small, the pool filtration system can be active create significant flow forces, and even a fountain can interfere with the positioning of the equipment.

As a possible solution, we suggest the use of external points of reference, which can communicate to the robot sensors. However, this solution makes the project more expensive and requires much user interaction: if the user places it inadequately, the robot will lose its efficiency, or not clean some pool areas.

Therefore, for low cost and simple equipment, we decided to just monitor the obstacles.

When our robot comes across a wall, for instance, it needs to recognize it and decide what to do. Thus, we need a sensor to tell the robot that there is an obstacle in front of it. Knowing this, the equipment will stop and discharge water from the floater. This will make it float.

When the robot arrives at the water surface, a special switch warns the machine. Then, it will receive water in the floater and submerge. This switch can consist of a lever with a very low specific mass material in the extremity. Thus, it will always have a greater thrust than weight when inside the water, but when it is exposed to the air, the opposite occurs and the weight moves

the lever down, signaling the robot has reached the surface.

When the equipment is submerging, the brushes can be turned off. Then, when the robot arrives at the floor, it will need a rear sensor to warn it. Therefore, with this sensor information, the robot needs to run backwards and make a curve, for instance, 150° arc of a circle. Then, the cycle is repeated.

However, the equipment can reach a wall on its lateral side. Since this is not its preferential direction, two lateral sensors must warn the robot of this. Consequently, there is one on each side. With this information, the equipment should make a curve and go ahead.

The robot should follow this algorithm for a limited time. This limit can be a pre-defined value of a group of values that the user can choose according to the pool dirtiness.

It could be selected near the energy source. Since we need some electric cables, we can place an adjustable timer, which permits the flux of current only for a limited time. As pre-defined times, we can offer, for instance, one hour for fast cleaning, one hour and a half for normal cleaning and two hours for heavy cleaning (see Figure 7).

Sizing and Selection

Now, we already have the basis of our project, so we have to build the systems and parts of our robot.

Our starting point is the filter and pump system. We chose a 1000 gallon per hour bilge pump, which has a relatively low price compared with others, and this flow can give to the robot the needed force. It costs between US$ 32 and US$ 48.

Our intake angle will be set as 90°, the input pipe has a 1 1/8" diameter, and the output pipe has a reduction, ending in a ½" diameter. With this data, we can calculate the forces.

The normal force that will press the equipment against the wall will be 7.5N and the force aligned to the movement will be 1.2N.

Figure 7. Control algorithm

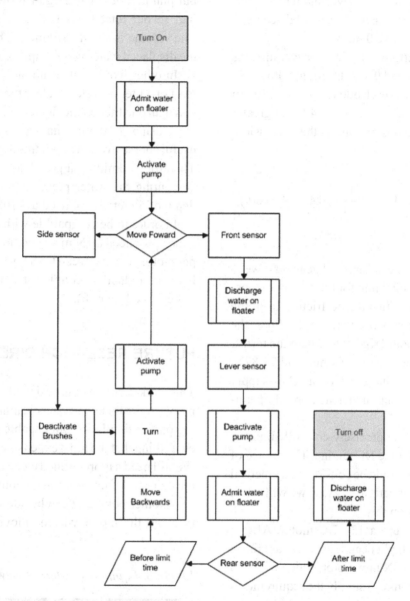

Additionally, we also have the filtration element. Since it is a prototype, we can use a commercial one, so, there is a 3600 liter per hour element used in stationary pool filters. This is approximately the flow capacity given by the pump and, therefore, is adequate to our needs.

Now we will choose the powertrain. Before any selection, we have to figure out the pool sizes where the equipment will run. Thinking of a home pool, a 10m x 10m pool area appears

to be the largest value the robot would have to clean. Therefore, the robot is designed to clean a 100m² area. Since our equipment has a width of 300mm, it will have to move approximately 333m. Thus, in order to cover this area in one hour as we discussed before, it will need a speed of approximately 0.1m/s.

However, we must also count the wall cleaning time, and remember, since our equipment does not map the environment, it must pass more than one time in the same place to cover the whole pool. Then, for this situation, a higher speed seems to be more adequate, like 0.5m/s.

The nylon friction coefficients on wet surfaces are between 0.04 and 0.3. Admitting that weight and thrust cancel out each other, by using a 30mm wheel radius and a crown radius 4 times greater than the pinion radius, we obtain the expression:

$$T_d =$$

$$\frac{r_p}{r_c} r\mu \left[M + m - \rho_w Vg + \frac{\rho_w V_o^2 A_o}{A_i} [A_i - A_o \cos(\theta)] \right]$$

The torque in the differential gear is between 2.6N.mm and 19.62N.mm for the maximum possible torque transmitted due to friction from the pool floor. If we suppose that, on the floor, the total weight is about 1Kg heavier than the thrust, those torques increase to 2.9N.mm and 21.87N.mm. Thus, the engine must be able to supply more than this amount under its normal operating condition.

Since there is no precisely controlled movement, we use a simple DC engine. The choice of the engine tension could be considered a practical issue, but the other parts are 12v, so we will select the engine based on this.

Then we will use a 12v DC motor. After a quick search of supplier catalogs, we could find a 2500rpm and 15.68N.mm motor, which with two four time reductions can supply the requirements and some extra power needed to move the brushes. This type of motor can be bought for US$ 3 to US$ 5, if purchased in large quantities.

For the directional system, a micro servo that releases 147N.mm is enough as it is used only to rotate the front wheels axis. This micro servo can be found for US$ 5 to US$ 10.

In order to dimension the size of the floating system, we need to estimate its weight. Based on the 3D model, we have approximately 8Kg, and

our pump system has 2Kg of water when working, so our total mass is approximately 10Kg. The volume is about 11E6mm³. This means that the displaced water weighs approximately 11Kg.

In other words, our equipment must have at least one liter of water to clear the pool floor and less than one liter on the floater to clean the walls.

If using a 100mm diameter and 128mm of useful course, we can admit this amount of water. Therefore, admitting a pre-charge of 50mm and computing the water pressure at the maximum depth of 5 meters, we need a 7.70KN/m spring.

In order to be compatible with the maximum course, we need a 1N.m stepper motor. The stepper motor costs between US$ 50 and US$ 55 and has a 4:1 reduction, coupled at a 10mm diameter pulley (see Figure 8).

FUTURE RESEARCH DIRECTIONS

This work described the design of an autonomous mobile robot for underwater cleaning, always aiming on the simplicity, but capable of performing everything that the most expensive robots can do. We still need a more refined electrical project and the control definitions must be improved.

Another point that can be addressed is trying to avoid the motor wheels, moving the equip-

Figure 8. Equipment external view

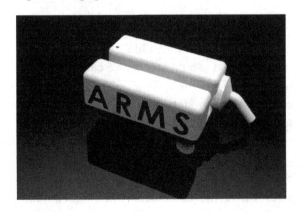

ment only with fluid force. This would make the equipment unable to run on dry surfaces, but it can remove the differential gear system.

The next step is to develop the detailed project, part by part, making it possible to build a prototype for experimental tests.

This is the most demanding phase of the project, first due to the costs (everything is expensive when not bought in quantity), but also because some processes which would provide us a better result in terms of functionality, do not make sense to use. For instance, for the careen, the use of injected abs is more adequate for this proposal, but the costs are prohibitive. Therefore, we have to opt for the molded abs.

Since the testing phase is iterative, this will guide subsequent research that provides new solutions for this proposal.

CONCLUSION

We tried to make the equipment as simple as possible, but some complexity cannot be avoided. As an engineering project, many of the systems reflect the designer's previous experience, which sometimes does not represent the best solution.

Anyway, this chapter intended to show the reader a project sequence to develop a robot and is transferable to robot development in general. The physical model can differ significantly, but the basic principles and considerations remain the same.

REFERENCES

Fox, R. W., & McDonald, A. T. (1981). *Introdução a mecânica dos fluidos*. Rio de Janeiro, Brazil: Guanabara Dois.

Siciliano, B., & Khatib, O. (2008). *Springer handbook of robotics*. Berlin, Germany: Springer-Verlag. doi:10.1007/978-3-540-30301-5

Siegwart, R., & Nourbakhsh, I. R. (2004). *Introduction to autonomous mobile robots*. Cambridge, MA: The MIT Press.

ADDITIONAL READING

Alves, J. A., & Mota, J. (2003). *Casas inteligentes*. Lisbon, Portugal: Centro Atlântico.

Amaral, R. D. C., & de Pina Filho, A. C. (2010). *Estudo e modelagem computacional de um robô aspirador de pó*. Paper presented at the IX Simpósio de Mecânica Computacional (SIMMEC). São João Del-Rei, Brazil.

Bianchi, R. A. C. (2007). *Curso de robótica móvel*. Rio de Janeiro, Brazil: Centro Universitário da FEI.

Bolzani, C. A. M. (2004). *Residências inteligentes*. Rio de Janeiro, Brazil: Editora Livraria da Física.

Budynas, R. G., & Nisbett, J. K. (2011). *Elementos de maquinas de shigley: Projeto de engenharia mecânica*. Porto Alegre, Brazil: AMGH Editora.

Craig, J. J. (2004). *Introduction to robotics: Mechanics and control* (3rd ed.). Upper Saddle River, NJ: Prentice Hall.

Doughty, S. (2001). *Mechanics of machines*. Lexington, KY: Lulu.

Dudek, G., & Jenkin, M. (2000). *Computational principles of mobile robotics*. Cambridge, UK: Cambridge University Press.

Gomide, F. A. C., & Gudwin, R. R. (1994). Modelagem, controle, sistemas e lógica fuzzy. *SBA Controle & Automação, 4*(3).

Groover, M. P., Weiss, M., Nagel, R. N., & Odrey, N. G. (1989). *Robótica: Tecnologia e programação*. São Paulo, Brazil: McGraw-Hill.

Indústrias Schneider, S. A. (2012). *Manual de hidráulica - Bombas*. Rio de Janeiro, Brazil: Joinvile.

Jones, J. L., & Flynn, A. M. (1993). *Mobile robots: Inspiration to implementation*. Reading, MA: A. K. Peters.

Lazinica, A. (2006). *Mobile robots - Toward new applications*. Vienna, Austria: Pro Literatur Verlag. doi:10.5772/33

Leonik, T. (2000). *Home automation basics: Practical applications using visual basic*. New York, NY: Delmar Learning.

Marques, A. L., de Pina Filho, A. C., & de Pina, A. A. (2008). *Usando ambientes virtuais no projeto de robôs móveis*. Paper presented at the XXIX Iberian Latin-American Congress on Computational Methods in Engineering (CILAMCE). Maceió, Brazil.

Marques, A. L., de Pina Filho, A. C., & Lima, F. R. (2008). *Modelagem de ambientes virtuais usando software livre*. Paper presented at the XII Congreso Internacional de Gráfica Digital (SIGraDi). Havana, Cuba.

Mateus, V. A. C. B., de Pina Filho, A. C., & de Pina, A. C. (2008). *Characteristics and concepts on mobile robots with wheels*. Paper presented at the 7th Brazilian Conference on Dynamics, Control and Applications (DINCON). São Paulo, Brazil.

Mateus, V. A. C. B., de Pina Filho, A. C., & de Pina, A. C. (2009). *Modeling of a mobile robot with wheels using CAD techniques*. Paper presented at the 30º Iberian-Latin-American Congress on Computational Methods in Engineering (CILAMCE). Rio de Janeiro, Brazil.

Meyer, G. (2004). *Smart home hacks: Tips and tools for automating your house*. New York, NY: O'Reilly Media, Inc.

Nehmzow, U. (2003). *Mobile robotics: A practical introduction*. New York, NY: Springer-Verlag. doi:10.1007/978-1-4471-0025-6

Norton, R. L. (2004). *Projeto de máquinas: Uma abordagem integrada*. Porto Alegre, Brazil: Bookman.

Pieri, E. R. (2002). *Curso de robótica móvel*. Santa Catarina, Brazil: Universidade Federal de Santa Catarina.

Pina Filho, A. C. de, & de Pina, A. C. (2009). *Methods and techniques in urban engineering*. Viena, Austria: I-Tech Education and Publishing.

Rivin, E. I. (1988). *Mechanical design of robots*. New York, NY: McGraw-Hill.

Silveira, P. R., & Santos, W. E. (1998). *Automação e controle discreto*. Viena, Austria: Editora Érica.

KEY TERMS AND DEFINITIONS

Autonomous System: Mechanism that can perform desired tasks in unstructured environments without continuous human guidance.

Domotics: Use of the automation technologies and computer science applied to home.

Mobile Robot: Automatic machine that is capable of movement in a given environment.

Powertrain: Set of parts that provide movement to the equipment.

Sensor: Device able to identify the surroundings.

Sizing: Choice of the adequate part that meets the requirements.

Thrust: Force due the interaction with a fluid.

Chapter 15
Mobile Robotics Education

Gustavo Ramírez Torres
Siteldi Solutions, Mexico

Pedro Magaña Espinoza
Siteldi Solutions, Mexico

Guillermo Adrián Rodríguez Barragán
Siteldi Solutions, Mexico

ABSTRACT

With the educational mobile robot Worm-Type Mobile Educational Robot (Robot Móvil Educativo tipo Oruga, or ROMEO, by its Spanish acronym), the authors offer three hierarchical levels of experimental learning, where the operator can develop as far as his/her ability or imagination permits, gaining knowledge about the basics of sensors, communications, and mechanical and robot programming. Due the lack of learning focused on robotics in Mexican educational institutions, the authors present this chapter, where an early stimulation to this topic could trigger curiosity to research that leads to technological advancement. ROMEO is a mobile wireless communication platform with different types of sensors: moisture, brightness, temperature, etc., as well as a compass and accelerometers with similar characteristics to industrial and commercial applications that allow us to experiment with communication algorithms, sampling, and autonomous and semiautonomous navigation.

INTRODUCTION

The images that spring to mind when we hear the word robot are from machines that do human work and imitate movements and behavior of living things. The development of robots for goods and services or to exploit natural resources is strongly influenced by technological development; this leads us from simple machines that magnify muscle power of men in physical work, to machines that are also capable of processing information, complementing or replacing man in intellectual activities such as the choice of better roads, monitoring, process control, etc..

Industry makes extensive use of robotics in tasks that are hazardous, tedious, and need physical performance, which gives us an enormous gamma of opportunities concerning robot development.

DOI: 10.4018/978-1-4666-2658-4.ch015

Copyright © 2013, IGI Global. Copying or distributing in print or electronic forms without written permission of IGI Global is prohibited.

Mobile robots are useful in human rescue, hazardous environments, and tight spaces, which are inaccessible to humans. In general, robotic vehicles move autonomously with sensors and computing resources that guide their movement (Bekey, 2008), e.g. "Mobile Robots for Offshore Inspection and Manipulation" (Bengel, 2012).

The input to the process of perception is usually twofold: (1) digital data of a number of sensors, (2) partial modeling environments that include status information of the robot and different external entities of the environmental (Siciliano, 2007).

This chapter aims has three primary goals: (1) to ensure that users understand the mechanism and mechanics of the robots in this case mobile robots, (2) to inform the user that robots use sensors that are analogous to human senses, and (3) to achieve that users understand the programming language required to make a robot function.

DEVELOPMENT

Educational Mobile Robot "ROMEO" (Figure 1) is a modular robot that uses an aluminum chassis adapted to motor-reducers and chain sprockets. It has a wireless communication platform with different types of humidity, light, distance, and temperature sensors, as well as a compass and

Figure 1. ROMEO

accelerometers with characteristics similar to industrial and commercial applications. It provides the operator with experience and provides its users the opportunity to formulate questions about the operation of wireless communication, programming, and sensors.

ROMEO has three hierarchical leaning levels, which we describe below.

Level 1: Simple Programming of Mobile Robots

For this level, we developed a platform with ZigBee wireless communication protocol (Carlos, 2007) with a XBEE module (Figure 2). The platform functions as a transceiver.

The elements of the communication platform are:

- Power circuits for the control of motor-reducers.
- Connection for control of a servo motor.
- Voltage regulators.
- Output ports of the microcontroller.
- Output voltage for the implementation of sensors.
- USB communication module.

We then developed an accessory remote control (Figure 3) for the wireless communication platform. We are able to remotely control ROMEO and we can program motion routines.

The elements of "accessory remote control" are:

- Two Joysticks.
- LCD Screen.
- Matrix Keyboard.

As mentioned above, ROMEO has two types of movement at this level, remote-controlled and programmed. The following section will describe these two types of movements.

Figure 2. Wireless communication platform

Figure 3. Accessory remote control

Remote Control Movement

Remote control movement (Wikipedia, 2012) is performed by using two communications platforms.

- The first platform is adapted with the "remote control accessory" to lead ROMEO with joysticks.
- The second platform is the brain of Romeo, which receives and permits the interpretation of movement commands.

Programmed Movement

We developed and recorded its simple programming in C, which focuses on the communications platform and uses the joystick accessory. As in remote control movement, we used two communication platforms (see Table 1).

- In the first platform, we adapt the "accessory remote control" from which we can program motion routines using a LCD screen and a matrix keyboard as an interface.
- The second platform is the brain of ROMEO, which receives instructions and allows ROMEO to interpret programmed routines.

Table 1. Simple programming vs. C language

Term	Simple language	C Language
Forward	F	-
Back	B	-
Right	R	-
Left	L	-
Waiting (1-60)	D(¿)	Delay

These commands perform the first steps in programming aimed in learning C.

This level is intended to guide the user to learn a little about what causes the movement of a robot and how to tell the robot to move.

Level 2: Mechanisms, Sensors, and Medium Programming of Mobile Robots

At this level, we made the following improvements to ROMEO level 1 (Figure 4), adding:

- A module forklift;
- A distance sensor;
- A light sensor;
- A temperature sensor;
- A color sensor.

We developed software with a graphical interface to a PC (Figure 5) that allows us to program and move "ROMEO" remotely. The connection between "ROMEO" and the PC is performed via wireless serial communication with an XBEE modulo.

The movement of ROMEO at this level is classified into two types: autonomous and remote control.

Figure 4. ROMEO forklift module

Autonomous Movement

We develop a medium programming language based on C to interpret sensor signals. We can program algorithms to move ROMEO and its forklift mechanism remotely. We can also monitor environmental conditions.

Remote Control Movement

With the addition of buttons in the graphic interface, we can give movement and direction to ROMEO and operate the forklift mechanism (see Table 2).

These commands perform simple movement algorithms and signal interpretation based on C language.

This level is designed for the user to understand the function of sensors and create simple programming algorithms such the following example (Figure 6).

Level 3: Mechanisms, Sensors, Advanced Programming, and Wireless Communication Protocols

For this level, we developed a communication platform (Figure 7) with the following:

- 512 Mb of memory;
- A 32-bit Microcontroller;
- 2 serial communication modules;
- A USB communication module;
- Inputs and outputs to the microcontroller dedicated to Messrs;
- An ARM architecture;
- 3.3 v power consumption;
- An 802.15.4 XBEE communication module;
- Programming with open source OS (PaRTiKle);
- A light sensor;
- A temperature sensor;
- A relative Humidity Sensor;

Figure 5. Graphic interface

Table 2. Medium programming vs. C language

Term	Medium Language	C Language
UP	Up	-
Down	Dn	-
forward	F	-
Back	B	-
Right	R	-
Left	L	-
Waiting (1-60)	E(¿)	Delay
Temperature (10-40)	T(¿)	-
Brightness (1/0)	B(¿)	-
Distance (1/30)	D(¿)	-
Color (B/W)	C(¿)	-
If (¿, ¿)	Si (¿, ¿)	If

Figure 6. Simple movement algorithm

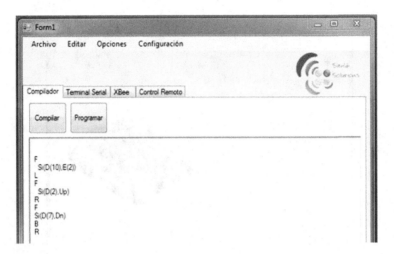

- An accelerometer;
- A digital compass;
- 2 Motor Control 3 A Drivers;
- 24-pin I / S GPIO;
- 5 5v outputs;
- 5 3.3v outputs;
- 3 I2C inputs;
- 5 inputs ADC.

For this level we use a gripper module and a container blocks module (Figure 8). We can manipulate blocks of 400 grams and 45mm wide. We developed algorithms for the collection of blocks.

Figure 7. Communication platform

We classify two types of movement for this level: remote control and autonomous:

Remote Control Movement

With the graphic interface developed at level 2, we can give movement and direction to ROMEO and operate the gripper mechanism.

Autonomous Movement

We developed reading sensors algorithms in C (Figure 9). We developed libraries to determine the orientation to the North Pole with a digital compass. We calculate the distance between robots measuring the power with which the signal is received.

To determine the distance between robots, we use the well-known formulas of path loss:

Free Space Path Loss:

$$L_p = 32.4 + 20 * log_{10} * f + 20 * log_{10} * d$$

f in Mhz

d in Km

Figure 8. Gripper module

Figure 9. Program algorithm in C

```
2   #fuses XT,NOWDT,NOPROTECT,NOLVP
3   #use delay(clock= 4000000)
4   #use standard_io(B)
5   #use standard_io(C)
6   #use standard_io(A)
7   #include <lcd.c>
8
9
10  int16 a=0,b=0,c=0;
11
12      void main() {
13  PORT_B_pullups(true);
14  lcd_init();
15   output_low(pin_c6);
16    output_low(pin_c0);
17     output_low(pin_c1);
18      output_low(pin_c2);
19
20
21      while (1)   {
22  output_low(pin_c6);
23   output_low(pin_c0);
24    output_low(pin_c1);
25     output_low(pin_c2);
26  inicio:
27  c=0;
28  a=0;
29  lcd_gotoxy(1,1);
30  printf(lcd_putc,"Seleciona");
31  lcd_gotoxy(1,2);
32  printf(lcd_putc,"Rutina");
```

Path Loss Model for indoor office:

$$L_p = 37 + 30 * log_{10} * R + 18.3 * n^{\left[\frac{n+2}{n+1}-0.46\right]}$$

R : *transceiver separation*

n : *number of floors on path*

Path Loss Model for outdoor to indoor:

$$L_p = 40log_{10}R + 30log_{10}f + 49$$

R : *receptor distance in km*

f : *frequency in MHz*

This level is intended for the user carry out sensor readings using algorithms programmed in C. It is designed for the user to employ his algorithms, orientation parameters, and distance between robots, using programming libraries.

This denotes the importance of the combination and implementation of different sensing systems that allow the robot to know its own state and that of its environment (Londono, 2000).

CONCLUSION

In this chapter, we proposed a mobile robot prototype to familiarize people—especially those in academics- to become familiar with robotics.

In the first level of this project, we hope users can become interested in the operation and the parts of the robot, regardless of whether the users are students or not. Furthermore, ROMEO offers users an introduction in programming. Users can design several routines for the mobile robot and enter it competitions to develop even better robot algorithms.

At the second level of this project, users understand the operation of sensors in a mobile robot. They interpret signals from the sensors and design algorithms for autonomous movement. We also noticed that people who have used the mobile robot are more inclined towards remote control operation, but success was achieved with the objective set for the second level.

At the third level of this project, users successfully design algorithms for mobile robot motion in C, but not all users have done it with the same pace or degree of comprehension. We noticed that users who enrolled at a secondary school level onwards were able to understand the programming of robots, even among users who were not enrolled in any academic program.

In general, we believe that the ROMEO prototype can spark interest in the subject of robotics, but we know beforehand that this topic is very broad and if someone wants to engage in it there is much more information to analyze.

REFERENCES

Bruno Siciliano, O. K. (2007). *Springer Handbook of Robotics*. Springer.

Carlos, V. R. (2007). *Universidad Nacional de Trujillo*. Retrieved from http://seccperu.org/files/ZigBee.pdf

George Bekey, R. A. (2008). *ROBOTICS STATE OF ART AND FUTURE CHALLENGES* (p. 11). Imperial College Press. doi:10.1142/9781848160071_0002

Matthias Bengel, K. P. (n.d.). *Mobile Robots for Offshore Inspection and Manipulation*.

N, L. (2000). La robótica: realidad y ficción. *Revista Silicio, 8*(11), 41-52.

WIKIPEDIA. (2012, April 10). Retrieved from http://es.wikipedia.org/wiki/Autom%C3%B3vil_teledirigido

Chapter 16
Ad Hoc Communications for Wireless Robots in Indoor Environments

Laura Victoria Escamilla Del Río
University of Colima, Mexico

Juan Michel García Díaz
University of Colima, Mexico

ABSTRACT

This chapter presents a theoretical and experimental comparison of electromagnetic propagation models for indoor robot communication using mobile ad-hoc IEEE802.11 and IEEE802.15.4. The analysis includes the behavior of the electromagnetic signal using the abovementioned standards in two scenarios, both located inside the building of the College of Telematics of the University of Colima. The results of the propagation of the electromagnetic signals in the two scenarios were then compared with the mathematical model.

INTRODUCTION

The mobility and convenience that wireless networks provide has contributed to the increased number of people using them; however, the required infrastructure cannot be provided in all scenarios(Luu, O'Brien, & Baran, 2007). An ad-hoc wireless network is a collection of mobile elements that form a temporary network without

an established infrastructure (Bracka, Midonnet, & Roussel, 2005). Advances in wireless technology and the miniaturization of robots have allowed the formation of ad hoc networks that can maintain communications in disaster and other important situations.

Ad hoc networks formed by mobile robots equipped which can communication among themselves through wireless transmission capabilities is mainly determined by the communication technology and the specific scenario.

DOI: 10.4018/978-1-4666-2658-4.ch016

Copyright © 2013, IGI Global. Copying or distributing in print or electronic forms without written permission of IGI Global is prohibited.

RELATED WORK

The mobility of wireless networks has been widely investigated. However, most of these papers focus more on random rather than controlled mobility (Jardosh, 2003). Controlled mobility is now a research area of mobile ad hoc networks (Bracka, et al., 2005). In the following research, the network communication has been addressed from different perspectives, without following any of these characteristics of mobility would be expected to have in our proposed scenarios.

In 2005, nine Lego Mindstorms robots equipped with an infrared interface used classic algorithms such as RIP or OSPF to form an ad hoc robot network, allowing robots to maintain communications with each other or with a fixed base (Bracka, et al., 2005). In 2006, the Army Research Laboratory of the United States (ARL) implemented a Mobile Ad Hoc Network (MANET) that allowed communication between an Operated Control Unit (OCU) and a robot. The ARL implementation employed IEEE 802.11g Linksys routers with 100mW internal amplifiers without high gain antennas. Test results show they were able to communication 400 meters, with the possibility of communicating longer distances (Luu, et al., 2007). In that same year, in a collaborative effort, the State of Mexico Technological Institute's Robotics Lab and the University of California at Santa Cruz's Internetworking Research Group formed a league of robots that allow multi-robot collaboration beyond the football field. The local robot control was managed by a single Digital Signal Processor (DSP) chip that was optimized for digital motor control. The DSP receives remote communications from the Artificial Intelligence system through a TX / RX Radiometrix radio[1] using the 914MHz and 869MHz frequency bands at a transmission rate of 64Kbits/seg. In this experiment, all robots were very close to each other, meaning that all the robots were within the transmission range of each other. This makes message routing between any two network nodes very simple as any robot can send a message to any other robot in a single transmission. However, for other applications would require a multi-protocol routing hop in order to maintain communication (Weitzenfeld, Martinez-Gomez, Francois, Levin-Pick, & Boice, 2006).

As mentioned above, another technology that has gained a field of application is based on IEEE 802.15.4. Due to its characteristics (low power consumption, cost and frequency), it is ideal for small applications where power consumption is minimized. In 2009, the Australian Research Council Center of Excellence for Autonomous Systems, in cooperation with the Australian Centre for Field Robotics (ACFR) and the University of Sydney, New South Wales, Australia, worked on an ad hoc wireless network based on ZigBee mesh. They employed JENNIC JN5139 as part of a prototype system consisting of 15 nodes. This device allowed them to run simple applications in the ZigBee module without an internal processor. Simulated and experimental tests concluded that ZigBee-based systems are good choices in networks consisting of many small robots (Fitch & Lal, 2009).

THE EVOLUTION OF ROBOT WIRELESS COMMUNICATION

Technology has advanced along with the electromagnetic propagation models used in mobile robots. An important design consideration for mobile ad hoc networks must consider not only the signal range, but also the power consumed by the devices and their size. Aspects such as these can be definitive in choosing a communication technology. Unlike a WAN where you have a robust physical infrastructure to support it, ad hoc networks are infrastructure less and its components must function both independently and in conjunction with each other during the packet routing procedure (Wang, 2003). Consequently, it is important to know how the signal propagates in scenarios where they plan to build the network. Table 1 lists the major wireless communications technologies used in mobile robotics.

A wireless communication link is characterized by a long-delay bandwidth, dynamic connectivity and link ages prone to errors. Because mobile robots are usually equipped with low-power wireless network interfaces and short ranges, a wireless communication link only allows direct communication with the nearest neighbors (Wang, 2003). Due to the above-mentioned reason, we propose an experiment using IEEE 802.11 and IEEE802.15.4 to communicate wireless network interfaces with mobile robots. The following section describes the scenarios and devices with which the experiments were conducted.

SCENARIOS

The experimental data collection was conducted in the building of the College of Telematics at the University of Colima, Mexico. The building has two floors, four standard classroom, four auditorium-type classrooms, and seven laboratories. Because Colima is prone to extreme seismicity, present-day constructions are reinforced concrete structures, as is the case of this building. As a first test scenario,

we used the main concourse of the building (see Figure 1), which offered line of sight between the transmitter and receiver throughout the test. As a second stage (Figure 2), measurements were made on the second floor where the classrooms are located. This was done to observe the signal decrease when obstacles were present between the network devices. For each of the experiment, we employed an ad hoc network consisting of two nodes. The test range was measured in both scenarios to test the two proposed technologies. Measurements were taken every meter, each for 30 seconds over a total distance of 35 meters. We monitored the Received Signal Strength Indicator (RSSI) and packet loss between the network devices. The following section provides a brief description of the technologies and devices used in the experiments.

THE IEEE 802.11n STANDARD

IEEE 802.11n is a modification to the IEEE802.11-2007 standard that improves network performance with increased transmission speeds of up to 600Mbps, depending on its settings. It defines changes to the physical layer and sub layer medium access control that enables persons to configure different operating modes that provide much

Table 1. Communication technologies applied to mobile robots

	RF band (Ghz)	Modulation	Date rate (Mbps)	Range (m)	Network structure	Power
Infrared			0.1-4	4	PPP	5mW
Bluetooth	2.4/5	FHSS	0.72 future10/20	10-100	Ad hoc	0.3mW-30mW
IEEE 802.11b	2.4/5	DSSS	5.5-11 (shared)	30-100	Infrastructure and ad hoc	2W
IEEE 802.11g	2.4	DSSS/OFDM	20-54	100	Infrastructure and ad hoc	1.5W
IEEE 802.11n	2.4/5	OFDM	200+	100+	Infrastructure and ad hoc	4.3W
802.15.4	0.868/0.915/ 2.4	O-QPSK /DSSS	0.02/0.04/ 0.25	10-100	Ad hoc	10mW-60mW

Figure 1. Esplanade of the College of Telematics

Figure 2. Second floor classroom area

higher performance and broader coverage than IEEE 802.11a/ b /g (IEEE, 2009).

To join a network employing the 802.11n technology used in these experiments, we used two computers with 11n MIMO Express high-power adapters with two external antennas on each team, which we call nodes. The technical specifications of 11n MIMO Express adapters are shown in Table 2.

Once the 802.11 network was created, we used two software tools to monitor the network: Chanalyzer for Windows and IPTraf for Linux.

IPTraf is a console-based application that works by collecting information about active TCP connections, such as statistics, activity of interfaces, and drops in TCP and UDP traffic (Java, 2001). Listed below are its main features, which include:

- Analyzing IP traffic, TCP flag information, packet and byte counts, and providing ICMP details and OSPF packet types.
- Providing information about general statistics, detailed IP interfaces, TCP, UDP, ICMP, IP packet counters and non-IP, interface activity.
- Monitoring and filtering TCP and UDP.
- Supporting Ethernet, FDDI, ISDN, SLIP, PPP, and loopback interfaces.
- Recognizing different protocols: IP, TCP, UDP, ICMP, IGMP, IGP, IGRP, OSPF, ARP, RARP.

Chanalyzer analysis software is a real-time radio frequency spectrum that allows users to monitor both 2.4GHz and 5.8GHz networks. This software requires a Wi-Spy 2.4x Wi-900x Spydbx or Wi-Spy. Chanalyzer can record and display data from a WiFi network (MetaGeek, 2011). Chanalyze offers the following:

- Recording Playback: allows a file to collect data from our network.
- Time frame: browses the timeline data from the network.
- Channels table: offers interesting features of reachable networks
- Multiple networks: provides a view of more than one network at a time
- Hardware configuration: sets the user's Wi-Spy to display networks on different frequencies.

THE IEEE 802.15.4 STANDARD

IEEE 802.15.4 standard is defined by its protocol and the interconnection of devices via radio communication in a Personal Area Network (PAN). This standard was completed in May 2003 and ratified on December 14, 2004, by the IEEE, but was not made public and available to universities for development until June 2005 (Acosta-Ponce, 2006).

Table 2. Technical specifications SR71-X (Networks, 2012)

SR71-X Hi-Power 802.11n Express Card	
Wireless Chipset	Atheros AR9280
Radio Operation	IEEE 802.11a/n, 5GHz, 802.11b/g/n 2.4GHz
Interface	Express Card PCMCIA
Antenna Ports	(2) MMCX for 2X2 MIMO Operation
Temperature Range	-30 to +75C
Avg. TX Power	24dBm, +/-2dB
Operation Voltage	3.3 VDC
AVG. TxConsumption	1.30A
Range Performance Indoors	Up to 200 meters
Range Performance Outdoors	Over 50km
Antenna	(2) swivel 6dBi omni for laptop
OS Supported	Windows Vista, XP, 2000

The standard uses the carrier sense multiple access with collision avoidance mechanism media access and is compatible with both star and point-to-point topologies. The media access is contention based; however, using the optional super frame structure, time intervals can be allocated by the PAN coordinator to devices with time critical data. Connectivity to higher performance networks is provided through a PAN coordinator. This standard specifies two PHYs:

- 869/915 MHz Direct Sequence Spread Spectrum (DSSS). PHY: Supports date rates over the air of 20 Kb/s and 40Kb/s.

- 2450 MHz DSSS, PHY: Supports a wireless rate of 250kb/s.

The PHY chosen depends on local regulations and user preferences (IEEE, 2003).

To test the 802.15.4 technology we used an XBP24-ACI-001 Max Stream X Bee module. This module operates at 2.4GHz and has an increased power output of 60mW, which supports both point-to-point and multi point connectivity, and is based on the Freescale chipset. Table 3 specifies the general characteristics of this module.

Table 3. Technical specifications XBee XBP24-ACI-001 (XBee.cl, 2009)

XBee XBP24-ACI-001	
RF Data Rate	250 kbps
Indoor/Urban Range	300 ft (100 m)
Outdoor/RF Line-of-Sight Range	1 mi (1.6 km)
Receiver Sensitivity (1% PER)	-100 dBm
Transmit Power	60 mW (+18 dBm)
Frequency Band	2.4 GHz
Serial Data Rate	1200 bps - 250 kbps

ELECTROMAGNETIC PROPAGATION MODELS

The electromagnetic propagation model defined in the simulation scenario is a simple but widely used model for wireless networks. The visibility between nodes can be with or without line of sight. In the first case, when there is no obstacle that interferes with communication between two nodes, we employ a medium communication type Line Of Sight (LOS). On the other hand, if there is an obstacle that causes attenuation between the two communicating nodes, a Non-Line-Of-Sight (NLOS) medium communication was used. This experiment considered both mediums.

Wireless communications are attenuated by multiple factors. For this work, we have defined variables affecting attenuation that interfere with communication between nodes: distance and obstacles. The following section describes each of these causes of attenuation and their mathematical model.

INDOORS ATTENUATION BY DISTANCE WITH IEEE 802.11n

Because mobile computing applications are becoming increasingly common in different scenarios, models of signal propagation and conducting feasibility studies of networks become increasingly necessary before initial deployment (Santos, 2009). For our case study, we compared the measurements taken with the MIMO 802.11n antenna technology empleando free space and indoor models. The power loss (Pl_{oss}) that might occur in a space free of obstacles or objects that can absorb the energy can be expressed logarithmically as:

$$P_{loss} = 32.4 + 20 \cdot \log_{10} \cdot d + 20 \cdot \log_{10} \cdot f..$$

(1)

Where:

- 32.4 = the reference loss constant
- d = distance in kilometers (km)
- f = frequency in Megahertz (MHz)

Equation (1) can be simplified for the exclusive use of the 2.4 GHz frequency band as:

$$P_{loss} = 40 + 20 \log_{10} d.$$

(2)

While Equation (2) gives us a look at data collected, the indoor model uses more environmental variables to more realistically approximate real-world data. The indoor model is based on the COST231 (ITU, 1997), which defines the following expression:

$$L = L_{fs} + L_c + \sum K_{wi} \cdot L_{wi} + n^{\left(\frac{n+2}{n+1} - b\right)} \cdot L_f.$$

(3)

Where:

L_{fs} : Free space loss between transmitter and receiver

L_c : Constant loss

K_{wi} : Number of penetrated walls of type i

n : Number of penetrated floors

L_{wi} : Loss of walls type i

L_f : Loss between adjacent floors

b : Empirical parameters

Equation (3) can be simplified into simple office environments as:

$$L = 37 + 30 \log_{10} R + 18.3 \cdot n^{\left(\frac{n+2}{n+1} - 0.46\right)}.$$

(4)

Where:

R : transmitter-receiver separation
n : number of floors in the path

Since our scenario included only a flat line of sight, the resulting equation is simplified as follows:

$$L = 37 + 30 \log_{10} R. \tag{5}$$

INDOOR ATTENUATION CAUSED BY OBSTACLES WITH THE IEEE 802.11n STANDARD

The indoor model presented in the previous section defined by Equation (3) shows variables such as walls and floors are barriers to signal dissemination. In the experiment we conducted, the scenario is carried out on a second floor for a total distance of 35 meters, with a maximum of five concrete walls obstructing the distance. Table 4 shows the different types of losses that may be considered depending the type of obstruction.

Since in our experiment the electromagnetic signal does not penetrate floors and the walls are made of concrete, we take a factor of 6.9 for L_{w2}. Expression as being:

$$L = L_{fs} + L_c + \sum K_{wi} \cdot L_{wi} \tag{6}$$

INDOOR ATTENUATION BY DISTANCE WITH IEEE 802.15.4

Induced attenuation distance is always present in any wireless communication model. For our case study, in the second experiment, we considered an ad-hoc network with IEEE802.15.4 technology and the propagation model exposed in the book *ZigBee Wireless Networks and Transceivers* (Farahani, 2008). The power loss would be expected in a smooth connection in a building can be expressed by the following equation:

$$P_d = P_0 - 20 \log_{10} (f) - 20 \log_{10} (d) + 27.56 \tag{7}$$

Where:

- P_d = signal power (dBm) at a distance d
- P_0 = signal power (dBm) at zero distance from the antenna
- f = signal frequency in MHz
- d = distance (meters) from the antenna

Table 5 illustrates the different values that the factor "n" can take in different environments (Farahani, 2008). Given the experimental stage, we consider a factor n=1.6 for line of sight.

Table 4. Weighted average for loss categories (ITU, 1997)

Loss category	Description	Factor(dB)
L_f	Typical floor structures (i.e. offices) • Hollow pot tiles • Reinforced concrete • Thickness type < 30 cm	18.3
L_{w1}	Light internal walls • Plasterboard • Walls with large numbers of holes • (e.g. windows)	3.4
L_{w2}	Internal walls • Concrete, brick • Minimum number of holes	6.9

Table 5. "n" factor for default settings

n	Environment
2.0	Free Space
1.6 -1.8	Indoors, line of sight
1.8	Supermarket
1.8	Factory
2.09	Conference room
2.2	Store
2 – 3	Factory indoors, without line of sight
2.8	Residence indoors
2.7 -4.3	Indoors office building, without line of sight

INDOORS ATTENUATION BY OBSTACLES WITH 802.15.4

The previous section mentioned the expected power loss for a smooth connection given by Farahani, so it took a factor n =1.6. Because our test scenario provided line of sight the first seven meters, which is where the first wall was located, but because it was a closed scenario, "n" is equal to 1.7.

In the next section shows the results of measurements made in experiments in both scenarios organized by technology. The graphs show the expected data given by the mathematical models compared to experimental data.

EXPERIMENTAL AND ANALYTICAL ANALYSIS OF THE WIRELESS PROPAGATION MODELS

Figure 3 shows the experimental results, analytical models, and indoor space. The free space model is one of the most widely used to predict the behavior of electromagnetic signals between two devices in a wireless network (Santos, 2009). However, we can see from our graph that the collected data fit the model better suited to interior line of sight that was presented in Equation (5). In Table 6, we see the overall statistics of packet loss that occurred during the test.

Figure 4 shows the model using indoor 802.11n technology barriers. MIMO antennas were used in this experiment so that the attenuation by obstacles does not behave as expected with the indoor model. While the theoretical data seem to approach the obtained data, the expected attenuation produced by signal penetration through the 5 walls is not presented in the experimental data, clearly showing a nearly constant signal throughout almost the entire experiment. Table 7 shows the packet loss ratio for this experiment, which is only 7.6%.

Figure 5 shows the theoretical and experimental results of the test line of sight using 802.15.4 technology nodes. As seen, the test only reaches 23 meters; after this, distance the nodes were not able to communicate each other. While the maximum distance attained in this test was 23 meters, the signal quality with respect to packet loss is also a factor. Figure 6 shows the percentage of packets received throughout this experiment.

The indoor model enclosed space is displayed in Figure 7. This experiment was intended to validate the propagation model with obstacles. However, the XBee nodes were not able to penetrate the concrete walls of the scenario so that the test had to be concluded with the first wall. The loss of packets in this test was greater than that obtained in the interior with line of sight, the percentage of successfully received packets is shown in Figure 8.

Table 6. Indoors line of sight (IEEE 802.11n): network statistics

Packets sent	Bytes sent	Packets received	Bytes received	% Lost Packets/ Bytes
974	95452	925	90650	5.0308

Table 7. Network statistics "indoors" without line of sight (IEEE 802.11n)

Packets sent	Bytes sent	Packets received	Bytes received	% Lost Packets/ Bytes
1135	111230	1048	102704	7.6651

Figure 3. Free space and indoors models with line of sight (IEEE 802.11n)

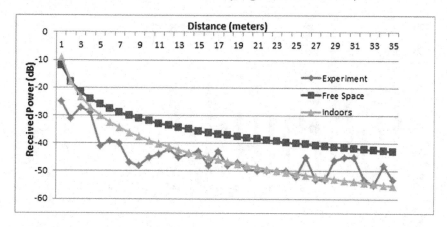

Figure 4. Indoor model with obstacles (IEEE 802.11 n)

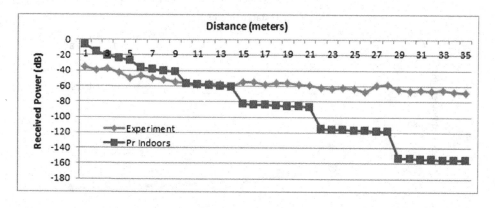

Figure 5. Indoor line of sight model (IEEE 802.15.4)

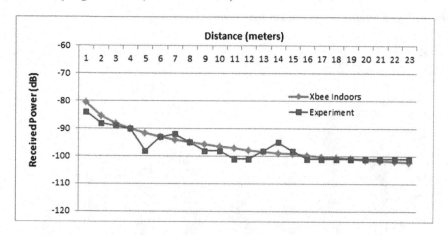

Figure 6. Indoor line of sight packet loss rate

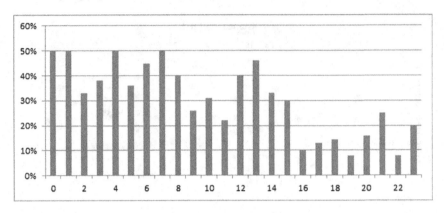

Figure 7. Indoor model in enclosed space

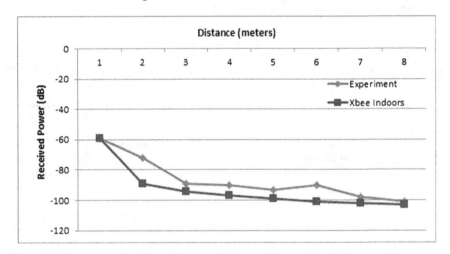

Figure 8. Indoor closed environment packet loss rate

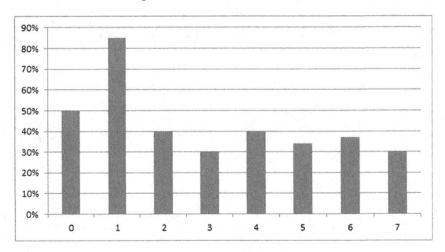

CONCLUSION

This chapter analyzed the behavior of electromagnetic signal propagation in two indoor scenarios with IEEE802.11n and IEEE802.15.4 technologies. We analyzed these two technologies as proposed wireless communication standards for ad hoc mobile robot communication. While IEEE 802.11n technology provides greater coverage, its high-energy consumption is not recommended for small mobile robots. In contrast, IEEE 802.15.4 does not possess a very wide coverage. Despite this, however, it is more convenient for small robot applications and concentrated networks due to its low power consumption. A secure routing protocol for packet reception could be the solution for packet loss that occurs. Another interesting option is the new XBee proposal, which uses RN-XVWifly technology, including IEEE802.11b/g with the same characteristics as traditional modules. This is particularly significant because it allows users to take advantage of this standard for devices with low energy consumption.

REFERENCES

Acosta-Ponce, M. C. (2006). *Estudio del estándar IEEE 802.15.4 "ZigBee" para comunicaciones inalámbricas de área personal de bajo consumo de energía y su comparación con el estándar IEEE 802.15.1 "Bluetooth"*. Quito, Ecuador: Escuela Politécnica Nacional.

Bracka, P., Midonnet, S., & Roussel, G. (2005). *Trayectory based communication in an ad hoc networl of robots*. Washington, DC: IEEE Press. doi:10.1109/WIMOB.2005.1512878

Farahani, S. (2008). *ZigBee wireless networks and transceivers*. London, UK: Newnes.

Fitch, R., & Lal, R. (2009). *Experiments with a ZigBee wireless communication system for self-reconfiguring modular robots*. Paper presented at the 2009 IEEE International Conference on Robotics and Automation. Kobe, Japan.

IEEE. (2003). *802.15.4 IEEE standard for information technology— Telecommunications and information exchange between systems— Local and metropolitan area networks— Specific requirements part 15.4: Wireless medium access control (MAC) and physical layer (PHY) specifications for low-rate wireless personal area networks (LR-WPANs)*. New York, NY: The Institute of Electrical and Electronics Engineers, Inc.

IEEE. (2009). *802.11n-2009 - IEEE standard for Information technology-- Local and metropolitan area networks-- Specific requirements-- Part 11: Wireless LAN medium access control (MAC) and physical layer (PHY) specifications amendment 5: Enhancements for higher throughput*. New York, NY: IEEE.

ITU. (1997). *Recommendation ITU-R M.1225 guidelines for evaluation of radio transmission technologies for IMT-2000*. Retrieved from http://www.itu.int

Jardosh, E. M. B.-R., Almeroth, K. C., & Suri, S. (2003). *Towards realistic mobility models for mobile ad hoc networks*. Paper presented at the 9th Annual International Conference on Mobile Computing and Networking. New York, NY.

Java, G. P. (2001). *IPTraf, IP network monitoring software (version 3.0.0)*. Retrieved from http://iptraf.seul.org/

Luu, B. B., O'Brien, B. J., & Baran, D. G. (2007). A soldier-robot ad hoc network. In *Proceedings of the Fifth Annual IEEE International Conference on Pervasive Computing and Communications Workshops (PerComW 2007)*. White Plains, NY: IEEE Press.

MetaGeek. L. (2011). *Chanalyzer (version 3.0)*. Retrieved from http://www.metageek.net

Networks, U. (2012). *Website.* Retrieved 2012, from http://site.microcom.us/sr71xdatasheet.pdf

Santos, R. A. G.-P., García-Ruiz, A., Edwards-Block, A., Rangel-Licea, V., & Villaseñor-González, L. A. (2009). Hybrid routing algorithm for emergency and rural wireless networks. *Electronics and Electrical Engineering, 1*(89), 6.

Wang, M. Z. A. N. A. (2003). *Ad-hoc robot wireless communication*. Paper presented at the Systems, Man and Cybernetics, 2003. New York, NY.

Weitzenfeld, A., Martinez-Gomez, L., Francois, J. P., Levin-Pick, A., & Boice, J. (2006). *Multi-robot systems: Extending RoboCup small-size architecture with local vision and ad-hoc networking*. Paper presented at the Robotics Symposium, 2006. Santiago, Chile.

XBee.cl. (2009). *Website.* Retrieved from http://www.xbee.cl/caracteristicas.html

KEY TERMS AND DEFINTIONS

Ad-Hoc Wireless Network: Collection of mobile elements that form a temporary network without an established infrastructure.

Attenuation: Is the gradual loss in intensity of any kind of flux through a medium. In electrical engineering and telecommunications, attenuation affects the propagation of waves and signals in electrical circuits, in optical fibers, as well as in air (radio waves).

Electromagnetic Spectrum: Range of all possible frequencies of electromagnetic radiation. The "electromagnetic spectrum" of an object is the characteristic distribution of electromagnetic radiation emitted or absorbed by that particular object.

IEEE 802.11: Set of standards for implementing Wireless Local Area Network (WLAN) computer communication in the 2.4, 3.6, and 5 GHz frequency bands.

IEEE 802.15.4: Standard that specifies the physical layer and media access control for Low-Rate Wireless Personal Area Networks (LR-WPANs).

Propagation Constant: An electromagnetic wave is a measure of the change undergone by the amplitude of the wave as it propagates in a given direction. The quantity being measured can be the voltage or current in a circuit or a field vector such as electric field strength or flux density. The propagation constant itself measures change per meter but is otherwise dimensionless.

Received Signal Strength Indicator (RSSI): Generic radio receiver technology metric of the power present in a received radio signal.

ENDNOTES

[1] Radiometrix: radio transmitter/receiver UHF frequencies 914MHz or 869MHz and a packet data controller of 64Kbit/s.

Compilation of References

Abolhasan, M., Wysocki, T., & Dutkiewicz, E. (2004). A review of routing protocols for mobile ad hoc networks. *Ad Hoc Networks*, *2*(1), 1–22. doi:10.1016/S1570-8705(03)00043-X

Aguiar, A. P., & Pascoal, A. (2000). Stabilization of the extended nonholonomic double integrator via logical based hybrid control: An application to point stabilization of mobile robots. In *Proceedings of the 6th International IFAC Symposium on Robot Control*. Vienna, Austria: IFAC.

Aicardi, M., Casalino, G., Bicchi, A., & Balestrino, A. (1995). Closed loop steering of unicycle-vehicles via Lyapunov techniques. *IEEE Robotics & Automation Magazine*, *2*(1), 27–35. doi:10.1109/100.388294

Akaya, K., & Younis, M. F. (2005). A survey on routing protocols for wireless sensor networks. *Ad Hoc Networks*, *3*(3), 325–349. doi:10.1016/j.adhoc.2003.09.010

Akyldiz, I. F., Su, W., Sankarasubramaniam, Y., & Cayirci, E. (2002). Wireless sensor network: A survey. *Computer Networks*, *38*, 393–422. doi:10.1016/S1389-1286(01)00302-4

Alami, R., Chatila, R., Fleury, S., Ghallab, M., & Ingrand, F. (2009). An architecture for autonomy. *The International Journal of Robotics Research*, *31*(11), 315–337.

Altera Corporation. (2011). *AHDL*. Retrieved from http://www.altera.com

Amaya, Y., & Ruiz, J. (2005). *Localización dinámica de móviles y obstáculos en una escena controlada para aplicaciones en robótica*. (Bachelor's Thesis). Industrial University of Santander. Santander, Colombia.

Anderson, M., & Papanikolopoulos, N. (2008). Implicit cooperation strategies for multi-robot search of unknown areas. *Journal of Intelligent and Robotic Systems: Theory and Applications*, *53*(4), 381–397. doi:10.1007/s10846-008-9242-5

Andrieu, C. (2003). An introduction to MCMC for machine learning. *Machine Learning*, *50*, 5–43. doi:10.1023/A:1020281327116

Antonelli, G., Arricchiello, F., Chiaverini, S., & Setola, R. (2006). Coordinated control of mobile antennas for ad hoc networks. *International Journal of Modelling. Identification and Control*, *1*(1), 63–71. doi:10.1504/IJMIC.2006.008649

Antonelli, G., Chiaverini, S., & Fusco, G. (2007). A fuzzy-logic-based approach for mobile robot path tracking. *IEEE Transactions on Fuzzy Systems*, *15*(2). doi:10.1109/TFUZZ.2006.879998

Antonelli, G., Chiaverini, S., Sarkar, N., & West, M. (1991). Adaptative control of an autonomous underwater vehicle: Experimental results on Odin. *IEEE Transactions on Control Systems Technology*, *9*(5), 756–765. doi:10.1109/87.944470

Arantes, G. M. A. Jr, França, F. M. G., & Martinhon, C. A. J. (2009). Randomized generation of acyclic orientations upon anonymous distributed systems. *Journal of Parallel and Distributed Computing*, *69*(3), 239–246. doi:10.1016/j.jpdc.2008.11.009

Astolfi, A. (1996). Discontinuous control of nonholonomic systems. *Systems & Control Letters*, *27*, 37–45. doi:10.1016/0167-6911(95)00041-0

Astrom, K. J., & Wittenmark, B. (1997). *Computer-controlled systems: Theory and design* (3rd ed.). Upper Saddle River, NJ: Prentice Hall.

Baccar, S. (2011). *Service-oriented architecture for wireless sensor networks: A RESTful approach.* (Unpublished Master Thesis Dissertation). National School of Engineering. Sfax, Tunisia.

Baccour, N., Koubâa, A., Mottola, L., Zuniga, M., Youssef, H., Boano, C., & Alves, M. (2012). Radio link quality estimation in wireless sensor networks: A survey. *ACM Transactions on Sensor Networks, 8*(4).

Bailey, T., & Durrant-Whyte, H. (2006). Simultaneous localisation and mapping (SLAM): Part II state of the art. *Robotics and Automation Magazine.* Retrieved from http://www.cs.berkeley.edu/~pabbeel/cs287-fa09/readings/Bailey_Durrant-Whyte_SLAM-tutorial-II.pdf

Balzarotti, M., & Ulivi, G. (1996). The fuzzy horizons obstacle avoidance method for mobile robots. In *Proceedings of the World Automation Conference, ISRAM 1996.* Montpellier, France: ISRAM.

Barbosa, V., & Gafni, E. (1989). Concurrency in heavily loaded neighborhood-constrained systems. *ACM Transactions on Programming Languages and Systems, 11*(4), 562–584. doi:10.1145/69558.69560

Barnes, M., Everett, H. R., & Rudakevych, P. (2005). ThrowBot: Design considerations for a man-portable throwable robot. In *Proceedings of SPIE-The International Society for Optical Engineering,* (pp. 511 - 520). SPIE.

Barnhard, D. H., Wimpey, B. J., & Potter, W. D. (2004). Odin and Hodur: Using Bluetooth communication for coordinated robotic search. In *Proceeding of the International Conference on Artificial Intelligence.* Las Vegas, NV: IEEE.

Barrett, A., Rabideau, G., Estlin, T., & Chien, S. (2007). Coordinated continual planning methods for cooperating rovers. *IEEE Aerospace and Electronic Systems Magazine, 22*(2), 27–33. doi:10.1109/MAES.2007.323296

Barros, E. A., & Soares, F. J. A. (1991). Desenvolvimento de um robô submarino de baixo custo. In *Proceedings of the XIV Congresso Brasileiro de Automática,* (vol. 4, pp. 2121-2126). Natal, Brazil: Federal University of Rio Grande do Norte.

Basye, K., Dean, T., Kirman, J., & Lejter, M. (1992). A decision-theoretic approach to planning, perception, and control. *IEEE Expert, 7*(4). doi:10.1109/64.153465

Batalin, M. A., & Sukhatme, G. S. (2004). Using a sensor network for distributed multi-robot task allocation. In *Proceedings of the IEEE International Conference on Robotics Automation (ICRA),* (pp. 158-164). IEEE Press.

Bellingham, J. G., & Rajan, K. (2007). Robotics in remote and hostile environments. *Science, 318*(5853), 1098–1102. doi:10.1126/science.1146230

Benton, C., Haag, M. M. M., Nitzel, R., Blidberg, D. R., Chappell, S. G., & Mupparapu, S. (2005). Advancements within the AUSNET protocol. In *Proceedings of the Fourteenth International Symposium on Unmanned Untethered Submersible Technology (UUST 2005).* Durham, NH: UUST.

Benton, C., Kenney, J., Nitzel, R., Blidberg, D. R., Chappell, S. G., & Mupparapu, S. (2004). Autonomous systems network (AUSNET) – Protocols to support ad-hoc AUV communications. In *Proceedings of IEEE/OES AUV 2004: A Workshop on Multiple Autonomous Underwater Vehicle Operations.* Sebasco Estates, ME: IEEE Press.

Berg, B. A. (2004). *Markov chain Monte Carlo simulations and their statistical analysis.* Singapore, Singapore: World Scientific. doi:10.1142/9789812700919_0001

Bhasker, J. (2010). *A VHDL primer.* Upper Saddle River, NJ: Prentice Hall.

Bhattacharya, P., & Gavrilova, M. L. (2008). Roadmap-based path planning - Using the Voronoi diagram for a clearance-based shortest path. *IEEE Robotics & Automation Magazine, 15,* 58–66. doi:10.1109/MRA.2008.921540

Binder, K. (1995). *The Monte Carlo method in condensed matter physics.* New York, NY: Springer. doi:10.1007/3-540-60174-0

Blanke, M., Lindegaard, K. P., & Fossen, T. I. (1991). Dynamic model for thrust generation of marine propellers. In *Proceedings of the IFAC Conference on Maneuvering and Control of Marine Craft,* (vol. 9). Aalborg, Denmark: IFAC.

Blankenship, J., & Mishal, S. (2010). *Enhancing the Pololu 3pi with RobotBASIC*. Retrieved from http://www.amazon.com/Enhancing-Pololu-RobotBASIC-John-Blankenship/dp/1438276052/ref=sr_1_6?ie=UTF8&qid=1346469953&sr=8-6&keywords=robotbasic

Blankenship, J., & Mishal, S. (2012). *The RROS User's Manual: A RobotBASIC Robot Operating System on a Chip*. Retrieved from http://www.robotbasic.org/6.html

Blankenship, J., & Mishal, S. (2008). *Robot Programmer's Bonanza*. New York, NY: McGraw-Hill.

Bloch, A. M., McClamroch, N. H., & Reyhanoglu, M. (1990). Controllability and stabilizability properties of a nonholonomic control system. In *Proceedings of the IEEE Conference on Decision and Control*. Honolulu, HI: IEEE.

Bloch, A. M. (Ed.). (2003). *Nonholonomic mechanics and control*. New York, NY: Springer. doi:10.1007/b97376

Bodhale, D., Afzulpurkar, N., & Thanh, N. T. (2008, February). *Path planning for a mobile robot in a dynamic environment*. Paper presented at IEEE International Conference on Robotics and Biomimetics, 2008. Bangkok, Thailand.

Bona, B., Indri, M., & Smaldone, N. (2006). Rapid prototyping of a model-based control with friction compensation for a direct-drive robot. *IEEE/ASME Transactions on Mechatronics*, *11*(5), 576–584. doi:10.1109/TMECH.2006.882989

Borenstein, J. H. R. (1996). *Navigation móbile robots: Systems and techniques*. New York, NY: AK Peters.

Box, G. E. P., Hunter, J. S., & Hunter, W. G. (2005). *Statistics for experimenters*. Hoboken, NJ: Wiley.

Bracka, P., Midonnet, S., & Roussel, G. (2005). *Trayectory based communication in an ad hoc networl of robots*. Washington, DC: IEEE Press. doi:10.1109/WIMOB.2005.1512878

Braitenberg, V. (1987). *Vehicles: Experiments in synthetic psychology*. Cambridge, MA: The MIT Press.

Brockett, R. W. (Ed.). (1983). Asymtotic stability and feedback stabilization. In R. W. Brockett, R. S. Millman, & H. S. Sussman (Eds.), *Differential Geometric Control Theory*, (pp. 181-191). Boston, MA: Birkhäuser.

Brooks, R. (1986). *Achieving artificial intelligence through building robots*. Cambridge, MA: The MIT Press.

Bruno Siciliano, O. K. (2007). *Springer Handbook of Robotics*. Springer.

Buniyamin, N., Sariff, N., Wan Ngah, W. A. J., & Mohamad, Z. (2011). Robot global path planning overview and a variation of ant colony system algorithm. *International Journal of Mathematics and Computers in Simulation*, *5*, 9–16.

Burgard, W., Moors, M., Stachniss, C., & Schneider, F. E. (2005). Coordinated multi-robot exploration. *IEEE Transactions on Robotics*, *21*(3), 376–386. doi:10.1109/TRO.2004.839232

Burke, J. L., & Murphy, R. R. (2004). Human-robot interaction in USAR technical search: Two heads are better than one. In *Proceedings of the IEEE International Workshop on Robot and Human Interactive Communication*, (pp. 307-312). Kurashiki, Japan: IEEE Press.

Burmitt, B. L., & Stentz, A. (1998). GRAMMPS: A generalized mission planner for multiple mobile robots in unstructured environments. In *Proceedings of the IEEE International Conference on Robotics and Automation*, (pp. 1564 - 1571). IEEE Press.

Byung, K., & Kang, S. (1984). An efficient minimum-time robot path planning under realistic conditions. In *Proceedings of the American Control Conference, 1984*, (pp. 296-303). IEEE.

Canudas de Wit, C., & Sørdalen, O. J. (1991). Exponential stabilization of mobile robots with nonholomic constraints. In *Proceedings of the IEEE Conference on Decision and Control*. Brighton, UK: IEEE Press.

Capi, G., & Kaneko, S. (2009). Evolution of neural controllers in real mobile robots for task switching behaviors. *International Journal of Innovative Computing, Information, & Control*, *5*(11), 4017–4024.

Carlone, L., Aragues, R., Castellanos, J. A., & Bona, B. (2011). A first-order solution to simultaneous localization and mapping with graphical models. In *Proceedings of the 2011 IEEE International Conference on Robotics and Automation (ICRA)*, (pp. 1764-1771). IEEE Press.

Carlos, V. R. (2007). *Universidad Nacional de Trujillo*. Retrieved from http://seccperu.org/files/ZigBee.pdf

Carvalho, D., Protti, F., Gregorio, M., & França, F. M. G. (2004). A novel distributed scheduling algorithm for resource sharing under near-heavy load. In *Proceedings of the International Conference on Principles of Distributed Systems (OPODIS)*, (Vol. 8). OPODIS.

Casper, J., Micire, M., & Murphy, R. R. (2000). Issues in intelligent robots for search and rescue. In *Proceedings of SPIE - The International Society for Optical Engineering*, (pp. 292 - 302). SPIE.

Casper, J., & Murphy, R. R. (2003). Human-robot interactions during the robot-assisted urban search and rescue response at the world trade center. *IEEE Transactions on Systems, Man, and Cybernetics. Part B, Cybernetics*, *33*(3), 367–385. doi:10.1109/TSMCB.2003.811794

Cassia, R. F., Alves, V. C., Besnard, F. G.-D., & França, F. M. G. (2009). Synchronous-to-asynchronous conversion of cryptographic circuits. *Journal of Circuits, Systems, and Computers*, *18*(2), 271–282. doi:10.1142/S0218126609005058

Chang, K. C. (2009). *Digital design and modeling with VHDL and synthesis*. Washington, DC: IEEE Computer Society Press.

Cheng, Y., Wang, X., & Lei, R. (2009). A fuzzy control system for path following of mobile robots. *ICIC Express Letters*, *3*(3), 403–408.

Choi, C., Hong, S.-G., Shin, J.-H., Jeong, I.-K., & Lee, J.-J. (1995). Dynamical path-planning algorithm of a mobile robot using chaotic neuron model. In *Proceedings of the 1995 IEEE/RSJ International Conference on Intelligent Robots and Systems: Human Robot Interaction and Cooperative Robots*, (Vol. 2, pp. 456 – 461). IEEE Press.

Chung, K. (1960). *Markov chains with stationary transition probabilities*. Berlin, Germany: Springer. doi:10.1007/978-3-642-49686-8

Clausen, T., & Jacquet, P. (2003). Optimized link state routing protocol (OLSR). *Network Working Group, 3626*.

Compton, K. (2009). Reconfigurable computing: A survey of systems and software. *ACM Computing Surveys*, *34*, 171–210. doi:10.1145/508352.508353

Conte, G., & Serrani, A. (1991). Modeling and simulation of underwater vehicles. In *Proceedings of the IEEE International Symposium on Computer-Aided Control System Design*, (pp. 62-67). Dearborn, MI: IEEE Press.

Coric, S., Leeser, M., & Miller, E. (2002). Parallel-bean backprojection: An FPGA implementation optimized for medical imaging. [Monterey, CA: FPGA.]. *Proceedings of FPGA*, *2002*, 217–226.

Coron, J. M. (1991). Global asymtotic stabilization for controllable systems without drift. *Mathematical Constrained System Design*, *15*, 295–312.

CoroWare Inc. (2011). *Website*. Retrieved from http://www.CoroWare.com

Cox, I. (1991). Blanche—An experiment in guidance and navigation of an autonomous robot vehicle. *IEEE Transactions on Robotics and Automation*, *7*(2). doi:10.1109/70.75902

Crenshaw, T. L., Hoke, S., Tirumala, A., & Caccamo, M. (2007). Robust implicit EDF: A wireless MAC protocol for collaborative real-time systems. *ACM Transactions on Embedded Computing Systems*, *6*(4).

Crowley, J. (1985). Navigation for an intelligent mobile robot. *IEEE Journal on Robotics and Automation*, *1*(1), 31–41. doi:10.1109/JRA.1985.1087002

Cunha, J. P. V. S. (1992). *Projeto e estudo de simulação de um sistema de controle a estrutura variável de um veículo submarino de operação remota*. (Unpublished Master Dissertation). Federal University of Rio de Janeiro. Rio de Janeiro, Brazil.

Daltorio, K., Gorb, S., Peressadko, A., Horchler, A. D., Ritzmann, R. E., & Quinn, R. D. (2006). A robot that climbs walls using micro-structured polymer feet. In *Climbing and Walking Robots* (pp. 131–138). Berlin, Germany: Springer. doi:10.1007/3-540-26415-9_15

Dayawansa, W. P., Martin, C. F., & Samuelson, S. (1995). Asymtotic stabilization of a generic class of 3-dimensional homogeneus quadratic systems. *Systems & Control Letters*, *24*(2), 115–123. doi:10.1016/0167-6911(94)00040-3

De Doná, J. A., Suryawan, F., Seron, M. M., & Lévine, J. (2009). A flatness-based iterative method for reference trajectory generation in constrained NMPC. In Magni, L., Raimondo, D. M., & Allgöwer, F. (Eds.), *Nonlinear Model Predictive Control: Towards New Challenging Applications* (pp. 325–333). Berlin, Germany: Springer-Verlag.

De Luca, A., Oriolo, G., & Vendittelli, M. (2001). Control of wheeled mobile robots: An experimental overview. In *RAMSETE: Articulated and Mobile Robots for Service and Technology*. Berlin, Germany: Springer.

Deligiannis, V., Davrazos, G., Manesis, S., & Arampatzis, T. (2006). Flatness conservation in the n-trailer system equipped with a sliding kingpin mechanism. *Journal of Intelligent & Robotic Systems, 46*(2), 151–162. doi:10.1007/s10846-006-9056-2

Demirkol, I., Ersoy, C., & Alagoz, F. (2006). MAC protocols for wireless sensor networks: A survey. *IEEE Communications Magazine, 44*(4), 115–121. doi:10.1109/MCOM.2006.1632658

Dias, B. M., Zlot, R., Kalra, N., & Stentz, A. (2006). Market-based multirobot coordination: A survey and analysis. *Proceedings of the IEEE, 94*, 1257–1270. doi:10.1109/JPROC.2006.876939

Diniz, P. C., & Park, J. (2002). Data reorganization engines for the next generation of system-on-a-chip FPGAs. In *Proceedings of FPGA*, (pp. 237-244). Monterey, CA: FPGA.

Dominguez, R. B. (1989). *Simulação e controle de um veículo submarino de operação remota.* (Unpublished Master Dissertation). Federal University of Rio de Janeiro. Rio de Janeiro, Brazil.

Dorigo, M., & Stützle, T. (2004). *Ant colony optimization.* Cambridge, MA: The MIT Press. doi:10.1007/b99492

Doroodgar, B., Ficocelli, M., Mobedi, B., & Negat, G. (2010). The search for survivors: Cooperative human-robot itneraction in search and rescue environments using semi-autonomous robots. In *Proceedings of the IEEE International Conference on Robotics and Automation*, (pp. 2858-5863). IEEE Press.

Dudek, G., Jenkin, M., Milios, E., & Wilkes, D. (1996). *Reflections on modeling a sonar range sensor.* Retrieved from http://www.cs.cmu.edu/~motionplanning/papers/sbp_papers/integrated2/dudek_snr_mdl.pdf

Dudek, G., & Jenkin, M. (2000). *Computational principles of mobile robotics.* Cambridge, UK: Cambridge University Press.

Durrant-White, H., & Bailey, T. (2006). Simultaneous localization and mapping: Part 1. *IEEE Robotics & Automation Magazine, 13*(2), 99–110. doi:10.1109/MRA.2006.1638022

Dzung, D., Apneseth, C., Endersen, J., & Frey, J. E. (2005). Design and implementation of a real-time wireless sensor/actuator communication system. In *Proceedings of the 10th IEEE Conference on Emerging Technologies and Factory Automation*, (pp. 443-452). IEEE Press.

Eberhart, Y., & Shi, Y. (2001). Particle swarm optimization: Developments, applications and resources in evolutionary computation. In *Proceedings of the 2001 Congress on Evolutionary Computation*, (pp. 81-86). IEEE.

Elfes, A. (1989). Using occupancy grids for mobile robot perception and navigation. *IEEE Computer, 22*(6), 46–57. doi:10.1109/2.30720

El-Hoiydi, A., & Decotignie, J. D. (2004). WiseMAC: An ultra low power mac protocol for the downlink of infrastructure wireless sensor networks. In *Proceedings of the Ninth IEEE Symposium on Communication, ISCC 2004*, (pp. 244-251). IEEE Press.

Ellips, M., & Sedighizadeh, D. (2007). Classic and heuristic approaches in robot motion planning – A chronological review. *Proceedings of World Academy of Science. Engineering and Technology, 23*, 101–106.

Elmogy, A. M., Khamis, A. M., & Karray, F. O. (2009). Dynamic complex task allocation in multisensor surveillance systems. In *Proceedings of the International Conference on Signals, Circuits and Systems (SCS)*, (pp. 1-6). SCS.

Fan, X., Luo, X., Yi, S., Yang, S., & Zhang, H. (2003). Optimal path planning for mobile robot based on intensified ant colony optimization algorithm. In *Proceedings of the IEEE International Conference on Robotics, Intelligent Systems and Signal Processing*, (Vol. 1, pp. 131-136). IEEE Press.

Farahani, S. (2008). *ZigBee wireless networks and transceivers*. London, UK: Newnes.

Feng, M., Fu, Y., Pan, B., & Liu, C. (2011). A medical robot system for celiac minimally invasive surgery. In *Proceedings of the 2011 IEEE International Conference on Information and Automation (ICIA)*, (pp. 33-38). IEEE Press.

Fielding, R. T. (2000). *Architectural styles and the design of network-based software architectures*. (Doctoral Dissertation). University of California. Irvine, CA.

Fischer, W., Tâche, F., & Siegwart, R. (2008). Magnetic wall climbing robot for thin surfaces with specific obstacles. In Laugier, C., & Siegwart, R. (Eds.), *Field and Service Robotics* (pp. 551–561). Berlin, Germany: Springer-Verlag. doi:10.1007/978-3-540-75404-6_53

Fitch, R., & Lal, R. (2009). *Experiments with a ZigBee wireless communication system for self-reconfiguring modular robots*. Paper presented at the 2009 IEEE International Conference on Robotics and Automation. Kobe, Japan.

Flickinger, M. D. (2007). *Motion planning and coordination of mobile robot behavior for medium scale distributed wireless network experiments*. (Unpublished Master Dissertation). University of Utah. Salt Lake City, UT.

Fliess, M., Lévine, J., & Rouchon, P. (1991). A simple approach of crane control via a generalized state-space model. In *Proceedings of IEEE Control and Decision Conference*, (pp. 736-741). IEEE Press.

Fliess, M., Levine, J., Martin, P., & Rouchon, P. (1994). *Flatness and defect of non liner systems: Introductory theory and examples*. CAS.

Flocchini, P., Nayak, A., & Schulz, A. (2005). Cleaning an arbitrary regular network with mobile agents. In G. Chakraborty (Ed.), *Distributed Computing and Internet Technology, Second International Conference*, (Vol. 3816, pp. 132-142). Berlin, Germany: Springer.

Fogel, D. (1994). An introduction to simulated evolutionary optimization. *IEEE Transactions on Neural Networks*, *5*(1). doi:10.1109/72.265956

Foka, A. F., & Trahanias, P. E. (2002). Predictive autonomous robot navigation. In *Proceedings of the IEEE/RSJ Internacional Conference on Intelligent Robots and Systems (IROS)*, (pp. 490-495). IEEE Press.

Fox, D., Burgard, W., & Thrun, S. (1999). Markov localization for robots in dynamic environments.

Fox, D., Burgard, W., Kruppa, H., & Thrun, S. (2000). A probabilistic approach to collaborative multi-robot localization. *Autonomous Robots*, *8*(3), 325–344. doi:10.1023/A:1008937911390

Fox, M. S. (1979). *Organization structuring: Designing large complex software*. Pittsburgh, PA: Carnegie Mellon University Press.

Fox, M. S. (1981). An organizational view of distributed systems. *IEEE Transactions on Systems, Man, and Cybernetics*, *11*(1), 70–80. doi:10.1109/TSMC.1981.4308580

Fox, R. W., & McDonald, A. T. (1981). *Introdução a mecânica dos fluidos*. Rio de Janeiro, Brazil: Guanabara Dois.

Funk, N. (2003). *A study of Kalman filter applied to visual tracking*. Retrieved from http://www.njfunk.com/research/courses/652-probability-report.pdf

Gaddour, O., Koubaa, A., Chaudhry, S., Tezeghdanti, M., Chaari, R., & Abid, M. (2012). Simulation and performance evaluation of DAG construction with RPL. In *Proceedings of the Third International Conference on Communication and Networking, COMNET*. COMNET.

Gao, M., Xu, J., & Tian, J. (2008). Mobile robot path planning based on improved augment ant colony algorithm. In *Proceedings of the Second International Conference on Genetic and Evolutionary Computing*, (pp. 273-276). IEEE.

Gasparri, A., Panzieri, S., & Pascucci, F. (2008). A fast conjunctive resampling particle filter for collaborative multi-robot localization. In *Proceedings of the Workshop on Formal Models and Methods for MultiRobot Systems*. AAMAS.

Gasparri, A., Panzieri, S., Pascucci, F., & Ulivi, G. (2009). An interlaced Kalman filter for sensors networks localization. *International Journal of Sensor Networks*, 5(3), 164–172. doi:10.1504/IJSNET.2009.026364

George Bekey, R. A. (2008). *ROBOTICS STATE OF ART AND FUTURE CHALLENGES* (p. 11). Imperial College Press. doi:10.1142/9781848160071_0002

Gerkey, B. P., Vaughan, R. T., Stoy, K., Howard, A., Sukhatme, G. S., & Mataric, M. J. (2001). Most valuable player: A robot device server for distributed control. In *Proceedings of IEEE/RSJ IROS 2001,* (pp. 1226-1231). Wailea, Hawaii: IEEE Press.

Gerkey, B., & Mataric, M. (2004). A formal analysis and taxonomy of task allocation in multi-robot systems. *The International Journal of Robotics Research*, 23(9), 939–954. doi:10.1177/0278364904045564

Glielmo, L., Setola, R., & Vasca, F. (1999). An interlaced extended Kalman filter. *IEEE Transactions on Automatic Control*, 44(8), 1546–1549. doi:10.1109/9.780418

Gobriel, S., Mosse, D., & Cleric, R. (2009). TDMA-ASAP: Sensor network TDMA scheduling with adaptive slot-stealing and parallelism. In *Proceedings of the 29th IEEE International Conference on Distributed Computing Systems, ICDCS 2009,* (pp. 458-465). IEEE Press.

Goldberg, D. E. (1989). *Genetic algorithms in search optimization, and machine learning.* Reading, MA: Addison-Wesley Publishing Company.

Goldman, J. A. (1994). Path planning problems and solutions. In *Proceedings of the IEEE 1994 National Aerospace and Electronics Conference, 1994,* (Vol. 1, pp. 105-108). IEEE Press.

Gonçalves, V. C. F., Lima, P. M. V., Maculan, N., & França, F. M. G. (2010). A distributed dynamics for webgraph decontamination. In T. Margaria & B. Steffen (Eds.), *4th International Symposium on Leveraging Applications of Formal Methods, Verification and Validation,* (Vol. 6415, pp. 462-472). Berlin, Germany: Springer.

Graf, B., Wandosell, J. M. H., & Schaeffer, C. (2001). Flexible path planning for nonholonomic mobile robots. [Stuttgart, Germany: Angenommenzur Eurobot.]. *Proceedings of Angenommenzur Eurobot, 2001,* 456–565.

Grigorescu, S. M., Macesanu, G., Cocias, T. T., & Moldoveanu, F. (2011). On the real-time modelling of a robotic scene perception and estimation system. In *Proceedings of the 15th International Conference on System Theory, Control, and Computing (ICSTCC),* (pp 1-4). ICSTCC.

Grisetti, G., Diego, G., Stachniss, C., Burgard, W., & Nardi, D. (2006). *Fast and accurate SLAM with Rao-Blackwellized particle filters.* Retrieved from http://www.informatik.uni-freiburg.de/~grisetti/pdf/grisetti06jras.pdf

Habib, M. K., & Yuta, S. (1989). Structuring free space as prime rectangular areas (PRAs) with on-line path planning and navigation for mobile robots. In *Proceedings of the IEEE International Conference on Systems, Man and Cybernetics,* (Vol. 2, pp. 557 – 565). IEEE Press.

Haight, J. M., & Kecojevic, V. (2005). Automation vs. human intervention: What is the best fit for the best performance? *Process Safety Progress*, 24(1), 45–51. doi:10.1002/prs.10050

Halberstam, E., Navarro-Serment, L., Conescu, R., Mau, S., Podnar, G., Guisewite, A. D., et al. (2006). A robot supervision architecture for safe and efficient space exploration and operation. In *Proceedings of the 10th Biennial International Conference on Engineering, Construction, and Operations in Challenging Environments,* (pp. 92-102). IEEE.

Hartl, D., & Jones, E. (1998). *Genetic principles and analysis* (4th ed.). New York, NY: Jones and Bartlett Publishers.

Harvey, I. (1996). Artificial evolution and real robots. In *Proceedings of International Symposium on Artificial Life and Robotics (AROB),* (pp. 138-141). Beppu, Japan: Masanori Sugisaka.

Hassanzadeh, I., Madani, K., & Badamchizadeh, M. A. (2010). *Mobile robot path planning based on shuffled frog leaping optimization algorithm.* Paper presented at the IEEE Conference on Automation Science and Engineering (CASE). Toronto, Canada.

Hauser, J., & Hindman, R. (1995). Maneuver regulation from trajectory tracking: Feedback linearizable systems. In *Proceeding World Congress of International Federation Automatic Control, Symposium of Nonlinear Control System Design,* (pp. 638-643). IEEE.

Hayes, M. H. (2008). *Schaum's outline of digital signal processing*. New York, NY: McGraw-Hill.

Haynes, G. C., Khripin, A., Lynch, G., Amory, J., Saunders, A., & Rizz, I. A., & Koditschek, D. E. (2009). Rapid pole climbing with a quadrupedal robot. In *Proceedings of the IEEE Conference on Robotics and Automation*. IEEE Press.

He, T., Stankovic, J. A., Lu, C., & Abdelzaher, T. (2003). SPEED: A stateless protocol for real-time communication in sensor networks. In *Proceedings of International Conference on Distributed Computing Systems*. Providence, RI: IEEE.

Heinzelman, W., Chandrakasan, A., & Balakrishnan, H. (2000). Energy-efficient communication protocol for wireless sensor networks. In *Proceedings of the Hawaii International Conference System Sciences*. IEEE.

Heinzelman, W., Kulik, J., & Balakrishnan, H. (1999). Adaptive protocols for information dissemination in wireless sensor networks. In *Proceedings of the 5th Annual ACM/IEEE International Conference on Mobile Computing and Networking (MobiCom 1999)*. Seattle, WA: ACM/IEEE Press.

Hofner, C., & Schmidt, G. (1994). Path planning and guidance techniques for an autonomous mobile cleaning robot. In *Proceedings of the IEEE/RSJ/GI International Conference on Intelligent Robots and Systems 1994: Advanced Robotic Systems and the Real World, IROS 1994*, (Vol. 1, pp. 610 – 617). IEEE Press.

Holland, J. (1992). Genetic algorithms. *Scientific American*, *267*(1), 66–72. doi:10.1038/scientificamerican0792-66

Hong, X., Xu, K., & Gerla, M. (2002). Scalable routing protocols for mobile ad hoc networks. *IEEE Network*, *16*, 11–21. doi:10.1109/MNET.2002.1020231

Horling, B., & Lesser, V. (2005). A survey of multi-agent organizational paradigms. *The Knowledge Engineering Review*, *19*(4), 281–386. doi:10.1017/S0269888905000317

Hsu, L., Cunha, J. P. V. S., Lizarralde, F., & Costa, R. R. (2000). Avaliação experimental e simulação da dinâmica de um veículo submarino de operação remota. *Controle & Automação*, *11*(2), 82–93.

Huang, S., & Dissanayake, G. (2007). Convergence and consistency analysis for extended Kalman filter based SLAM. *IEEE Transactions on Robotics*, *23*, 1036–1049. doi:10.1109/TRO.2007.903811

Hui, M., & Yu-Chu, T. (2008). Robot path planning in dynamic environments using a simulated annealing based approach. In *Proceedings of the 10th International Conference on Control, Automation, Robotics and Vision, ICARCV 2008*, (pp. 1253-1258). ICARCV.

Hui, J. W., & Culler, D. E. (2008). Extending IP to low-power, wireless personal area networks. *IEEE Internet Computing*, *12*(4), 37–45. doi:10.1109/MIC.2008.79

IEEE. (2003). *802.15.4 IEEE standard for information technology— Telecommunications and information exchange between systems— Local and metropolitan area networks— Specific requirements part 15.4: Wireless medium access control (MAC) and physical layer (PHY) specifications for low-rate wireless personal area networks (LR-WPANs)*. New York, NY: The Institute of Electrical and Electronics Engineers, Inc.

IEEE. (2009). *802.11n-2009 - IEEE standard for Information technology-- Local and metropolitan area networks-- Specific requirements-- Part 11: Wireless LAN medium access control (MAC) and physical layer (PHY) specifications amendment 5: Enhancements for higher throughput*. New York, NY: IEEE.

Indiveri, G., Aicardi, M., & Casalino, G. (2000). Robust global stabilization of an underactuated marine vehicle on a linear course by smooth time-invariant feedback. In *Proceedings of the Conference on Decision and Control, CDC 2000*, (pp. 2156-2161). Sydney, Australia: CDC.

Intanagonwiwat, C., Govindan, R., & Estrin, D. (2000). Directed diffusion: A scalable and robust communication paradigm for sensor networks. In *Proceedings of the 6th Annual ACM/IEEE International Conference on Mobile Computing and Networking (MobiCom 2000)*. Boston, MA: ACM/IEEE Press.

Ito, S. A., & Carro, L. (2000). A comparison of microcontrollers targeted to {FPGA}-based embedded applications. In *Proceedings of the IEEE 13th Symposium on Integrated Circuits and Systems Design*, (pp. 397-402). Manaus, Brazil: IEEE Press.

ITU. (1997). *Recommendation ITU-R M.1225 guidelines for evaluation of radio transmission technologies for IMT-2000*. Retrieved from http://www.itu.int

Jamieson, K., Balakrishnan, H., & Tay, Y. C. (2006). sift: a mac protocol for event-driven wireless sensor networks. In *Proceedings of the Third European Workshop on Wireless Sensor Networks (EWSN)*, (pp. 260-275). EWSN.

Jan, G. E., Ki, Y. C., & Parberry, I. (2008). Optimal path planning for mobile robot navigation. *IEEE/ASME Transactions on Mechatronics, 13*, 451–460. doi:10.1109/TMECH.2008.2000822

Jardosh, E. M. B.-R., Almeroth, K. C., & Suri, S. (2003). *Towards realistic mobility models for mobile ad hoc networks*. Paper presented at the 9th Annual International Conference on Mobile Computing and Networking. New York, NY.

Jardosh, S., & Ranjan, P. (2008). A survey: Topology control for wireless sensor networks. In *Proceedings of the International Conference on Signal Processing, Communications and Networking*, (pp. 422-427). IEEE.

Java, G. P. (2001). *IP Traf, IP network monitoring software (version 3.0.0)*. Retrieved from http://iptraf.seul.org/

Jian-She, J., Jing, J., Yong-Hui, W., Ke, Z., & Jia-Jun, H. (2008). Development of remote-controlled home automation system with wireless sensor network. In *Proceedings of the Fifth IEEE International Symposium on Embedded Computing*, (pp 169-173). IEEE Press.

Johnson, D. B., Maltz, D. A., & Hu, Y.-C. (2003). *The dynamic source routing protocol for mobile ad hoc networks (DSR)*. Unpublished.

Joon-Woo, L., & Ju-Jang, L. (2010). Novel ant colony optimization algorithm with path crossover and heterogeneous ants for path planning. In *Proceedings of the IEEE International Conference on Industrial Technology*, (pp. 559-564). IEEE Press.

Jung, S. J., Kwon, T. H., & Chung, W. Y. (2009). A new approach to design ambient sensor network for real time healthcare monitoring system. [IEEE Press.]. *Proceedings of IEEE Sensors, 2009*, 576–580.

Kadous, M. W., Sheh, R. K., & Sammut, C. (2006). Controlling heterogeneous semi-autonomous rescue robot teams. In *Proceedings of the IEEE International Conference on Systems, Man, and Cybernetics*, (pp. 3204-3209). Taipei, Taiwan: IEEE Press.

Kahn, J. M., & Barry, J. R. (1997). Wireless infrared communication. *Proceedings of the IEEE, 85*(2). doi:10.1109/5.554222

Kalman, R. E. (1960). A new approach to linear filtering and prediction problems. *Transactions of the ASME – Journal of Basic Engineering, 82*, 35–45.

Kamiriski, B., Wejrzanowski, K., & Koczara, W. (2004). An application of PSIM simulation software for rapid prototyping of DSP based power electronics control systems. In *Proceedings of the 2004 IEEE 35th Annual Power Electronics Specialists Conference PESC 2004*, (Vol. 1, pp. 336-341). IEEE Press.

Kensington, L. I., & Englewood, D. R. (2003). *Vortex attractor*. United States Patent No 6.595.753. Washington, DC: US Patent Office.

Khamis, A. M., Elmogy, A. M., & Karray, F. O. (2011). Complex task allocation in mobile surveillance systems. *Journal of Intelligent & Robotic Systems, 64*(1), 33–55. doi:10.1007/s10846-010-9536-2

Khoshnevis, B., & Bekey, G. (1998). Centralized sensing and control of multilpe mobile robots. *Computers & Industrial Engineering, 35*(3-4), 503–506. doi:10.1016/S0360-8352(98)00144-2

Kim, B. K., Miyazaki, M., Ohba, K., Hirai, S., & Tanie, K. (2005). Web services based robot control platform for ubiquitous functions. In *Proceedings of the International Conference on Robotics and Automation*, (pp. 690-696). IEEE.

Kim, K., Park, S., Chakeres, I., & Perkins, C. (2007). *Dynamic MANET on-demand for 6LoWPAN (DYMO-low) routing*. Retrieved from http://tools.ietf.org/html/draft-montenegro-6lowpan-dymo-low-routing-03

Kim, K., Yoo, S., Park, J., Park, S. D., & Lee, J. (2005). *Hierarchical routing over 6LoWPAN (HiLow)*. Retrieved from http://tools.ietf.org/html/draft-daniel-6lowpan-hilow-hierarchical-routing-01

Kim, S., Spenko, M., Trujillo, S., Heyneman, B., Santos, D., & Cutkosky, M. R. (2008). Smooth vertical surface climbing with directional adhesion. *IEEE Transactions on Robotics, 24*(1).

Kolmanovky, I., & McClamroch, N. H. (1996). Stabilization of wheeled vehicles by hybrid nonlinear time-varying feedback laws. In *Proceedings of the IEEE Conference on Control Applications*, (pp. 66-72). IEEE Press.

Kolmanovky, I., & McClamroch, N. H. (1995). Developments in nonholomic control problems. *IEEE Control Systems Magazine, 15*, 20–36. doi:10.1109/37.476384

Koubaa, A., Alves, M., & Tovar, E. (2007). IEEE 802.15.4: A federating communication protocol for time-sensitive wireless sensor networks. In *Sensor Networks and Configurations: Fundamentals, Techniques, Platforms, and Experiments* (pp. 19–49). Berlin, Germany: Springer-Verlag.

Koutsopoulos, I., & Tassiulas, L. (2007). Joint optimal access point selection and channel assignment in wireless networks. *IEEE/ACM Transactions on Networking, 15*(3), 521–532. doi:10.1109/TNET.2007.893237

Kurazume, R., Nagata, S., & Hirose, S. (1994). Cooperative positioning with multiple robots. [Los Alamitos, CA: IEEE Press.]. *Proceedings of IEEE ICRA, 1994*, 1250–1257.

Lamiraux, F., & Laumond, J. (2000). Flatness and small-time controllability of multibody mobile robots: Application to motion planning. *IEEE Transactions on Automatic Control, 45*(10), 1878–1881. doi:10.1109/TAC.2000.880989

Landauer, T. (1995). *The trouble with computers*. Cambridge, MA: MIT Press.

Latombe, J. C. (1991). *Robot motion planning*. Norwell, MA: Kluwer Academic Publishers. doi:10.1007/978-1-4615-4022-9

Lee, E. A. (2009). *The problem with threads*. Retrieved from http://www.eecs.berkeley.edu/Pubs/TechRpts/2006/EECS-2006-1.pdf

Lengerke, O., Carvalho, D., Lima, P. M. V., Dutra, M. S., Mora-Camino, F., & França, F. M. G. (2008). Controle distribuído de sistemas JOB SHOP usando escalonamento por reversão de arestas. In *Proceedings of the XIV Congreso Latino Ibero Americano de Investigación de Operaciones*, (pp. 1-3). IEEE.

Lengerke, O., Dutra, M., França, F., & Tavera, M. (2008). Automated guided vehicles (AGV): Searching a path in the flexible manufacturing system. *Journal of Konbin, 8*(1), 113–124. doi:10.2478/v10040-008-0106-7

Leonard, J. J., & Durrant-Whyte, H. F. (1991). Mobile robot localization by tracking geometric beacons. *IEEE Transactions on Robotics and Automation, 7*(3), 376–382. doi:10.1109/70.88147

Levine, J., & Vickers, L. (2001). Robots controlled through web services. *Technogenesis Research Project*. Retrieved from http://reference.kfupm.edu.sa/content/l/e/learning_robots_____91__a_class_63381.pdf

Li, J., Gao, X., Fan, N., Li, K., & Jiang, Z. (2009). BIT climber: A centrifugal impeller-based wall. In *Proceedings of the 2009 IEEE International Conference on Mechatronics and Automation*. IEEE Press.

Liang, K., & Jian-Jun, W. (2010). *The solution for mobile robot path planning based on partial differential equations method*. Paper presented at the International Conference on Electrical and Control Engineering (ICECE) 2010. Wuhan, China.

Lingelbach, F. (2004). Path planning for mobile manipulation using probabilistic cell decomposition. In *Proceedings of the IEEE/RSJ International Conference on Intelligent Robots and Systems*, (Vol. 3, pp. 2807-2812). Stockholm, Sweden: Centre for Autonomous Systems.

Liu, S., Sun, D., & Zhu, C. (2011). Coordinated motion planning for multiple mobile robots along designed paths with formation requirement. *IEEE/ASME Transactions on Mechatronics, 16*, 1021–1031. doi:10.1109/TMECH.2010.2070843

Ljung, L. (1987). *System identification: Theory for the user*. Englewood Cliffs, NJ: Prentice Hall.

Longo, D., & Muscato, G. (2008). Adhesion techniques for climbing robots: state of the art and experimental considerations. In *Proceedings of CLAWAR 2008*. CLAWAR.

Longo, D., Muscato, G., & Sessa, S. (2005). Simulation and locomotion control for the Alicia3 climbing robot. In *Proceedings of the 22nd International Symposium on Automation and Robotics in Construction, ISARC 2005.* ISARC.

Lozano-Perez, T. (1983). Spatial planning: A configuration space approach. *IEEE Transactions on Computers, C-32,* 108–120. doi:10.1109/TC.1983.1676196

Luccio, F., & Pagli, L. (2007). Web marshals fighting curly links farms. In P. Crescenzi, G. Prencipe, & G. Pucci (Eds.), *Fun with Algorithms, 4th International Conference,* (Vol. 4475, pp. 240-248). Berlin, Germany: Springer.

Lumelsky, V., & Stepanov, A. (1984). Effect of uncertainty on continuous path planning for an autonomous vehicle. In *Proceedings of the 23rd IEEE Conference on Decision and Control,* (Vol. 23, pp. 1616-1621). IEEE Press.

Luu, B. B., O'Brien, B. J., & Baran, D. G. (2007). A soldier-robot ad hoc network. In *Proceedings of the Fifth Annual IEEE International Conference on Pervasive Computing and Communications Workshops (PerComW 2007).* White Plains, NY: IEEE Press.

Ma, Y. J., & Hou, W. J. (2010). Path planning method based on hierarchical hybrid algorithm. In *Proceedings of the International Conference on Computer, Mechatronics, Control and Electronic Engineering,* (Vol. 1, pp. 74-77). IEEE.

MacQueen, J. B. (1967). Some methods for classification and analysis of multivariate observations. In *Proceedings of the 5th Berkeley Symposium on Mathematical Statistics and Probability,* (pp. 281-297). IEEE.

Maes, P. (1991). A bottom-up mechanism for behavior selection in an artificial creature. In *Proceedings of the First International Conference on Simulation of Adaptive Behavior (SAB90),* (pp. 238-246). The MIT Press.

Manjeshwar, A., & Agrawal, D. P. (2001). TEEN: A protocol for enhanced efficiency in wireless sensor networks. In *Proceedings of the 1st International Workshop on Parallel and Distributed Computing Issues in Wireless Networks and Mobile Computing.* San Francisco, CA: IEEE.

Manjeshwar, A., & Agrawal, D. P. (2002). APTEEN: A hybrid protocol for efficient routing and comprehensive information retrieval in wireless sensor networks. In *Proceedings of the 2nd International Workshop on Parallel and Distributed Computing Issues in Wireless Networks and Mobile Computing.* Ft. Lauderdale, FL: IEEE.

Marques, L., Almeida, A., Tokhi, M. O., & Virk, G. S. (2008). *Advances in mobile robotics.* Singapore, Singapore: World Scientific.

Martin, P., Murray, R., & Rouchon, P. (1997). *Flat systems.* Paper presented at Mini-Course European Control Conference. Brussels, Belgium.

Martinelli, A., & Siegwart, R. (2005). Observability analysis for mobile robot localization. In *Proceedings of the IEEE/RSJ IROS 2005,* (pp. 1264-1269). Edmonton, Canada: IEEE Press.

Martinelli, A., Pont, F., & Siegwart, R. (2005). Multi-robot localization using relative observation. In *Proceedings of IEEE ICRA 2005.* Barcelona, Spain: IEEE Press.

Masehian, E., & Amin-Naseri, M. R. (2006). A tabu search-based approach for online motion planning. In *Proceedings of the IEEE International Conference on Industrial Technology,* (pp. 2756-2761). Tehran, Iran: IEEE Press.

Mataric, M. J. (1992). Designing emergent behavior: From local interactions to collective intelligence, from animals to animat 2. In *Proceedings of the Second International Conference on Simulation of Adaptive Behavior (SAB92),* (pp. 432-441). Cambridge, MA: The MIT Press.

Matthias Bengel, K. P. (n.d.). *Mobile Robots for Offshore Inspection and Manipulation.*

Maturana, F., Shen, W., & Norrie, D. (1999). MetaMorph: An adaptive agent-based architecture for intelligent manufacturing. *International Journal of Production Research, 37*(10), 2159–2173. doi:10.1080/002075499190699

Melo, L. F., & Rosario, J. M. (2006). A Proposal for a hybrid opened architecture with hardware reconfigurable control applied in mobile robots. Proceedings of IEEE International Conference on Robotic and Bionemetics - ROBIO 2006, v. 1.pp. 1101-1106, China.

Melo, L. F., & Rosario, J. M. (2007). A hybrid opened architecture with hardware reconfigurable control applied in nonholonomic mobile robots. In *Proceedings of the 19th International Congress of Mechanical Engineering - COBEM 2007*, (pp. 100-108). COBEM.

Melo, L. F., Lima, C. R. E., & Rosario, J. M. (2005). A reconfigurable control architecture for mobile robots. In *Proceedings of 2nd International Symposium on Mutibody and Mechatronics-MuSMe 2005*, (vol. 1, pp. 1-8). MuSMe.

MetaGeek. L. (2011). *Chanalyzer (version 3.0)*. Retrieved from http://www.metageek.net

Microchip Technology Inc. (2006). *dsPIC30F family reference manual*. New York, NY: Microchip Technology, Inc.

Miller, D. P. (1996). Design of a small cheap UUV for under-ship inspection and salvage. In *Proceedings of the IEEE Symposium on Autonomous Underwater Vehicle Technology*, (pp. 18-20). Monterey, CA: IEEE Press.

Mita, M., Mizuno, T., Ataka, M., & Toshiyoshi, H. (2004). A 2-axis MEMS scanner for the landing laser radar of the space explorer. In *Proceedings of the 30th Annual Conference of IEEE*, (pp. 2497-2501). IEEE Press.

Mitchell, M. (1998). *An introduction to genetic algorithms*. Cambridge, MA: MIT Press.

Mitsuo, G., & Runwei, C. (2000). *Genetic algorithms & engineering optimization*. New York, NY: Wiley.

Miyake, T., Ishihara, H., & Yoshimura, M. (2007). Application of wet vacuum-based adhesion system for wall climbing mechanism. In *Proceedings of Micro-Nano-Mechatronics Symposium*, (pp. 532-537). IEEE Press.

Miyazaki, T. (1998). Reconfigurable systems. In *Proceedings of the Design Automation Conference, ASP-DAC 1998*, (pp. 447-452). IEEE Press.

Mobile Robots, Inc. (2011). *Pioneer 3D datasheet*. Retrieved from http://www.mobilerobots.com/ResearchRobots/PioneerP3DX.aspx

Mokarizadeh, S., Grosso, A., Matskin, M., Kungas, P., & Haseeb, A. (2009). Applying semantic web service composition for action planning in multi-robot systems. In *Proceedings of the Fourth International Conference on Internet and Web Applications and Services, 2009, ICIW 2009*, (pp. 370-376). ICIW.

Mondada, F., Bonani, M., Raemy, X., Pugh, J., Cianci, C., & Klaptocz, A. … Martinoli, A. (2009). The e-puck, a robot designed for education in engineering. In *Proceedings of the 9th Conference on Autonomous Robot Systems and Competitions*, (Vol. 1, pp. 59-65). IEEE.

Montemerlo, M., Roy, N., & Thrun, S. (2003). Perspectives on standardization in mobile robot programming: The Carnegie Mellon navigation (CARMEN) toolkit. In *Proceedings of the International Conference on Intelligent Robots and Systems*. Las Vegas, Nevada: IEEE.

Montenegro, G., Kushalnagar, N., Hui, J., & Culler, D. (2007). *Transmission of IPv6 packets over IEEE 802.15.4 networks*. Retrieved from http://tools.ietf.org/html/draft-montenegro-lowpan-ipv6-over-802.15.4-02

Moritz, G., Zeeb, E., Pruter, S., Golatowski, F., Timmermann, D., & Stoll, R. (2010). Devices profile for web services and the REST. In *Proceedings of the 2010 8th IEEE International Conference on Industrial Informatics (INDIN)*, (pp. 584-591). IEEE Press.

Moscarini, M., Petreschi, R., & Szwarcfiter, J. L. (1998). On node searching and starlike graphs. In *Proceedings of Congressus Numerantion, 131*, 75–84.

Mupparapu, S. S., Bartos, R., & Haag, M. M. (2005). Performance evaluation of ad hoc protocols for underwater networks. In *Fourteenth International Symposium on Unmanned Untethered Submersible Technology (UUST 2005)*. Durham, NH: UUST.

Murphy, R. R. (2004). Human-robot interactions in rescue robotics. *IEEE Transactions on Systems, Man and Cybernetics. Part C, Applications and Reviews, 34*(2), 138–153. doi:10.1109/TSMCC.2004.826267

Murphy, R., Blitch, J., & Casper, J. (2002). AAAI/RoboCup-2001 urban search and rescue events: Reality and competitions. *AI Magazine, 23*(1), 37–42.

Murray, R. (1995). Non linear control of mechanical systems: A Lagrangian perspective. In *Proceedings of the IFAC Symposium on Nonlinear Control Systems Design (NOLCOS)*. IFAC.

Murray, R., Rathinam, M., & Sluis, W. (1995). Differential flatness of mechanical control systems: A catalog of prototype systems. In *Proceedings of the ASME International Mechanical Engineering Congress and Expo*. San Francisco, CA: ASME.

N, L. (2000). La robótica: realidad y ficción. *Revista Silicio, 8*(11), 41-52.

Nagaoka, K., Kubota, T., Otsuki, M., & Tanaka, S. (2009). Robotic screw explorer for lunar subsurface investigation: Dynamics modelling and experimental validation. In *Proceedings of the International Conference on Advanced Robotics*, (pp. 1-6). IEEE.

Nagib, G., & Gharieb, W. (2004). Path planning for a mobile robot using genetic algorithms. *IEEE International Conference on Electrical, Electronic and Computer Engineering*, (pp. 185-189). IEEE Press.

Navas, O., & Ortiz, N. (2006). *Algoritmos genéticos aplicados al planeamiento de trayectorias de un robot móvil.* (Bachelor`s Thesis). Industrial University of Santander. Santander, Colombia.

Navas, O., Ortiz, N., Tibaduiza, D. A., & Martinez, R. (2005). *Algoritmos genéticos aplicados al planeamiento de trayectorias de un robot móvil.* Paper presented at I Simposio Regional de Electrónica y Aplicaciones Industriales. Ibagué, Colombia.

Networks, U. (2012). *Website.* Retrieved 2012, from http://site.microcom.us/sr71xdatasheet.pdf

Nguyen, H. G., Pezeshkian, M. R., Raymond, M., Gupta, A., & Spector, J. M. (2003). Autonomous communication relays for tactical robots. In *Proceedings of the 11th International Conference on Advanced Robotics.* Coimbra, Portugal: IEEE.

Normey-Ricoa, J. E., Gómez-Ortegab, J., & Camachob, E. F. (1999). A Smith-predictor-based generalized predictive controller for mobile robot path-tracking. *Control Engineering Practice, 7*(6), 729–740. doi:10.1016/S0967-0661(99)00025-8

Norton, J. P. (1986). *An introduction to identification.* New York, NY: Academic Press.

Ogata, K. (2009). *Modern control engineering* (5th ed.). Upper Saddle River, NJ: Prentice Hall.

Ögren, P., & Leonard, N. (2005). A convergent dynamic window approach to obstacle avoidance. *IEEE Transactions on Robotics, 21*(2), 188–195. doi:10.1109/TRO.2004.838008

Ollero Baturone, A. (2007). *Robótica, manipuladores y robots móviles.* Barcelona, Spain: Alfaomega-Marcombo.

Oxford. (2012). *An improved particle filter for non-linear problems.* Oxford, UK: Oxford University Press.

Pacheco, L., Luo, N., Ferrer, I., Cufí, X., & Arbusé, R. (2012). Gaining control knowledge through an applied mobile robotics course. In *Mobile Robots Current Trends* (pp. 69–86). Rijeka, Croatia: InTech. doi:10.5772/27994

Panzieri, S., Pascucci, F., & Setola, R. (2005). Simultaneous localization and map building algorithm for real-time applications. In *Proceedings of the 16th IFAC World Congress.* Praha, Czech Republic: IFAC.

Panzieri, S., Pascucci, F., & Setola, R. (2006). Multirobot localisation using interlaced extended Kalman filter. In *Proceedings of the IEEE/RSJ IROS 2006.* Beijing, China: IEEE Press.

Parker, L. E. (1998). ALLIANCE: An architecture for fault tolerant multirobot cooperation. *IEEE Transactions on Robotics and Automation, 14*(2), 220–240. doi:10.1109/70.681242

Parker, L. E. (2008). Multiple mobile robot systems. In Siciliano, B., & Khatib, O. (Eds.), *Springer Handbook of Robotics* (pp. 921–941). Berlin, Germany: Springer. doi:10.1007/978-3-540-30301-5_41

Perkins, C. E., & Bhagwat, P. (1994). Highly dynamic destination-sequenced distance-vector routing (DSDV) for mobile computers. In *Proceedings of the ACM Special Interest Group on Data Communications (SIGCOMM)*, (pp. 234-244). ACM Press.

Perkins, C. E., Royer, E. M., & Das, S. R. (2003). *Ad hoc on-demand distance vector (AODV) routing.* Retrieved from http://www.ietf.org/rfc/rfc3561.txt

Pimentel, J. C., & Le-Huy, H. (2000). A VHDL-based methodology to develop high performance servo drivers. In *Proceedings of the Industry Applications Conference*, (pp. 1505-1512). IEEE Press.

Pinna, J., Masson, F., & Guivant, J. (2008). *Planeamiento de trayectoria en mapas extendidos mediante programación dinámica de dos niveles*. Paper presented at V Jornadas Argentinas de Robotica-JAR 2008. Bahía Blanca, Argentina.

Pizzocaro, D., & Preece, A. (2009). Towards a taxonomy of task allocation in sensor networks. In *Proceedings of the IEEE International Conference on Computer Communications Workshops*, (pp. 413-414). IEEE Press.

Polastre, J., Hill, J., & Culler, D. (2004). Versatile low power media access for wireless sensor networks. [ACM Press.]. *Proceedings of the SenSys, 2004*, 3–5.

Porta Garcia, M. A., Montiel, O., Castillo, O., Sepulveda, R., & Melin, P. (2009). Path planning for autonomous mobile robot navigation with ant colony optimization and fuzzy cost function evaluation. *Journal of Applied Soft Computing, 9*(3), 1102–1110. doi:10.1016/j.asoc.2009.02.014

Pourboughrat, F., & Karlsson, M. P. (2002). Adaptive control of dynamic mobile robots with nonholomic constraints. *Computers & Electrical Engineering, 28*, 241–253. doi:10.1016/S0045-7906(00)00053-7

Prahlad, H., Pelrine, R., Stanford, S., Marlow, J., & Kornbluh, R. (2008). Electroadhesive robots-wall climbing robots enabled by a novel, robust, and electrically controllable adhesion technology. In *Proceedings of the IEEE International Conference on Robotics and Automation*. IEEE Press.

Prakashgoud, P., Vidya, H., Shreedevi, P., & Umakant, K. (2011). Wireless sensor network for precision agriculture. In *Proceedings of the 2011 International Conference on Computational Intelligence and Communication Networks (CICN)*, (pp. 763-766). CICN.

Qixin, C., Yanwen, H., & Jingliang, Z. (2006). *An evolutionary artificial potential field algorithm for dynamic path planning of mobile robot*. Paper presented at the International Conference on Intelligent Robots and Systems. Beijing, China.

Rajendran, V., Obraczka, K., & Garcia Luna Aceves, J. J. (2003). Energy efficient, collision free medium access control for wireless sensor networks. In *Proceedings of SenSys 2003*. ACM Press. doi:10.1145/958491.958513

Ramesh, J., Kastur, R., & Shunck, B. (1995). *Machine vision*. New York, NY: McGraw Hill Book Co. Guo, J., Liu, L., Liu, Q., & Qu, Y. (2009). An improvement of D* algorithm for mobile robot path planning in partial unknown environment. In *Proceedings of the Second International Conference on Intelligent Computation Technology and Automation, 2009*, (Vol. 3, pp. 394-397). ICICTA.

Rangarajan, H., & Garcia-Luna-Aceves, J. J. (2004). Using labeled paths for loop-free on-demand routing in ad hoc networks. In *Proceedings of the 5th ACMMOBIHOC*, (pp. 43-54). ACM Press.

Reeds, J. A., & Shepp, L. A. (1990). Optimal paths for a car that goes forwards and backwards. *Pacific Journal of Mathematics, 145*.

Reif, J. H. (1979). Complexity of the mover's problem and generalizations. In *Proceedings of the 20th Annual Symposium on Foundations of Computer Science*, (pp 421-427). ACM.

Rekleitis, I. M., Dudek, G., & Milios, E. E. (2002). Multi-robot cooperative localization: A study of trade-offs between efficiency and accuracy. In *Proceedings of the IEEE/RSJ IROS 2002*, (pp. 2690-2696). Lausanne, Switzerland: IEEE Press.

Rekleitis, I. M., Dudek, G., & Milios, E. E. (2000). Multi-robot collaboration for robust exploration. [San Francisco, CA: IEEE Press.]. *Proceedings of the IEEE ICRA, 2000*, 3164–3169.

Rekleitis, I. M., Dudek, G., & Milios, E. E. (2003). Probabilistic cooperative localization and mapping in practice. [Taipei, Taiwan: IEEE Press.]. *Proceedings of the IEEE ICRA, 2003*, 1907–1912.

Renner, M., Hoffmann, K. J., Markert, K., & Glesner, M. (2002). Desing methodology of application specific integrated circuit for mechatronic systems. In *Microprocessors and Microsystems* (pp. 95–102). London, UK: Elselvier Science Ltd.

Ridao, P., Batlle, J., & Carreras, M. (2001). *Dynamic model of an underwater robotic vehicle. Research Report*. Girona, Spain: University of Girona.

Rimon, E., & Koditschek, D. (1992). Exact robot navigation using artificial potential functions. *IEEE Transactions on Robotics and Automation, 8*(5), 501–518. doi:10.1109/70.163777

Ristic, B., Arulampalam, S., & Gordon, N. (2004). *Beyond the Kalman filter*. London, UK: Artech House.

Rosier, L. (1992). Homogeneous Lyapunov functions for homogeneous continuous vector field. *Systems & Control Letters*, *19*(6), 467–473. doi:10.1016/0167-6911(92)90078-7

Rotella, F., & Zambettakis, I. (2008). Comande des systèmes par platitude. *Techniques de l'ingénieur, 450*.

Rouached, M., Chaudhry, S., & Koubaa, A. (2011). Service-oriented architecture meets LowPANs: A survey. *The International Journal of Ubiquitous Systems and Pervasive Networks*, *1*(1).

Rouchon, P., Fliess, M., Lévine, J., & Martin, P. (1993). Flatness and motion planning: The car with n trailers. In *Proceedings European Control Conference*, (pp. 1-6). ECC.

Roumeliotis, S. I., & Bekey, G. A. (2000). Synergetic localization for groups of mobile robots. [Sidney, Australia: IEEE Press.]. *Proceedings of the IEEE CDC, 2000*, 3477–3482.

Roumeliotis, S. I., & Bekey, G. A. (2002). Distribuited multi-robot localization. *IEEE Transactions on Robotics and Automation*, *18*(5), 781–795. doi:10.1109/TRA.2002.803461

Rowe, A., Mangharam, R., & Rajkumar, R. (2006). RT-link: A time-synchronized link protocol for energy-constrained multi-hop wireless networks. In *Proceedings of the 3rd Annual IEEE Communications Society on Sensor and Ad Hoc Communications and Networks, 2006*, (pp. 402-411). IEEE Press.

Russell, S. J., & Norvig, P. (1995). *Artificial intelligence: A modern approach*. Upper Saddle River, NJ: Prentice Hall.

Ryu, J. C., Franch, J., & Agrawal, S. K. (2008). Motion planning and control of a tractor with a steerable trailer using differential flatness. *Journal of Computation and Nonlinear Dynamics*, *3*(3), 1–8.

Sang Hyuk, L., Soobin, L., Heecheol, S., & Hwang Soo, L. (2009). Wireless sensor network design for tactical military applications: Remote large-scale environments. In *Proceedings of the Military Communications Conference, 2009*, (pp. 1-7). IEEE Press.

Santos, J., & Duro, R. (2005). *Evolución artificial y robótica autónoma*. Mexico City, México: Alfaomega - Ra - Ma.

Santos, R. A. G.-P., García-Ruiz, A., Edwards-Block, A., Rangel-Licea, V., & Villaseñor-González, L. A. (2009). Hybrid routing algorithm for emergency and rural wireless networks. *Electronics and Electrical Engineering*, *1*(89), 6.

Sariff, N. B., & Buniyamin, N. (2009). Comparative study of genetic algorithm and ant colony optimization algorithm performances for robot path planning in global static environments of different complexities. In *Proceedings of the IEEE International Symposium on Computational Intelligence in Robotics and Automation (CIRA)*, (pp.132-137). Mara, Malaysia: IEEE Press.

Scerri, P., Velagapudi, P., Sycara, K., Wang, H., Chien, S. Y., & Lewis, M. (2010). Towards an understanding of the impact of autonomous path planning on victim search in USAR. In *Proceedings of the IEEE/RSJ International Conference on Intelligent Robots and Systems*. Taipei, Taiwan: IEEE Press.

Schreckenghost, D., Fong, T. W., & Milam, T. (2008). Human supervision of robotic site survey. In *Proceedings of the AIP Conference*, (pp. 776-783). AIP.

Scillato, A. E., Colón, D. L., & Balbuena, J. E. (2012). *Tesis de grado para obtener el grado de ingeniero electrónico*. Neuquén, Argentina: IEEE.

Seib, V., Gossow, D., Vetter, S., & Paulus, D. (2011). Hierarchical multi-robot coordination. *Lecture Notes in Computer Science*, *6556*, 314–323. doi:10.1007/978-3-642-20217-9_27

Shim, H.-S., Kim, J.-H., & Koh, K. (1995). Variable structure control of nonholonomic wheeled mobile robot. In *Proceedings of IEEE International Conference on Robotics and Automation*, (vol. 2, pp. 1694-1699). IEEE Press.

Shiroma, N., Chiu, Y. H., Sato, N., & Matsuno, F. (2005). Cooperative task execution of a search and rescue mission by a multi-robot team. *Advanced Robotics*, *19*(3), 311–329. doi:10.1163/1568553053583670

Siciliano, B., Bicchi, A., & Valigi, P. (Eds.). (2001). Control of wheeled mobile robots: An experimental overview. *Lecture Notes in Computer Information Science, 270*, 181-226.

Siciliano, B., & Khatib, O. (2008). *Springer handbook of robotics*. Berlin, Germany: Springer-Verlag. doi:10.1007/978-3-540-30301-5

Siciliano, B., Sciavicco, L., Villani, L., & Oriolo, G. (2009). *Robotics, modelling, planning and control*. London, UK: Springer-Verlag.

Siegwart, R., & Nourbakhsh, I. (2004). *Introduction to autonomous mobile robots*. Cambridge, MA: MIT Press.

Siegwart, R., & Nourbakhsh, I. R. (2004). *Introduction to autonomous mobile robots*. Cambridge, MA: The MIT Press.

Simon, X. Y., & Meiig, M. (1999). Real-time collision-free path planning of robot manipulators using neural network approaches. In *Proceedings of the IEEE International Symposium on Computational Intelligence in Robotics and Automation*, (pp. 47-52). Guelph, Canada: IEEE Press.

Sohn, S. Y., & Kim, M. J. (2008). Innovative strategies for intelligent robot industry in Korea. In *Proceedings of the IEEE International Conference on Industrial Engineering and Engineering Management*, (pp. 101-106). IEEE Press.

Song, B., Tian, G., Li, G., Zhou, F., & Liu, D. (2011). *ZigBee based wireless sensor networks for service robot intelligent space*. Paper presented at the International Conference on Information Science and Technology. Nanjing, China.

Sørdalen, O. J., & Canudas de Wit, C. (1993). Exponential control law for a mobile robot: Extension to path following. *IEEE Transactions on Robotics and Automation, 9*(6), 837–842. doi:10.1109/70.265927

Souza, E., & Maruyama, N. (2002). An investigation of dynamic positioning strategies for unmanned underwater vehicles. In *Proceedings of the XIV Congresso Brasileiro de Automática, XIV Congresso Brasileiro de Automática*, (pp. 1273-1278). Natal, Brazil: Federal University of Rio Grande do Norte.

Stopforth, R., Holtzhausen, S., Bright, G., Tlale, N. S., & Kumile, C. M. (2008). Robots for search and rescue purposes in urban and underwater environment- A survey and comparison. [IEEE.]. *Proceedings of the Mechatronics and Machine Vision in Practice, 2008*, 476–480.

Struemper, H., & Krishnaprasad, P. S. (1997). On approximate tracking for systems on three-dimensional matrix lie groups via feedback nilpotentization. In *Proceedings of the International Federation Automatic Control Symposium on Robot Control*. IEEE.

Sugihara, M., & Suzuki, I. (1990). Distributed motion coordination of multiple mobile robots. In *Proceedings of the IEEE International Symposium on Intelligent Control*, (pp. 138-143). Philadelphia, PA: IEEE Press.

Sugihara, K., & Smith, J. (1999). Genetic algorithms for adaptive planning of path and trajectory of a mobile robot in 2D terrains. *IEICE Transactions on Information and Systems, 82*(1), 309–317.

Suguihara, K. (1998). GA-based on-line path planning for SAUVIM. *Lecture Notes in Computer Science, 1416*, 329–338. doi:10.1007/3-540-64574-8_419

Sweeney, J. D., Li, H., Grupen, R. A., & Ramamritham, K. (2003). Scalability and schedulability in large, coordinated distributed robot systems. In *Proceedings of the IEEE International Conference on Robotics and Automation*, (pp. 4074-4079). IEEE Press.

Tadokoro, S. (2009). Earthquake disaster and expectation for robotics. In Tadokoro, S. (Ed.), *Rescue Robotics: DDT Project on Robots and Systems for Urban Search and Rescue* (pp. 1–16). London, UK: Springer.

Tang, K. S., Man, K. F., & Kwong, S. (1996). Genetic algorithms and their applications. *IEEE Signal Processing Magazine, 13*, 22–37. doi:10.1109/79.543973

Tavakoli, A. (2009). *HYDRO: A hybrid routing protocol for Lossy and low power networks*. Retrieved from http://tools.ietf.org/html/draft-tavakoli-hydro-01

Tavera, M., Dutra, M., Veslin, E., & Lengerke, O. (2009). *Implementation of chaotic behavior on a fire fighting robot*. Paper presented at the 20th International Congress of Mechanical Engineering. Gramados, Brasil.

TelosB mote, Inc. (2011). *TelosB mote datasheet*. Retrieved from http://www.willow.co.uk/TelosB_Datasheet.pdf

Tewolde, G. S., & Weihua, S. (2008). Robot path integration in manufacturing processes: Genetic algorithm versus ant colony optimization. *IEEE Transactions on Systems, Man, and Cybernetics. Part A, Systems and Humans, 38*, 278–287. doi:10.1109/TSMCA.2007.914769

Thorpe, C., & Matthies, L. (1984). Path relaxation: Path planning for a mobile robot. [OCEANS.]. *Proceedings of OCEANS, 1984*, 576–581.

Thrun, S., Fox, D., Burgard, W., & Dellaert, F. (2001). Robust Monte Carlo localization for mobile robots. *Artificial Intelligence, 128*(1-2), 99–141. doi:10.1016/S0004-3702(01)00069-8

TI. (2009). *DSP*. Retrieved from http://www.dsp.ti.com

Tibaduiza, D. A. (2006). *Planeamiento de trayectorias de un robot móvil*. (Master Thesis). Industrial University of Santander. Santander, Colombia.

Tibaduiza, D. A., Martínez, R., & Barrero, J. (2005). *Planificación de trayectorias para robots móviles*. Paper presented at the Simposio Regional de Electrónica y Aplicaciones Industriales. Ibagué, Colombia.

Tibaduiza, D. A., Martinez, R., & Barrero, J. (2006). *Algoritmos de planificación de trayectorias para un robot móvil*. Paper presented at the 2nd IEEE Colombian Workshop on Robotics and Automation. Bogotá, Colombia.

Tibaduiza, D. A., Amaya, Y., Ruiz, J., & Martínez, R. (2007). Localización dinámica de móviles y obstáculos para aplicaciones en robótica. *Colombian Journal of Computation, 8*, 93–120.

Tibaduiza, D. A., & Anaya, M. (2007). Campos de potencial aplicados al planeamiento de trayectorias de robots móviles. *Revista de Ingeniería, 1*, 23–34.

Tibaduiza, D. A., & Chio, N. (2008). Metodologías de campos de potencial para el planeamiento de trayectorias de robots móviles. *Colombian Journal of Computation, 9*, 104–119.

Torres, C. H., & Mendoza, E. Y. (2006). *Control de dos móviles en un entorno dinámico*. (Bachelor's Thesis). Industrial University of Santander. Santander, Colombia.

Trawny, N., Roumeliotis, S. I., & Giannakis, G. B. (2005). Cooperative multi-robot localization under communication constraints. In *Proceedings of the IEEE ICRA 2005*. Kobe, Japan: IEEE Press.

Unver, O., Uneri, A., Aydemir, A., & Sitti, M. (2006). Geckobot: A gecko inspired climbing robot using elastomer adhesives. In *Proceedings of ICRA*, (pp. 2329–2335). IEEE Press.

Vachtsevanos, G., & Hexmoor, H. (1986). A fuzzy logic approach to robotic path planning with obstacle avoidance. In *Proceedings of the 25th IEEE Conference on Decision and Control*, (Vol. 25, pp. 1262-1264). IEEE Press.

van Nieuwstadt, M. (1997). *Trajectory generation for nonlinear control systems. Technical Report CS 96-011*. Pasadena, CA: California Institute of Technology.

Veslin, E., Slama, J., Dutra, M., Lengerke, O., & Morales, M. (2011). Trajectory tracking for robot manipulators using differential flatness. *Ingeniería e Investigación, 31*(2), 84–90.

Viguria, A., Maza, I., & Ollero, A. (2008). S+T: An algorithm for distributed multirobot task allocation based on services for improving robot cooperation. In *Proceedings of the IEEE International Conference on Robotics and Automation (ICRA 2008)*, (pp. 3163-3168). IEEE Press.

Vincent, R., Fox, D., Ko, J., Konolige, K., Limketkai, B., & Morisset, B. (2008). Distributed multirobot exploration, mapping and task allocation. *Annals of Mathematics and Artificial Intelligence, 52*(2-4), 229–255. doi:10.1007/s10472-009-9124-y

Vishay Inc. (1999). *TCRT1000 datasheet*. New York, NY: Vishay, Inc.

Wagner, M., Chen, X., Nayyerloo, M., Wang, W., & Chasepp, J. G. (2008). A novel wall climbing robot based on Bernoulli effect. In *Proceedings of Mechatronics and Embedded Systems and Applications* (pp. 210–215). IEEE Press. doi:10.1109/MESA.2008.4735656

Wang, M. Z. A. N. A. (2003). *Ad-hoc robot wireless communication.* Paper presented at the Systems, Man and Cybernetics, 2003. New York, NY.

Wang, X., Yi, Z., Gong, Y., & Wang, Z. (2009). Optimum dynamic modeling of a wall climbing robot for ship rust removal. [Berlin, Germany: Springer-Verlag.]. *Proceedings of ICIRA, 2009,* 623–631.

Wang, Z., Liu, L., & Zhou, M. (2005). Protocols and applications of ad-hoc robot wireless communication networks: An overview. *International Journal of Intelligent Control and Systems, 10,* 296–303.

Warren, C. W. (1989). Global path planning using artificial potential fields. In *Proceedings of the IEEE International Conference on Robotics and Automation,* (Vol. 1, pp. 316-321). IEEE Press.

Weitzenfeld, A., Martinez-Gomez, L., Francois, J. P., Levin-Pick, A., & Boice, J. (2006). *Multi-robot systems: Extending RoboCup small-size architecture with local vision and ad-hoc networking.* Paper presented at the Robotics Symposium, 2006. Santiago, Chile.

Werger, B. B., & Matarić, M. J. (2000). Broadcast of local eligibility: Behavior-based control for strongly cooperative robot teams. In *Proceedings of the International Conference on Autonomous Agents,* (pp. 21 - 22). IEEE.

Wickens, C. D., & Hollands, G. J. (2000). *Engineering psychology and human performance* (3rd ed.). Upper Saddle River, NJ: Prentice Hall.

WifiBot Inc. (2011). *Website.* Retrieved from http://www.wifibot.com

Wijetunge, S., Gunawardana, U., & Liyanapathirana, R. (2010). Wireless sensor networks for structural health monitoring: Considerations for communication protocol design. In *Proceedings of the 2010 IEEE 17th International Conference on Telecommunications (ICT),* (pp. 694-699). IEEE Press.

WIKIPEDIA. (2012, April 10). Retrieved from http://es.wikipedia.org/wiki/Autom%C3%B3vil_teledirigido

Winter, T., & Thubert, P. (2010). *RPL: IPv6 routing protocol for low power and Lossy networks.* Retrieved from http://tools.ietf.org/html/draft-ietf-roll-rpl-19

Wong, C. Y., Seet, G., & Sim, S. K. (2011). Multiple-robot systems for USAR: Key design attributes and deployment issues. *International Journal of Advanced Robotic Systems, 2*(3), 85–101.

Wu, T., Duan, Z. H., & Wang, J. (2010). The design of industry mobile robot based on LL WIN function blocks language and embedded system. In *Proceedings of the 2nd International Conference on Computer Engineering and Technology,* (pp. 622-625). IEEE.

XBee.cl. (2009). *Website.* Retrieved from http://www.xbee.cl/caracteristicas.html

Xu, Y., Heidemann, J., & Estrin, D. (2001). Geography-informed energy conservation for ad hoc routing. In *Proceedings of the 7th Annual ACM/IEEE International Conference on Mobile Computing and Networking (MobiCom 2001).* Rome, Italy: ACM/IEEE Press.

Yamauchi, B. (2004). PackBot: A versatile platform for military robotics. *The International Society for Optical Engineering, 5422,* 228–237.

Yanco, H. A., Drury, J. L., & Scholtz, J. (2004). Beyond usability evaluation: Analysis of human-robot interaction at a major robotics competition. *Human-Computer Interaction, 19,* 117–149. doi:10.1207/s15327051hci1901&2_6

Ye, W., Heidemann, J., & Estrin, D. (2002). An energy-efficient MAC protocol for wireless sensor networks. In *Proceedings of the IEEE INFOCOM,* (pp. 1567-1576). IEEE Press.

Ye, W., Heidemann, J., & Estrin, D. (2004). Medium access control with coordinated, adaptive sleeping for wireless sensor networks. *IEEE/ACM Transactions on Networking, 12*(3), 493–506. doi:10.1109/TNET.2004.828953

Yogeswaran, M., & Ponnambalam, S. G. (2009). An extensive review of research in swarm robotics. In *Proceedings of the World Congress on Nature & Biologically Inspired Computing, 2009,* (pp. 140-145). IEEE.

Youngseok, L., Kyoungae, K., & Yanghee, C. (2002). Optimization of AP placement and channel assignment in wireless LANs. In *Proceedings of the 27th Annual IEEE Conference on Local Computer Networks, 2002,* (pp. 831-836). IEEE Press.

Younis, M., Youssef, M., & Arisha, K. (2002). Energy-aware routing in cluster-based sensor networks. In *Proceedings of the 10th IEEE/ACM International Symposium on Modeling, Analysis and Simulation of Computer and Telecommunication Systems (MASCOTS 2002)*. Fort Worth, TX: IEEE/ACM Press.

Yuh, J. (2000). Design and control of autonomous underwater robots: A survey. *International Journal of Autonomous Robots, 1*(3), 1–14.

Yu, Y., Estrin, D., & Govindan, R. (2001). *Geographical and energy aware routing: A recursive data dissemination protocol for wireless sensor networks. UCLA Computer Science Department Technical Report, UCLA-CSD TR-01-0023*. Los Angeles, CA: UCLA.

Zhang, H. D., Dong, B.-H., Cen, Y.-W., Zheng, R., Hashimoto, S., & Saegusa, R. (2007). *Path planning algorithm for mobile robot based on path grids encoding novel mechanism*. Paper presented at the Third International Conference on Natural Computation, 2007. Haikou, China.

Zhang, Y., Zhang, L., & Zhang, X. (2008). Mobile robot path planning base on the hybrid genetic algorithm in unknown environment. In *Proceedings of Eight International Conference on Intelligent Systems Design and Applications, 2008*, (Vol. 2. pp. 661-665). IEEE.

Zhang, H., Zhang, J., Wei Wang, W., Liu, R., & Zong, G. (2007). A series of pneumatic glass-wall cleaning robots for high-rise buildings. *Industrial Robot: An International Journal, 34*(2), 150–160. doi:10.1108/01439910710727504

Zhao, J., Zhu, L., Liu, G., & Han, Z. (2009). A modified genetic algorithm for global path planning of searching robot in mine disasters. In *Proceedings of the IEEE International Conference on Mechatronics and Automation*, (pp. 4936-4940). IEEE Press.

Zhao, Y., & BeMent, S. L. (1992). Kinematics, dynamics and control of wheeled mobile robots. In *Proceedings of IEEE International Conference on Robotics and Automation*, (vol. 1, pp. 91-96). IEEE Press.

Zheng, T., Li, R., Guo, W., & Yang, L. (2008). Multi-robot cooperative task processing in great environment. In *Proceedings of the IEEE Conference on Robotics, Automation and Mechatronics*, (pp. 1113-1117). IEEE Press.

Zhuang, Y., Gu, M. W., Wang, W., & Yu, H. Y. (2010). Multi-robot cooperative localization based on autonomous motion state estimation and laser data interaction. *Science China –. Information Sciences, 53*(11), 2240–2250.

About the Contributors

Raúl Aquino Santos graduated from the University of Colima with a BE in Electrical Engineering and received his MS degree in Telecommunications from the Centre for Scientific Research and Higher Education in Ensenada, Mexico, in 1990. He holds a PhD from the Department of Electrical and Electronic Engineering of the University of Sheffield, England. Since 2005, he has been with the College of Telematics, at the University of Colima, where he is currently a Research-Professor in telecommunications networks. His current research interests include wireless and sensor networks.

Omar Lengerke received his Engineering Degree in Computational Systems Engineering from the Universidad Autónoma de Bucaramanga – UNAB, Colombia, in 1999, and a M.Sc. in Control and Automation of Flexible Manufacturing Systems from the Monterrey Institute of Technology and Higher Education – ITESM-CEM, Mexico, in 2002. In 2010, he received a Ph.D. in Mechanical Engineering from the Federal University of Rio de Janeiro – Brazil – COPPE/UFRJ, Brazil, with the thesis "Mechatronics Architecture for Navigation for Automated Guided Vehicles with Trailers on Environments of Flexible Manufacturing Systems." Dr. Lengerke realized a postdoctoral stay in Mechatronics Systems and Robotics at the Robotics Laboratory from the Federal University of Rio de Janeiro – Brazil – COPPE/UFRJ. Presently, he is titular Professor in the Mechatronics Engineering Department at UNAB, Colombia, Editor of the *Colombian Journal of Computation* and Secretary of Government of Santander (Colombia) of Information and Communication Technologies. Dr. Lengerke has published more than 50 works in national and international congresses and indexed journals in the last 5 years. Currently, his areas of interest are mechatronics design, Wireless Sensor Networks (WSNs), robotics, AGVs, ICT, industrial automation.

Arthur Edwards received his Master's degree in Education from the University of Houston in 1985. He has been a Researcher-Professor at the University of Colima since 1985, where he has served in various capacities. He has been with the School of Telematics since 1998. His primary areas of research are Computer Assisted Language Learning (CALL), distance learning, collaborative learning, multimodal leaning, and mobile learning. The primary focus of his research is presently in the area of mobile collaborative learning.

* * *

Hernando Gonzalez Acevedo received his B.Sc. degree in Electronic Engineering from the Industrial University of Santander (UIS), Bucaramanga, Colombia, in 2000, and the M.Sc. degree in Electronic Engineering from UIS in 2007. Currently, he is a Professor with the Mechatronic Engineering School at Universidad Autónoma de Bucaramanga (UNAB), Bucaramanga, Colombia. His areas of interest are control, robotics, artificial intelligence, and wind power generation. He is a member of GICYM – Research Group on Control and Mechatronic.

Hernán González Acuña was born in 1982, received his B.S degree in Mechatronics Engineering from the Universidad Autónoma de Bucaramanga (UNAB), Colombia, in 2005. HE is a Specialist in Industrial Automation from UNAB in 2006. He received his M.Sc degree in Mechanical Engineering from the Universidad Federal do Rio de Janeiro – COPPE/UFRJ, Brazil, in 2009. He is currently Professor of the Mechatronic Engineering program at Universidad Autónoma de Bucaramanga. His main research interests are climbing robots, mobile robots, industrial robots, industrial automation, intelligent control, and hydraulics and pneumatics systems. He is a member of GICYM – Research Group on Control and Mechatronics.

Auzuir R. Alexandria was born in Fortaleza, Brazil, in 1970. He is currently an Associate Professor with the Department of Automation and Control Engineering, Federal Institute for Education, Science, and Technology of Ceará (IFCE), Fortaleza, Brazil. He received the Ph.D. degree from the Federal University of Ceará (UFC), Fortaleza, Brazil, in 2011. His research interests include embedded systems, industrial automation, artificial vision, and robotics.

Jairo de Jesús Montes Alvarez received his B.S degree in Mechatronics Engineering from the Universidad Autónoma de Bucaramanga (UNAB), Colombia, in 2012. He worked in design of Mechanical Parts at Dana Transejes S.A., Bucaramanga. Currently he works in projects development. His areas of interest are wireless communication, control, hydraulic, JAVA platform, image processing applications to quality control, and robotic systems.

Daniel S. F. Alves graduated in Computer Science (magna cum laude) at Universidad Federal do Rio de Janeiro (UFRJ), in 2012. Currently, he is a MSc student in Computer Science at COPPE-UFRJ. He has interests in artificial intelligence, swarm robotics, and logic.

Edicarla P. Andrade was born in Fortaleza, Brazil, in 1988. She is a student in Automation and Control Engineering from the Federal Institute for Education, Science, and Technology of Ceará (IFCE), Fortaleza, Brazil. He is currently a researcher at the Laboratory of Technological Innovation (LIT) of IFCE. His research interests include embedded systems, automation, control, and electronics.

André L. C. Araújo was born in Fortaleza, Brazil, in 1973. He is currently an Associate Professor with the Department of Telecommunications Engineering, Federal Institute for Education, Science, and Technology of Ceará (IFCE), Fortaleza, Brazil. He works toward the Ph.D. degree in Robust Control Theory with the Federal University of Ceará (UFC), Fortaleza, Brazil. His research interests include telemetry, wireless sensor networks, embedded systems, and data communications.

José Victor C. Azevedo was born in São Paulo, Brazil, in 1986. He is a student in Automation and Control Engineering from the Federal Institute for Education, Science, and Technology of Ceará (IFCE), Fortaleza, Brazil, in 1999. He is currently a Researcher at the Laboratory of Technological Innovation (LIT) of IFCE. His research interests include embedded systems, automation, control, and electronics.

Guillermo Adrián Rodríguez Barragán was born in Tuxpan, Jalisco, Mexico, in 1989, but currently lives in Villa de Alvarez, Colima, Mexico. He received his degree in Mechatronics in the Instituto Tecnologico de Colima in 2012. Since then, he has been developing projects related with robotics, vending applications, and control systems in domotics environments, based on medium-level microcontrollers from microchip and ARM microcontrollers. In addition, he is interested on electronic and mechanical design using CAD software.

Ítalo J. L. Batista was born in Fortaleza, Brazil, in 1983. He is currently an Associated Professor with the Department of Automation and Control Engineering, University of Fortaleza (UNIFOR), Fortaleza, Brazil. He working toward the Ph.D. degree in Predictive Control Theory applied to AUV and Mobile Robots with the Federal University of Ceará (UFC), Fortaleza, Brazil. His research interests include predictive control, automation, embedded systems, and robotic systems.

John Blankenship taught Computer and Electronics Technology for 33 years at DeVry University. He has also worked as an Engineer and as an Independent Consultant. He received a B.S in Electronic Engineering from Virginia Polytechnic State University, a Master in Electronic Engineering Technology from Southern Polytechnic State University, and an M.B.A. from Georgia State University. He spends his retirement in Florida developing robotics hardware and software.

Diego Alexander Tibaduiza Burgos received his B.S. degree in Electronic Engineering from the Industrial University of Santander, Colombia, in 2003, and the M.S degree in 2006 at the Industrial University of Santander, Colombia. Between 2005 and 2009, he worked as Assistant Professor with the Industrial University of Santander and Autonomous University of Bucaramanga in Bucaramanga-Colombia in research and teaching courses of robotics, control, and electronic systems. He is currently working as Researcher and finishing his Ph.D. at the Universitat Politècnica de Catalunya with a contract of Agència de Gestió d'Ajuts Universitaris i de Recerca of the Generalitat de Catalunya in Barcelona, Spain. His research interests include automation, robotics, structural health monitoring, and control systems.

Guilherme P. S. de Carvalho graduated in Control and Automation Engineering (cum laude) at Universidade Federal do Rio de Janeiro (UFRJ) in 2012. He is currently working at Chemtech, Siemems Business, and a Brazilian leading consulting company in engineering solutions to the process industry.

Silvia Galvão de Souza Cervantes has a degree in Electrical Engineering from Paulista State University of Julio de Mesquita Filho (1992), Masters in Electrical Engineering from Federal University of Santa Catarina (1995), and Ph.D. in Electrical Engineering from Federal University of Santa Catarina (2005). She is currently Professor at State University of Londrina. She has experience in Electrical Engineering with emphasis on automation, mainly in the following areas: control systems, traffic control, optimization, artificial intelligence, power systems, and synchronous machine.

Imen Chaâri is a PhD Researcher candidate in the Computer and Embedded System Laboratory (CES) at the National School of Engineers of Sfax (ENIS), in Tunisia. She has also been involved in the RTRACK research project as a Master student researcher. Her research interest deals with path planning. Imen received her Master degree in Computer Science from the National School of Engineering of Sfax in February 2012, and her Engineering degree in Computer Science from National School of Engineers of Sfax (ENIS), Tunisia, in June 2009.

Marcos Vinícius B. do Couto is an undergraduate student (will graduate during 2012) in Control and Automation Engineering at Universidade Federal do Rio de Janeiro (UFRJ). He currently works with precision control and instrumentation of test equipment and iterative learning cyclic control of motion systems.

Elkin Yesid Veslin Díaz graduated as Mechatronics Engineer from the Universidad Autónoma de Bucaramanga – UNAB, Colombia (2005), obtained a Master title in Sciences of Mechanical Engineering in the Universidade Federal do Rio de Janeiro – COPPE/UFRJ, Brazil (2010). He worked in the Tecnológico de Monterrey, Campus Estado de México, as a Researcher in Teaching Methodologies on the Industrial Automation area. At present, he is the Program Director of Mechatronics Engineering in the Universidad de Boyacá. His investigation areas are focused in robotics, applications of artificial intelligence, industrial automation, and control based in flatness.

Elkin Yesid Veslin Diaz received his B.S degree in Mechatronics Engineering from the Universidad Autónoma de Bucaramanga (UNAB), Colombia, in 2005, obtained a Master title in Sciences of Mechanical Engineering in the Universidade Federal do Rio de Janeiro – COPPE/UFRJ, Brazil (2010). Worked at the Tecnológico de Monterrey, Campus Estado de México, as a Researcher in Teaching Methodologies on the Industrial Automation area. At present, he is the Program Head of Mechatronics Engineering program at the Universidad de Boyacá. His investigation areas are focused in robotics, applications of artificial intelligence, industrial automation, and control based in flatness.

Juan Michel Garcia Díaz was born in Guadalajara, Jalisco, Mexico, on January 12, 1986, and lives in Villa de Alvarez, Colima, Mexico. He graduated in 2008 as a Computer Systems Engineer at the University of Colima, where he is now studying the Master of Computer Science. He is currently developing projects related to vehicular networks and multimodal interfaces. He is also interested in wireless communications networks.

Max Suell Dutra received the title of Mechanical Engineer from the Universidad Federal Fluminenese in the year of 1987, obtained a Master degree in Mechanical Engineering from the Universidad Federal do Rio de Janeiro in the year of 1990, and a PhD in Mechanical Engineering of the Gerhard Mercator en Universität Duisburg, Germany, in the year of 1995. At present, he is Associate Professor of the Universidad Federal do Rio de Janeiro COPPE/UFRJ, Brazil. His research areas are mechanical engineering with emphasis in robotics, working mainly in the following topics: mechatronic design, robotics, machine design, biomechanical, nonlinear dynamics, multi body systems, domotics, and industrial automation.

Pedro Magaña Espinoza was born in Tamazula de Gordiano City, Jalisco, Mexico, in 1990, but currently lives in Colima, Mexico. He received his degree in Telematics from the University of Colima in 2012. He works in the area of wireless sensor networks, communications of mobile robots, remote control with actuators, and micro controllers of medium level.

Armando Carlos de Pina Filho D.Sc. in Mechanical Engineering in Robotics (2005, COPPE/UFRJ). M.Sc. in Mechanical Engineering in Robotics (2001, COPPE/UFRJ). Industrial Mechanical Engineer and Technician (1995, CEFET-RJ). Professor at the POLI/UFRJ, heading the research group ARMS (Automation, Robotics, and Modeling of Systems). Member of Brazilian Society of Mechanical Sciences and Engineering (ABCM). One book about humanoid robots edited in 2007, and one book about biped robots edited in 2011. More than 30 works published in national and international congresses and journals in the last 3 years.

Felipe M. G. França was born in Rio de Janeiro, Brazil. He received his Electronics Engineer degree from the Universidad Federal do Rio de Janeiro (UFRJ), in 1981, the M.Sc. in Computer Science from COPPE/UFRJ, in 1987, and his PhD from the Department of Electrical and Electronics Engineering of the Imperial College London, in 1994. Since 1996, he has been with the Systems Engineering and Computer Science (Graduate) Program, COPPE/UFRJ, as Associate Professor, and he has research and teaching interests in asynchronous circuits, computational intelligence, computer architecture, cryptography, distributed algorithms, and other aspects of parallel and distributed computing.

Renan S. de Freitas graduated in Control and Automation Engineering at Universidad Federal do Rio de Janeiro (UFRJ) in 2012. He has interests in all aspects of robotics.

Tiago L. Garcia was born in Fortaleza, Brazil, in 1985. He received the degree in Mechatronics from the Federal Institute for Education, Science, and Technology of Ceará (IFCE), Fortaleza, Brazil, in 2009. He is currently a Researcher at the Laboratory of Technological Innovation (LIT) of IFCE. His research interests include process automation in electric power systems and energy measurement.

Vanessa C. F. Gonçalves graduated in Computer Science (magna cum laude) at Universidade Federal do Rio de Janeiro (UFRJ). She received her MSc in Computer Science from COPPE-UFRJ, in 2011. She currently works at M. I. Montreal Informática, a company that provides consulting services in information technology for Petroléo Brasileiro S.A. – Petrobrás.

Ricardo Ortiz Guerrero: received his B.S degree in Mechatronics Engineering from the Universidad Autónoma de Bucaramanga (UNAB), Colombia, in 2011. Currently, he is a student of Project Management of Engineering at the Universidad del Norte. He works in Projects Development at TEAMS S.A.

Alejandro Hossian is a Civil Engineer (Universidad Católica Argentina), specializing in expert systems – Instituto Técnico Buenos Aires (ITBA). Master in Engineering of Software (Universidad

Politécnica de Madrid). Professor of Universidad Tecnológica Nacional – Facultad Regional Neuquén (UTN-FRN). The Research group's Director of Robotics and Automation. His topics of interest include robotics, artificial neural networks, and artificial intelligence.

Lafaete Creomar Lima Junior Mechanical Engineering student (POLI/UFRJ). Member of the research group ARMS (Automation, Robotics, and Modeling of Systems). One paper published in Iberian Latin American Congress on Computational Methods in Engineering.

Sim Siang Kok joined the Nanyang Technological Institute (now Nanyang Technological University) as Lecturer in 1986 and was promoted to Associate Professor of School of Mechanical and Aerospace Engineering from 1991 to June 2010, when he retired from academia. Prior to the teaching appointment, he was engaged as a Naval Architect involved in the design of naval vessels. He is actively involved in research relating the application of artificial intelligence to design, robotics, and manufacturing. He has published over 60 papers in referred journals and conferences and in book chapters. He read Naval Architecture and Shipbuilding at the University of Newcastle upon Tyne, in 1974 with BSc (1st class Honours). Under the Academic Staff Development Scholarship and Foreign Commonwealth Overseas Scholarships, he read MSc (in Knowledge-Based Systems) at the University of Essex in 1987. He was awarded a PhD in the year 2000 at the University of Strathclyde.

Anis Koubaa is an Associate Professor in Computer Science at Al-Imam Mohamed bin Saudi University (Riyadh, Saudi Arabia) and the leader of COINS Research group. He received his habilitation in Computer Science from the National School of Engineering of Sfax in Tunisia, in 2011, and his PhD from the National Polytechnic Institute of Lorraine (INPL) and INRIA/LORIA in 2004. He was a post-doctoral researcher in 2005/2006 at CISTER Research Unit. Currently, his research interest spans over several areas related to wireless sensor networks, robotics, localization and tracking, standard protocols, link quality estimation, etc. Anis Koubaa has been involved in several research projects and international working groups, in particular chair of the TinyOS ZigBee Working Group. He has also been actively involved in the design and development of several open source toolset for experimentation with wireless sensor networks and IEEE 802.15.4/ZigBee standard protocols, such as RadiaLE, open-ZB, Z-Monitor, and others. He has more than 100 publications in several high-quality international journals and conferences. He has received several awards including the Research Excellence Prize from Al-Imam University in 2010, SensorNets Best Demo Awards in 2009, ECRTS 2007 Best Paper Award in 2007. His H-Index is 18.

Alfonso René Quintero Lara is currently a student of Mechatronics Engineering program at the Universidad Autónoma de Bucaramanga (UNAB). His areas of interest are mobile robots and industrial automation.

Priscila M. V. Lima graduated in Computer Science (magna cum laude) at Universidad Federal do Rio de Janeiro (UFRJ). She received her MSc in Computer Science from COPPE-UFRJ, in 1987, and her PhD in Computing (Artificial Intelligence) from Imperial College London, in 2000. She is currently with the Department of Mathematics, Universidad Federal Rural do Rio de Janeiro. Among her research interests are logical reasoning with artificial neural networks, applications of artificial and computational intelligence, and distributed systems.

Ningsu Luo (Full Professor, Spain). Prof. Luo is the Coordinator and Principal Investigator of three national projects and the UdG Responsible Investigator of three European projects. He is the author and coauthor of 214 publications: 65 international journal papers (38 in journals indexed in the ISI JCR), one scientific book, 15 book chapters, 133 international conference papers, and two plenary lectures. He has extensive experience in advanced control technologies and applications, especially in modeling and the identification of nonlinear dynamics using parametric and nonparametric (NN and wavelets analysis) methods, and in the design of robust, predictive, sliding mode, backstepping, QFT, and mixed H2/Hinf controllers for systems with complex dynamics (parametric uncertainties, nonlinearities, unknown disturbances, time-delays, and dynamic couplings), with applications in mechatronic systems, networked control systems, mobile robotics, civil engineering structures, and automotive and aeronautic suspension systems.

Nelson Maculan is Professor in the Department of Systems Engineering and Computer Science of the Graduate School and Research in Engineering (COPPE), Federal University of Rio de Janeiro (UFRJ), Brazil. He has supervised over 60 PhD theses and 150 MSc dissertations; was former president of the Federal University of Rio de Janeiro (1990-1994); served as National Secretary for Higher Education (2004-2007) and as State Secretary for Education in Rio de Janeiro (2007-2008). He was awarded several honorary degrees, such as: Chevalier dans l'Ordre National du Mérite, France; Grand-Croix National Order of Scientific Merit, Brazil; and Docteur Honoris Causa de l'Université Paris 13, France. His membership includes the Brazilian Academy of Sciences, the National Academy of Engineering (Brazil), and the European Academy of Arts, Sciences, and Humanities (Paris, France).

Leonimer Flávio de Melo is graduated in Electrical Engineering from the State University of Campinas (1985), Masters in Electrical Engineering from State University of Campinas (2002), and Ph.D. in Mechanical Engineering from the State University of Campinas (2007). He is currently a Professor at State University of Londrina (UEL). He has experience in the Electrical Engineering and Mechatronic Engineering, working mainly in the following areas: industrial automation, manufacturing engineering, embedded control systems, robotics, industrial mechatronic systems, reconfigurable architecture, distance education, and traffic engineering.

Rafael Mathias de Mendonça is a graduate student. He received his Engineering degree in Electronics in State University of Rio de Janeiro. He is pursuing his Master in Electronics Engineering and his interests are in swarm robots in particular and for intelligent systems in general.

Samuel Mishal is a Software Engineer and Systems Analyst. He has worked as a consultant for major government departments and businesses around the world. He taught Mathematics and Computer Technology at the college level. He received a B.S. in Electronic Engineering Technology from DeVry University, a Bachelor in Computer Science from the University of Western Australia, a Master in Engineering Science from Oxford University, and a Master in Structural Engineering from Imperial College.

Gustavo Eduardo Monte. Electrical Engineering with orientation in Electronics (Universidad Nacional de Mar del Plata). Master of Science, State University of New York at Stony Brook. Professor and Researcher of Universidad Tecnológica Nacional Facultad Regional del Neuquén (UTN-FRN). His topics of interest include digital signal processing, intelligent sensors, and embedded systems. The IES Standards Committee of the IEEE's member.

Luiza de Macedo Mourelle is an Associate Professor in the Department of System Engineering and Computation at the Faculty of Engineering, State University of Rio de Janeiro, Brazil. Her research interests include computer architecture, embedded systems design, hardware/software codesign, and reconfigurable hardware. Mourelle received her PhD in Computation from the University of Manchester – Institute of Science and Technology (UMIST), England, her MSc in System Engineering and Computation from the Federal University of Rio de Janeiro (UFRJ), Brazil, and her Engineering degree in Electronics also from UFRJ, Brazil.

Nadia Nedjah is an Associate Professor in the Department of Electronics Engineering and Telecommunications at the Faculty of Engineering, State University of Rio de Janeiro, Brazil. Her research interests include computer architecture, embedded systems, and reconfigurable hardware design as well as distributed algorithms for swarm robots control. Nedjah received her Ph.D. in Computation from the University of Manchester – Institute of Science and Technology (UMIST), England, her M.Sc. in System Engineering and Computation from the University of Annaba, Algeria, and her Engineering degree in Computer Science also from the University of Annaba, Algeria.

Epitácio K. F. Neto was born in Fortaleza, Brazil, in 1986. He received the B.Sc. degree in Automation and Control Engineering from the Federal Institute for Education, Science, and Technology of Ceará (IFCE), Fortaleza, Brazil, in 2012. He is currently a Researcher at the Laboratory of Technological Innovation (LIT) of IFCE. His research interests include embedded systems, automation, control, and electronics.

Verónica Olivera. Advanced student of Engineering Electronics Universidad Tecnológica Nacional – Facultad Regional Neuquén (UTN-FRN). She participated in the preparation of articles on cognitive robotics and automation. Teaching Assistant of "Technology of Neural Networks" in UTN-FRN. Her topics of interest include cognitive robotics, automation, signals and systems, artificial neural networks, and artificial intelligence.

Lluís Pacheco (Associate Professor, Spain). Dr. Pacheco has excellent knowledge of electronic engineering, computer architecture and technology, and model-based control techniques with applications in applied mobile robotics. He has extensive experience in the construction of experimental platforms for innovative research and academic activities. He is co-author of 1 scientific book, 15 publications in peer-reviewed international journals and book chapters, and 30 publications in conference proceedings. His areas of interest include robotics, computer vision, and off-shore wind turbines.

Wee Ching Pang was born in Fortaleza, Brazil, in 1983. He is currently an Associate Professor with the Department of Automation and Control Engineering, University of Fortaleza (UNIFOR), Fortaleza, Brazil. He is working toward the Ph.D. degree in Predictive Control Theory applied to AUV and Mobile Robots with the Federal University of Ceará (UFC), Fortaleza, Brazil. His research interests include predictive control, automation, embedded systems, and robotic systems.

Wee Ching Pang is currently a Research Associate with the School of Mechanical and Aerospace Engineering, Nanyang Technological University, Singapore. She graduated with a Bachelor's degree from the School of Computer Engineering, Nanyang Technological University, in 2005, and was a key programmer for the UniSeeker robot, which took part in the 2008 Tech-X Challenge organized by the Defence Science and Technology Agency. In 2010, she graduated with a MSc in Digital Media Technology from the School of Computer Engineering, Nanyang Technological University. Her research interests include telepresence robots, mobile, aerial, as well as underwater robots.

Federica Pascucci (1975) received the "Laurea" degree in Computer Science from the University of "Roma Tre" and the Research Doctorate degree in Systems Engineering from the University of Rome "La Sapienza" in 2000 and 2004, respectively. She is with the Department of Computer Science and Automation of the University of "Roma Tre," where she is an Assistant Professor since 2005. Her research interests are in the field of industrial control systems, robotics, sensor fusion, and Critical Infrastructure Protection (CIP). Several published papers in the robotics field are in the area of mobile robotic localization in unstructured environment. Many techniques derived from fuzzy logic, Bayesian estimation, and Dempster-Shafer theory have been developed and applied to the problem of mapping building and vision-based localization. More recently, she has been interested in search and rescue localization in highly dynamic environments using sensor networks.

Stefano Panzieri (1963) received the "Laurea" degree in Electronic Engineering in 1989 and the Ph.D. in Systems Engineering in 1994, both from the University of Roma "La Sapienza." Since February 1996, he is with the Dipartimento di Informatica e Automazione (DIA) of University of "Roma Tre," as Associate Professor, where he is the director of the Automatic Laboratory. His research interests are in the field of industrial control systems, robotics, and sensor fusion. In the area of mobile robots, some attention has been given to the problem of navigation in structured and unstructured environments with a special attention to the problem of sensor-based navigation and sensor fusion. Many techniques derived from fuzzy logic, Bayesian estimation (Kalman filtering), and Dempster-Shafer theory have been developed and applied to the problem of mapping, building, and vision-based localization.

Aloísio Carlos de Pina D.Sc. in Systems Engineering and Computer Science in Artificial Intelligence (2008, COPPE/UFRJ). M.Sc. in Systems Engineering and Computer Science in Artificial Intelligence (2003, COPPE/UFRJ). B.Sc. in Computer Science (2000, UFRJ). Professor at the IM/UFRJ, postdoctoral researcher at the Laboratory of Computational Methods and Offshore Systems (COPPE/UFRJ), researcher in the group ARMS (Automation, Robotics, and Modeling of Systems). More than 20 works published in national and international congresses and journals in the last 3 years.

Laura Victoria Escamilla del Río was born in Guadalajara, Jalisco, México, in 1985, and currently lives in Villa de Álvarez, Colima, México. She graduated in 2010 as a Telematics Engineer at the University of Colima, where she is now studying a Master in Computer Science. Currently, she is developing projects related to control system in domotics environments and wireless communication networks.

João Maurício Rosario is Professor at Faculty of Mechanical Engineering of the University of Campinas, UNICAMP, responsible of the Laboratory of Automation and Robotics and Coordinator of Robotics and Automation in Brazilian Manufacturing Network. From 1998-2002, he was the Head of Department graduate course of Automation and Control Engineering (Mechatronics). Currently, he develops various industrial research projects on the national and international level in the areas of industrial automation, control design of mechatronics systems, and biomechanics. He was educated at University of Campinas, São Paulo, Brazil, receiving the B Sc. degree in Mechanical Engineering in 1981 and the MSc degree in Systems and Control in 1983, and Specialization Degree in Production and Automation Systems in 1986 at Nancy University, France. He was awarded the PhD degree in 1990 by Ecole Centrale – Paris, France, for research into Automation and Robotics. He worked briefly as a control engineer and robotics in Spain, Switzerland, and France, and underwater robotics at GKSS, Germany. Currently, he is an invited Professor at Automation and Control department, at SUPELEC, France.

Thiago M. Santos graduated in Control and Automation Engineering (cum laude) at Universidad Federal do Rio de Janeiro (UFRJ) in 2012. He is currently working at Kongsberg Maritime, a Norwegian technology enterprise, as a Marine Automation Service Engineer.

Lorenzo Sciavicco (1938) received the Laurea degree in Electronic Engineering from the University of Rome in 1963. From 1968 to 1995, he worked at the Faculty of Engineering of the University of Naples. He is currently Professor of Robotics in the Department of Computer Engineering and Automation of the University of "Roma Tre." His research interests include automatic control theory and applications, manipulator inverse kinematics techniques, redundant manipulator control, force/motion control of manipulators, and cooperative robot manipulation. He has published more than 80 journal and conference papers, and he is co-author of the book *Modeling and Control of Robot Manipulators* (McGraw-Hill, 1996). Professor Sciavicco has been one of the pioneers of robot control research in Italy, and has been awarded numerous research grants for his robotics group. He has served as a referee for industrial and academic research projects on robotics and automation in Italy.

Gerald Seet is currently an Associate Professor with the School of Mechanical and Aerospace Engineering, Nanyang Technological University, Singapore. He is a Chartered Engineer and a Fellow of the Institution of Mechanical Engineers, United Kingdom. He lectures in mechatronics, engineering design, and real-time systems. He is Director of the Robotics Research Centre, and also Director of the Mechatronics Programme of the School of Mechanical and Aerospace Engineering. His main research interests are in mechatronics, man-machine systems, and field robotics, with specific interest in underwater mobile robotics and fluid power systems.

Roberto Setola (1969) obtained the Master Degree in Electronic Engineering at the University of Naples Federico II (1992) and the PhD in Electronic Engineering and Computer Science (1996). Currently, he is Associate Professor at University CAMPUS BioMedico, where he is the head of the Complex Systems and Security Lab and the Director of the Second Level Master in Homeland Security. He was the coordinator of the EU DG JLS project SecuFood, and he was involved in several national and international projects related to critical infrastructure protection and homeland security. Since 1992, he carried out researches on many topics related to modeling, simulation, and control of complex networks and systems, and about the protection of critical infrastructures. He is co-author of 3 books, editor of 3 books, guest editor of 3 special issues of international journals, editor in chief of 2 magazines, and co-author of about 130 scientific publications.

Jules G. Slama received the title of Electronique Electrotechnique Automatique from the Université De Provence – Faculté De St Jerome in the year of 1970, received a Master degree in Acoustics, DEA d'Acoustique, from the Université de Provence in the year of 1971, and a PhD degree in Acoustique et Dynamique des Vibrations from the Université d'Aix Marseille II in the year of 1988 in France. At present, he is Associate Professor in the Universidade Federal do Rio de Janeiro – COPPE/UFRJ, in Brazil. Has researches in mechanical engineering in the areas of acoustics and vibrations, architecture, and urbanism, with emphasis in planning and design of urban space, working mostly in the following themes: environmental acoustic, airport noise, urban acoustic, and architectural acoustic.

Eduardo Elael de M. Soares is an undergraduate student (will graduate during 2012) in Control and Automation Engineering at Universidade Federal do Rio de Janeiro (UFRJ).

Daniel H. da Silva was born in Fortaleza, Brazil, in 1987. He received the degree in Mechatronic from the Federal Institute for Education, Science, and Technology of Ceará (IFCE), Fortaleza, Brazil, in 2009. He is currently a Researcher at the Laboratory of Technological Innovation (LIT) of IFCE. His research interests include embedded systems, automation, control, and electronics.

Guilherme C. Strachan graduated in Control and Automation Engineering at Universidade Federal do Rio de Janeiro (UFRJ) in 2012. He will start his Master Degree in Telematics at Graz University of Technology, Austria, in the second semester of 2012. His interests are in artificial intelligence and embedded systems.

Gustavo Ramírez Torres was born in Colima, México, in 1976, and currently lives in Villa de Alvarez, Colima, México. He graduated in the year of 2012 as Mechatronic Engineer in the Technological Institute of Colima, has developed some projects related to robotics, has been a consultant in the field of robotics for students from the career of telematics at the University of Colima, and has worked in the design and manufacturing of mechanical parts. He is also interested in the development of aerial vehicles, especially in monitoring quadcopters.

Sahar Trigui is a PhD Researcher Candidate in the Computer and Embedded System Laboratory (CES) at the National School of Engineers of Sfax (ENIS), in Tunisia. She has been involved in the RTRACK Research Project as a Master Student Researcher. Her research interest deals with multi-robot task allocation. Sahar received her Master degree in Computer Science from the National School of Engineering of Sfax in December 2011, and her Bachelor degree in Computer Science from Faculty of Sciences of Sfax, Tunisia, in June 2009.

Antonio T. Varela was born in Crato, Brazil, in 1968. He is currently an Associated Professor with the Department of Automation and Control Engineering, Federal Institute for Education, Science, and Technology of Ceará (IFCE), Fortaleza, Brazil. He received the Ph.D. degree from the Federal University of Ceará (UFC), Fortaleza, Brazil, in 1999. His research interests include embedded systems, automation, acquisition, telemetry, control, and robotic systems.

Maribel Anaya Vejar received her B.S. degree in Electronic Engineering from the Industrial University of Santander, Colombia, in 2005, and the M.S degree in 2011 at the Industrial University of Santander, Colombia. In 2005, she worked as Assistant Professor with Industrial University of Santander teaching control course in Bucaramanga-Colombia. Between 2009 and 2010, she worked as Assistant Professor with Autonomous University of Bucaramanga in Bucaramanga-Colombia teaching courses of analogue and digital electronics. In 2011, she worked in Universitat Politécnica de Catalunya as Technical Support Level Media Research. She is currently working on her Ph.D. degree in Earthquake and Structural Dynamic at the Universitat Politècnica de Catalunya. Her research interest includes structural health monitoring, control, robotics, and biomedical systems.

Amina Waqar is a Pakistani Engineer who made significant achievements in the studies of Telecom engineering and its allied disciplines. Amina was born in Lahore, Pakistan, in 1988. She lived in a family of middle socio-economic status and has a father who served in the Pakistan Air Force. Her brother, Omer Waqar, was two years older than she was and both she and Omer studied together in the same educational institutions. She and her brother pursued similar academic careers and graduated from National University of Computers and Emerging Sciences (Foundation for Advancement of Science and Technology). On completing her graduate studies, Amina decided to continue her engineering education in the same university and is at present a student of MS (Electrical Engineering) in National University of Computers and Emerging Sciences. Throughout her stay in the university, she has been more inclined towards research-oriented pursuits. Besides her routine projects, she has been published in the World Academy of Science "Mobile Robot Navigation through Monte Carlo Method." In addition, Amina is married to Ahmed Humayon Mirza, who is also an Electrical Engineer. Amina and Ahmed both plan on continuing their education to achieve doctorate degrees.

Choon Yue Wong is currently pursuing a PhD with the Robotics Research Centre, Nanyang Technological University, Singapore. In 2003, he graduated with a Bachelor's degree from the the School of Mechanical and Production Engineering, Nanyang Technological University, Singapore. In 2007, he graduated with a MSc. in Smart Product Design from the School of Mechanical and Aerospace Engineering, Nanyang Technological University. His current research interests include unmanned aerial robots, multiple-robot deployment, and the associated human-factor issues.

Marco Fernandes S. Xaud graduated in Control and Automation Engineering at Universidad Federal do Rio de Janeiro (UFRJ) in 2012. He is currently working at Chemtech, Siemems Business, a Brazilian leading consulting company in engineering solutions to the process industry.

Index